DESIGN, EVALUATION, AND ANALYSIS OF QUESTIONNAIRES FOR SURVEY RESEARCH

DESIGN, EVALUATION, AND ANALYSIS OF QUESTIONNAIRES FOR SURVEY RESEARCH

Second Edition

WILLEM E. SARIS AND IRMTRAUD N. GALLHOFER
Research and Expertise Centre for Survey Methodology
Universitat Pompeu Fabra
Barcelona, Spain

Published by John Wiley & Sons, Inc., Hoboken, New Jersey
Published simultaneously in Canada

For general information on our other products and services or for technical support, please contact our
Customer Care Department within the United States at (800) 762-2974, outside the United States at
(317) 572-3993 or fax (317) 572-4002.

Wiley also publishes its books in a variety of electronic formats. Some content that appears in print may
not be available in electronic formats. For more information about Wiley products, visit our web site at
www.wiley.com.

Library of Congress Cataloging-in-Publication Data:

Saris, Willem E.
  Design, evaluation, and analysis of questionnaires for survey research / Willem
E. Saris, Irmtraud Gallhofer. – Second Edition.
     pages   cm
  Includes bibliographical references and index.
  ISBN 978-1-118-63461-5 (cloth)
1. Social surveys.   2. Social surveys–Methodology.   3. Questionnaires.
4. Interviewing.   I. Title.
  HN29.S29 2014
  300.72′3–dc23

                                                              2013042094

ISBN: 9781118634615

10 9 8 7 6 5 4 3 2 1

CONTENTS

PREFACE TO THE SECOND EDITION

The most innovative contribution of the first edition of the book was the introduction of a computer program (SQP) for predicting the quality of survey questions, created on the basis of analyses of 87 multitrait–multimethod (MTMM) experiments. At that time (2007), this analysis was based on 1067 questions formulated in three different languages: English, German, and Dutch. The predictions were therefore also limited to questions in these three languages.

The most important rationale for this new edition of the book is the existence of a new SQP 2.0 program that provides predictions of the quality of questions in more than 22 countries based on a database of more than 3000 extra questions that were evaluated in MTMM experiments to determine the quality of the questions. The new data was collected within the European Social Survey (ESS). This research has been carried out since 2002 every two years in 36 countries. In each round, four to six experiments were undertaken to estimate the quality of approximately 50 questions in all countries and in their respective languages. This means that the new program has far more possibilities to predict the quality of questions in different languages than its predecessor, which was introduced in the first edition of the book.

Another very important reason for a new edition of the book is also related to the new program. Whereas the earlier version had to be downloaded and used on the same PC, the new one is an Internet program with a connected database of survey questions. These contain all questions used in the old experiments as well as the new experiments, but equally, all questions asked to date in the ESS. This means that the SQP database contains more than 60,000 questions in all languages used in the ESS and elsewhere. The number of questions will grow in three ways: (1) by way of the new studies done by the ESS, which adds another 280 questions phrased in all of its working languages used in each round; (2) as a result of the new studies added to the database by other large-scale cross-national surveys; and (3) thanks to the introduction of new questions by researchers who use the program in order to evaluate the quality of their questions. In this way, the SQP program is *a continuously growing database of survey questions in most European languages with information about the quality of the questions and about the possibility for evaluating the quality of questions that have not yet been evaluated.* The program will thus be a permanently growing source of information about survey questions and their quality. To our knowledge, there is no other program that exists to date that offers the same possibilities.

We have used this opportunity to improve two chapters based on the comments we have received from program users. This is especially true for Chapter 1 and Chapter 15. Furthermore, we decided to adjust Chapters 12 and 16 on the basis of new developments in the field.

WILLEM E. SARIS
IRMTRAUD GALLHOFER

PREFACE

Designing a survey involves many more decisions than most researchers realize. Survey specialists, therefore, speak of the art of designing survey questions (Payne 1951). However, this book introduces the methods and procedures that can make questionnaire design a scientific activity. This requires knowledge of the consequences of the many decisions that researchers take in survey design and how these decisions affect the quality of the questions.

It is desirable to be able to evaluate the quality of the candidate questions of the questionnaire before collecting the data. However, it is very tedious to manually evaluate each question separately on all characteristics mentioned in the scientific literature that predicts the quality of the questions. It may even be said that it is impossible to evaluate the effect of the combination of all of these characteristics. This would require special tools that did not exist so far. A computer program capable of evaluating all the questions in a questionnaire according to a number of characteristics and providing an estimate of the quality of the questions based on the coded question characteristics would be very helpful. This program could be a tool for the survey designer in determining, on the basis of the computer output, which questions in the survey require further study in order to improve the quality of the data collected.

Furthermore, after a survey is completed, it is useful to have information about the quality of the data collected in order to correct for errors in the data. Therefore, there is a need for a computer program that can evaluate all questions of a questionnaire based on a number of characteristics and provide an estimate of the quality of the questions. Such information can be used to improve the quality of the data analysis.

In order to further such an approach, we have

1. Developed a system for coding characteristics of survey questions and the more general survey procedure;
2. Assembled a large set of studies that used multitrait–multimethod (MTMM) experiments to estimate the reliability and validity of questions;
3. Carried out a meta-analysis that relates these question characteristics to the reliability and validity estimates of the questions;
4. Developed a semiautomatic program that predicts the validity and reliability of new questions based on the information available from the meta-analysis of MTMM experiments.

We think that these four steps are necessary to change the development of questionnaires from an "art" into a scientific activity.

While this approach helps to optimize the formulation of a single question, it does not necessarily improve the quality of survey measures. Often, researchers use complex concepts in research that cannot be measured by a single question. Several indicators are therefore used. Moving from complex concepts to a set of questions that together may provide a good measure for the concept is called operationalization. In order to develop a scientific approach for questionnaire design, we have also provided suggestions for the *operationalization* of complex concepts.

The purpose of the book is, first, to specify a three-step procedure that will generate questions to measure the complex concept defined by the researcher. The approach of operationalization is discussed in Part I of the book.

The second purpose of the book is to introduce to survey researchers the different choices they can make and are making while designing survey questionnaires, which is covered in Part II of the book.

Part III discusses quality criteria for survey questions, the way these criteria have been evaluated in experimental research, and the results of a meta-analysis over many of such experiments that allow researchers to determine the size of the effects of the different decisions on the quality of the questions.

Part IV indicates how all this information can be used efficiently in the design and analysis of surveys. Therefore, the first chapter introduces a program called "survey quality predictor" (SQP), which can be used for the prediction of the quality of survey items on the basis of cumulative information concerning the effect of different characteristics of the different components of survey items on the data quality. The discussion of the program will be specific enough so that the reader can use it to improve his/her own questionnaires.

The information about data quality can and should also be used after a survey has been completed. Measurement error is unavoidable, and this information is useful for how to correct it. The exact mechanics of it are illustrated in several chapters of Part IV. We start out by demonstrating how this information can be applied to estimate the quality of measures of complex concepts, followed by a discussion on how to correct for measurement error in survey research. In the last chapter, we discuss how one can cope with measurement error in cross-cultural research.

In general, we hope to contribute to the scientific approach of questionnaire design and the overall improvement of survey research with the book.

ACKNOWLEDGMENTS

This second edition of the book would not have been possible without the dedicated cooperation in the data collection by the national coordinators of the ESS in the different countries and the careful work of our colleagues in the central coordinating team of the ESS.

All the collected data has been analyzed by a team of dedicated researchers of the Research and Expertise Centre for Survey Methodology, especially Daniel Oberski, Melanie Revilla, Diana Zavala Rojas, and our visiting scholar Laur Lilleoja. We can only hope that they will continue their careful work in order to improve the predictions of SQP even more in the future. The program would not have been created without the work of two programmers Daniel Oberski and Tom Grüner.

Finally, we would like to thank our publisher Wiley for giving us the opportunity to realize the second edition of the book. A very important role was also played by Maricia Fischer-Souan who was able to transform some of our awkward English phrases into proper ones.

Last but not least, we would like to thank the many scholars who have commented on the different versions of the book and the program. Without their stimulating support and criticism, the book would not have been written.

INTRODUCTION

In order to emphasize the importance of survey research for the social, economic, and behavioral fields, we have elaborated on a study done by Stanley Presser, originally published in 1984. In this study, Presser performed an analysis of papers published in the most prestigious journals within the scientific disciplines of economics, sociology, political science, social psychology, and public opinion (or communication) research. His aim was to investigate to what extent these papers were based on data collected in surveys.

Presser did his study by coding the data collection procedures used in the papers that appeared in the following journals. For the economics field, he used the *American Economic Review*, the *Journal of Political Economy*, and the *Review of Economics and Statistics*. To represent the sociology field, he used the *American Sociological Review*, the *American Journal of Sociology*, and *Social Forces* and, for the political sciences, the *American Journal of Political Science*, the *American Political Science Review*, and the *Journal of Politics*. For the field of social psychology, he chose the *Journal of Personality and Social Psychology* (a journal that alone contains as many papers as each of the other sciences taken together). Finally, for public opinion research, the *Public Opinion Quarterly* was elected. For each selected journal, all papers published in the years 1949–1950, 1964–1965, and 1979–1980 were analyzed.

We have updated Presser's analysis of the same journals for the period of 1994–1995, a period that is consistent with the interval of 15 years to the preceding measurement. Presser (1984: 95) suggested using the following definition of a survey:

Design, Evaluation, and Analysis of Questionnaires for Survey Research, Second Edition.
Willem E. Saris and Irmtraud N. Gallhofer.
© 2014 John Wiley & Sons, Inc. Published 2014 by John Wiley & Sons, Inc.

TABLE I.1 Percentage of articles using survey data by discipline and year (number of articles excluding data from statistical offices in parentheses)

Discipline	Period			
	1949–1950	1964–1965	1979–1980	1994–1995
Economics	5.7% (141)	32.9% (155)	28.7% (317)	(20.0%) 42.3% (461)
Sociology	24.1% (282)	54.8% (259)	55.8% (285)	(47.4%) 69.7% (287)
Political science	2.6% (114)	19.4% (160)	35.4% (203)	(27.4%) 41.9% (303)
Social psychology	22.0% (59)	14.6% (233)	21.0% (377)	(49.0%) 49.9% (347)
Public opinion	43.0% (86)	55.7% (61)	90.6% (53)	(90.3%) 90.3% (46)

...any data collection operation that gathers information from human respondents by means of a standardized questionnaire in which the interest is in aggregates rather than particular individuals. (...) Operations conducted as an integral part of laboratory experiments are not included as surveys, since it seems useful to distinguish between the two methodologies. The definition is silent, however, about the method of respondent selection and the mode of data collection. Thus, convenience samples as well as census, self-administered questionnaires as well as face-to-face interviews, may count as surveys.

The results obtained by Presser, and completed by us for the years 1994–1995, are presented in Table I.1. For completing the data, we stayed consistent with the procedure used by Presser except in one point: we did not automatically subsume studies performed by organizations for official statistics (statistical bureaus) under the category "surveys." Our reason was that at least part of the data collected by statistical bureaus is based on administrative records and not collected by survey research as defined by Presser. Therefore, it is difficult to decide on the basis of the description of the data in the papers whether surveys have been used. For this reason, we have not automatically placed this set of papers, based on studies by statistical bureaus, in the class of survey research.

The difference in treating studies from statistical bureaus is reflected in the last column of Table I.1, relating to the years 1994–1995. We first present (within parentheses) the percentage of studies using survey methods based on samples (our own classification). Next, we present the percentages that would be obtained if all studies conducted by statistical bureaus were automatically subsumed under the category survey (Presser's approach).

Depending on how the studies of the statistical offices are coded, the proportion of survey research has increased, or slightly decreased, over the years in economics, sociology, and political science. Not surprisingly, the use of surveys in public opinion research is still very high and stable.

TABLE I.2 Use of different data collection methods in different disciplines as found in the major journals in 1994–1995 expressed in percentages with respect to the total number of empirical studies published in these years

	Disciplines				
Method	Economics	Sociology	Political science	Psychology	Public opinion
Survey	39.4	59.6	28.9	48.7	95.0
Experimental	6.0	1.7	5.4	45.6	5.0
Observational	3.2	0.6	31.9	4.1	0.0
Text analysis	6.0	4.6	7.2	0.6	0.0
Statistical data	45.4	33.5	26.6	9.0	0.0

Most remarkable is the increase of survey research in social psychology: the proportion of papers using survey data has more than doubled over the last 15-year interval. Surprisingly, this outcome contradicts Presser's assumption that the limit of the survey research growth in the field of social psychology might already have been reached by the end of the 1970s, due to the "field's embracing the laboratory/ experimental methodology as the true path to knowledge."

Presser did not refer to any other method used in the papers he investigated, except for the experimental research of psychologists. For the papers published in 1994–1995, we, however, also categorized nonsurvey methods of the papers. Moreover, we checked whether any empirical data were employed in the same papers.

In economics, sociology, and political science, many papers are published that are purely theoretical, that is, formulating verbal or mathematical theories or discussing methods. In economics, this holds for 36% of the papers; in sociology, this figure is 26%; and in political science, it is 34%. In the journals representing the other disciplines, such papers have not been found for the period analyzed.

Given the large number of theoretical papers, it makes sense to correct the percentages of Table I.1 by ignoring the purely theoretical papers and considering only empirical studies. The results of this correction for 1994–1995 are presented in Table I.2.

Table I.2 shows the overwhelming importance of the survey research methodology for public opinion research but also for sociology and even for social psychology. For social psychology, the survey method is at least as important as the experimental design, while hardly any other method is employed. In economics and sociology, existing statistical data also are frequently used, but it has to be considered that these data sets themselves are often collected through survey methods.

The situation in political science in the period of 1994–1995 is somewhat different, although political scientists also use quite a number of surveys and statistical data sets based on surveys; they also make observations in many papers of the voting behavior of representatives.

We can conclude that survey research has become even more important than it was 15 years ago, as shown by Presser. All other data collection methods are only used infrequently with the exception of what we have called "statistical data." These data are collected by statistical bureaus and are at least partially based on survey research and on administrative records. Observations, in turn, are used especially in the political sciences for researching voting behavior of different representative bodies, but hardly in any other science. The psychologists naturally use experiments but with less frequency than was expected from previous data. In communication science, experiments are also utilized on a small scale. All in all, this study clearly demonstrates the importance of survey research for the fields of the social and behavioral sciences.

I.1   DESIGNING A SURVEY

As a survey is a rather complex procedure to obtain data for research, in this section, we will briefly discuss a number of decisions a researcher has to take in order to design a survey.

I.1.1   Choice of a Topic

The first choice to be made concerns the substantive research in question. There are many possibilities, depending on the state of the research in a given field what kind of research problem will be identified. Basic choices are whether one would like to do a *descriptive* or *explanatory* study and in the latter case whether one would like to do *experimental* research or *nonexperimental* research.

Survey research is often used for descriptive research. For example, in newspapers and also in scientific journals like *Public Opinion Quarterly*, many studies can be found that merely give the distribution of responses of people on some specific questions such as satisfaction with the economy, government, and functioning of the democracy. Many polls are done to determine the popularity of politicians, to name just a few examples.

On the other hand, studies can also be done to determine the reasons for the satisfaction with the government or the popularity of a politician. Such research is called *explanatory research*. The class of explanatory studies includes nonexperimental as well as experimental studies in a laboratory. Normally, we classify research as *survey research* if large groups of a population are asked questions about a topic. Therefore, even though laboratory experiments employ questionnaires, they are not treated as surveys in this book. However, nowadays experimental research can also be done with survey research. In particular, computer-assisted data collection facilitates this kind of research by random assignment procedures (De Pijper and Saris 1986; Piazza and Sniderman 1991), and such research is included here as survey research. The difference between the two experimental designs is where the emphasis is placed, either on the data of individuals or small groups or on the data of some specified population.

I.1.2   Choice of the Most Important Variables

The second choice is that of the variables to be measured. In the case of a descriptive study, the choice is rather simple. It is directly determined by the purpose of the study. For example, if a study is measuring the satisfaction of the population

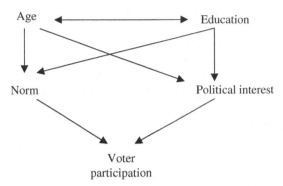

FIGURE I.1   A model for the explanation of participation in elections by voting.

with the government, it is clear that questions should be asked about the "satisfaction with the government."

On the other hand, to study what the effects of different variables are on participation in elections, the choice is not so clear. In this case, it makes sense to develop an inventory of possible causes and to develop from that list a preliminary model that indicates the relationships between the variables of interest. An example is given in Figure I.1. We suppose that two variables have a *direct effect* on "participation in elections" (voter participation): "political interest" and "the adherence to the norm that one should vote."

Furthermore, we hypothesize that "age" and "education" have a direct influence on these two variables but only an *indirect effect* on "participation in elections." One may wonder why the variables age and education are necessary in such a study if they have no direct effect on "voter participation." The reason is that these variables cause a relationship between the "norm" and "voter participation" and, in turn, between "political interest" and "voter participation." Therefore, if we use the correlation between, for example, "political interest" and "voter participation" as the estimate of the effect of "political interest," we would overestimate the size of the effect because part of this relationship is a "spurious correlation" due to "age" and "education."

For more details on this issue, we recommend the following books on causal modeling by Blalock (1964), Duncan (1975), and Saris and Stronkhorst (1984). Therefore, in this research, one not only has to introduce the variables "voter participation," "political interest," and "adherence to the norm" but also "age" and "education" as well as all other variables that generate spurious correlation between the variables of interest.

I.1.3   Choice of a Data Collection Method

The third choice to be made concerns the data collection method. This is an important choice related to costs, question formulation, and quality of data. Several years ago, the only choices available were between personal interviews (face-to-face interviews), telephone interviews, and mail surveys, all using paper questionnaires. A major difference in these methods was the presence of the interviewer in the data collection process. In personal interviews, the interviewer is physically present; in telephone interviewing, the interviewer is at a distance and the contact is by phone; while in mail surveys, the interviewer is not present at all. Nowadays, each of these modes of data collection can also be computerized by computer-assisted personal interviewing (CAPI), computer-assisted telephone interviewing (CATI), and computer-assisted self-interviewing (CASI) or Web surveys.

As was mentioned, these modes of data collection differ in their cost of data collection, where personal interviewing is the most expensive, telephone interviewing is less expensive, and mail interviewing is the cheapest. This holds true even with the aid of the computer. The same ordering can be specified for the response that one can expect from the respondents although different procedures have been developed to reduce the nonresponse (Dillman 2000).

Besides the aforementioned differences, there is a significant amount of literature on the variances in data quality obtained from these distinct modes of data collection. We will come back to this issue later in the book, but what should be clear is that the different modes require a corresponding formulation of the questions, and due to these differences in formulation, differences in responses can also be expected. Therefore, the choice of the mode of data collection is of critical importance not only for the resulting data quality but also for the formulation of the questions, which is the fourth decision to be made while designing a survey.

I.1.4   Choice of Operationalization

Operationalization is the translation of the concepts to the questions. Most people who are not familiar with designing questionnaires think that making questionnaires is very simple. This is a common and serious error. To demonstrate our point, let us look at some very simple examples of questions:

    I.1   Do you like football?

Most women probably answered the question: *Do you like to watch football on TV?*

    Most young men will answer the question: *Do you like to play football?*

    Some older men will answer the former question, some others the latter one, depending on whether they are still playing football.

    This example shows that the interpretation of the question changes for the age and gender of the respondents.

    Let us look at another example of a question that was frequently asked in 2003:

    I.2a   Was the invasion of Iraq in 2003 a success?

In general, the answer to this question is probably "yes." President Bush declared the war over in a relatively short time. But the reaction would have been quite different in 2004 if it had been asked:

    I.2b   Is the invasion of Iraq in 2003 a success?

Probably, the answer would be "no" for most people because after the end of the war, the initial problem was not solved.

While there is only a one word difference in these questions, the responses of the people would have been fundamentally different because in the first question (I.2a), people answer a question about the invasion, but in the second question (I.2b), they shift the object to evaluating the consequences of the invasion at that later point in time.

Given that such simple questions can already create a problem, survey specialists speak of "the art of asking questions" (Payne 1951; Dillman 2000: 78). We think that there is a third position on this issue: that it is possible to develop scientific methods for questionnaire design. In designing a question, many decisions are made. If we know the consequences of these decisions on the quality of the responses, then we can design *optimal questions* using a scientific method.

Now, let us consider some decisions that have to be made while designing a question.

Decision 1: *Subject and Dimension*
A researcher has to choose a subject and a dimension on which to evaluate the subject of the question. Let us expand on examples I.2a and I.2b:

    I.2c   Was the invasion a success?
    I.2d   Was the invasion justified?
    I.2e   Was the invasion important?

For examples I.2c–I.2e, there are many more choices possible, but what is done here is that the *subject* (the invasion) has been kept the same and the *dimension* on which people have to express their answer (*concept* asked) changes. The researcher has to make the choice of the dimension or concept depending on the purpose of the study.

Decision 2: *Formulation of the Question*
Many different formulations of the same question are also possible. For example:

    I.2f   Was the invasion a success?
    I.2g   Please tell me if the invasion was a success.
    I.2h   Now, I would like to ask you whether the invasion was a success.
    I.2i   Do you agree or not with the statement: the invasion was a success.

Again, there are many more formulation choices possible, as we will show later.

Decision 3: *The Response Categories*
The next decision is choosing an appropriate response scale. Here, again are some examples:

I.2j	*Was the invasion a success?*	*Yes/no*
I.2k	*How successful was the invasion?*	*Very much/quite/a bit/not at all*
I.2l	*How successful was the invasion?*	*Express your opinion with a number between 0 and 100 where 0 = no success at all and 100 = complete success*

Again, there are many more formulation options, as we will discuss later in the book.

Decision 4: *Additional Text*
Besides the question and answer categories, it is also possible to add:

- An introduction
- Extra information
- Definitions
- Instructions
- A motivation to answer

It is clear that the formulation of a single question has many possibilities. The study of these decisions and their consequences on the quality of the responses will be the main topic of this book. But before we discuss this issue, we will continue with the decisions that have to be made while designing a survey study.

I.1.5   Test of the Quality of the Questionnaire

The next step in designing a survey study is to conduct a check of the quality of the questionnaire. Some relevant checks are:

- Check on face validity
- Control of the routing in the questionnaire
- Prediction of quality of the questions with some instrument
- Use of a pilot study to test the questionnaire

It is always necessary to ask yourself and other people whether the concepts you want to measure are really measured by the way the questions are formulated. It is also necessary to control for the correctness of all routings in the questionnaire. This is especially important in computer-assisted data collection because otherwise the respondent or interviewer can be guided completely in the wrong direction, which normally leads to incomplete·responses.

There are also several approaches developed to control the quality of questions. This can be done by an expert panel (Presser and Blair 1994) or on the basis of a coding scheme (Forsyth et al. 1992; Van der Zouwen 2000) or by using a computer program (Graesser et al. 2000a, b). Another approach that is now rather popular is to present respondents with different formulations of a survey item in a laboratory setting in order to understand the effect of wording changes (Esposito et al. 1991; Esposito and Rothgeb 1997). For an overview of the different possible cognitive approaches to the evaluation of questions, we recommend Sudman et al. (1996).

In this book, we will provide our own tool, namely, survey quality predictor (SQP), which can be used to predict the quality of questions before they are used in practice.

I.1.6 Formulation of the Final Questionnaire

After corrections in the questionnaire have been made, the ideal scenario would be to test the new version again. With respect to the routing of computer-assisted data collection, that is certainly the case because of the serious consequences if something is off route. Another is to ensure that people actually understand a question better after correction. However, it will be clear that there is a limit to the iteration of tests and improvements.

Another issue is that the final layout of the questionnaire has to be decided on. This holds equally for both the paper-and-pencil approach and for questionnaires designed for computer-assisted data collection. However, research has only started on the effects of the layout on quality of the responses. For further analysis of the issue, see Dillman (2000).

After all these activities, the questionnaires can be printed if necessary to follow through with the data collection.

So far, we have concentrated on the design of the questionnaire. There is, however, another line of work that also has to be done. This concerns the selection of a population and sampling design and organization of the fieldwork, which will be discussed in the subsequent sections.

I.1.7 Choice of Population and Sample Design

With all survey research, a decision about what *population* to report on has to be made. One possible issue to consider is whether to report about the population of the country as a whole or about a specific subgroup. This decision is important because without it a sampling design cannot be specified. *Sampling* is a procedure to select a limited number of units from a population in order to describe this population. From this definition, it is clear that a population has to be selected first.

The sampling should be done in such a way that the researcher has no influence on the selection of the respondents; otherwise, the researcher can influence the results. The recommended procedure to satisfy this requirement is to select the respondents at random. Such samples based on a selection at random are called *random samples*.

If a random sampling procedure is used with a known selection probability for all respondents (not zero and not necessarily equal for all people), then it is in principle possible to *generalize* from the sample results to the population. The precision of the statements one can make about the population depends on the design of the sample and the size of the sample.

In order to draw a sample from a population, a *sampling frame* such as a list of names and addresses of potential respondents is needed. This can be a problem for specific populations, but if such a list is missing, there are also procedures to create a sampling frame. For further details, we refer to the standard literature in the area (Kish 1965; Cochran 1977; Kalton 1983). It should, however, be clear that this is a very important part of the design of the survey instrument that has to be worked out very carefully and on the basis of sufficient knowledge of the topic.

I.1.8   Decide about the Fieldwork

At least as important as the design of the sample is the design of the fieldwork. This stage determines the amount of cooperation and refusals from respondents and the quality of the work of the interviewers. In order to generate an idea of the complexity of this task, we provide an overview of the decisions that have to be made:

- Number of interviews for each interviewer
- Number of interviewers
- Recruitment of interviewers: where, when, and how
- How much to pay: per hour/per interview
- Instruction: kind of contacts, number of contacts, when to stop, and administration
- Control procedures: interviews done/not done
- Registration of incoming forms
- Coding of forms
- Necessary staff

All these decisions are rather complex and require special attention in survey research, which are beyond the scope of this book.

I.1.9   What We Know about These Decisions

In his paper mentioned at the beginning of this introduction, Presser (1984) complained that, in contrast with the importance of the survey method, methodological research was directed mainly at statistical analysis and not at the methods of data collection itself. That his observation still holds can be seen if one looks at the high proportion of statistical papers published in *Sociological Methodology* and in *Political Analysis*, the two most prestigious methodological outlets in the social sciences. However, we think that the situation has improved over the last 15 years in that research has been done directed at the quality of the survey method. The following section will be a brief review of this research.

In psychology, large sets of questions are used to measure a concept. The quality of these so-called tests are normally evaluated using factor analysis, classical test theory models, and reliability measures like Cronbach's α or item response theory (IRT) models. In survey research, such large sets of questions are not commonly used. Heise (1969) presented his position for a different approach. He argued that the questions used by sociologists and political scientists cannot be seen as alternative measures for the same concept as in psychology. Each question measures a different concept, and therefore, a different approach for the evaluation of data quality is needed. He suggested the use of the quasi-simplex models, evaluating the quality of a single question in a design using panel studies. Saris (1981) showed that different questions commonly used for the measurement of "job satisfaction" cannot be seen as indicators of the same concept. Independently of these theoretical arguments, survey researchers are frequently using single questions as indicators for the concepts they want to measure.

In line with this research tradition, many studies have been done to evaluate the quality of single survey questions. Alwin and Krosnick (1991) followed the suggestion by Heise and used the quasi-simplex model to evaluate the quality of survey questions. They suggested that on average approximately 50% of the variance in survey research variables is due to random measurement error. Split-ballot experiments are directed at determining bias due to question format (Schuman and Presser 1981; Tourangeau et al. 2000; Krosnick and Fabrigar forthcoming). Nonexperimental research has been done to study the effect of question characteristics on nonresponse and bias (Molenaar 1986). Multitrait–multimethod (MTMM) studies have been done to evaluate the effects of design characteristics on reliability and validity (Andrews 1984; Költringer 1995; Scherpenzeel 1995; Scherpenzeel and Saris 1997). Saris et al. (2004) have suggested the use of split-ballot MTMM experiments to evaluate single questions with respect to reliability and validity. Cognitive studies concentrate on the aspects of questions that lead to problems in the understanding, retrieval, evaluation, and response of the respondent (Belson 1981; Schwarz and Sudman 1996; Sudman et al. 1996). Several studies have been done to determine the positions of category labels in metric scales (Lodge et al. 1976; Lodge 1981). Interaction analysis has been done to study the problems that certain question formats and question wordings may cause with the interaction between the interviewer and the respondent (Van der Zouwen et al. 1991; Van der Zouwen and Dijkstra 1996). If no interviewer is used, the respondent can also have problems with the questions. This has been studied using keystroke analyses and response latency analyses (Couper et al. 1997).

A lot of attention has also been given to sampling. Kish (1965) and Cochran (1977) have published standard works in this context. More detailed information about new developments can be found in journals like the *Journal of Official Statistics* (JOS) and *Survey Methodology*. More recently, nonresponse has become a serious problem and has given a lot of attention in publication as in the *Journal of Public Opinion Quarterly* and in books by Groves (1989), de Heer (1999), Groves and Couper (1998), Voogt (2003), and Stoop (2005).

As this brief review has demonstrated, the literature on survey research is expanding rapidly. In fact, the literature is so expansive that the whole process cannot be discussed without being superficial. Therefore, we have decided to concentrate this book on the process of designing questionnaires. For the more statistical aspects like sampling and the nonresponse problems, we refer the reader to other books.

I.1.10 Summary

In this chapter, we have described the different choices of survey design. It is a complex process that requires different kinds of expertise. A lot of information about sampling and nonresponse problems can be found in the statistical literature. Organization of fieldwork requires a different kind of expertise. The fieldwork organizations know more about this aspect of survey research. Designing survey questions is again a distinct kind of work, and we do not recommend relying on the statistical literature or the expertise of fieldwork organizations. The design of survey questions is the typical task and responsibility of the researcher. Therefore, we will concentrate in this book on questionnaire design. For other aspects of survey research, we refer the reader to the standard literature on sampling and fieldwork. This does not mean that we will not use statistics. In the third part of this book, we will discuss the evaluation of the quality of questions through statistical models and analysis. The fourth part of this book will make use of statistical models to show how information about data quality can be employed to improve the analysis of survey data.

EXERCISES

1. Choose a research topic that you would like to study. What are the most important concepts for this topic? Why are they important?

2. Try to make a first questionnaire to study this topic.

3. Go through the different steps of the survey design mentioned in this chapter and make your choices for the research on which you have chosen to work.

PART I

THE THREE-STEP PROCEDURE TO DESIGN REQUESTS FOR ANSWERS

In this part, we explain a three-step procedure for the design of questions or as we call it requests for answers. We distinguish between concepts-by-intuition, for which obvious questions can be formulated, and concepts-by-postulation, which are formulated on the basis of concepts-by-intuition. A common mistake made by researchers is that they do not indicate explicitly how the concepts-by-postulation they use are operationalized in concepts-by-intuition. They immediately formulate questions they think are proper ones. For this reason, many survey instruments are not clear in their operationalization or even do not measure what they are supposed to measure.

In this part, we suggest a three-step approach that, if properly applied, will always lead to a measurement instrument that measures what is supposed to be measured.

The three steps are:

1. Specification of the concept-by-postulation in concepts-by-intuition (Chapter 1)
2. Transformation of concepts-by-intuition in statements indicating the requested concept (Chapter 2)
3. Transformation of statements into questions (Chapter 3)

Design, Evaluation, and Analysis of Questionnaires for Survey Research, Second Edition.
Willem E. Saris and Irmtraud N. Gallhofer.
© 2014 John Wiley & Sons, Inc. Published 2014 by John Wiley & Sons, Inc.

1

CONCEPTS-BY-POSTULATION AND CONCEPTS-BY-INTUITION

In this chapter, we will first discuss the difference between concepts-by-intuition and the concepts-by-postulation. After that we will illustrate the different ways in which concepts-by-postulation can be defined by concepts-by-intuition. In doing so, we will make a distinction between concepts-by-postulation, namely between concepts with reflective and formative indicators. These illustrations make it clear that there are many different ways to define concepts-by-postulation.

The effects that the wording of survey questions can have on their responses have been studied in depth by Sudman and Bradburn (1983), Schuman and Presser (1981), Andrews (1984), Alwin and Krosnick (1991), Molenaar (1986), Költringer (1993), Scherpenzeel and Saris (1997), and Saris and Gallhofer (2007b). In contrast, very little attention has been given to the problem of translating concepts into questions (De Groot and Medendorp 1986; Hox 1997). Blalock (1990) and Northrop (1947) distinguish between concepts-by-intuition and concepts-by-postulation.

1.1 CONCEPTS-BY-INTUITION AND CONCEPTS-BY-POSTULATION

Regarding the differentiation between concepts of intuition and concepts of postulation, Blalock (1990: 34) asserts the following:

> Concepts-by-postulation receive their meaning from the deductive theory in which they are embedded. Ideally, such concepts would be taken either as primitive or undefined or

Design, Evaluation, and Analysis of Questionnaires for Survey Research, Second Edition.
Willem E. Saris and Irmtraud N. Gallhofer.
© 2014 John Wiley & Sons, Inc. Published 2014 by John Wiley & Sons, Inc.

as defined by postulation strictly in terms of other concepts that were already understood. Thus, having defined mass and distance, a physicist defines density as mass divided by volume (distance cube). The second kind of concepts distinguished by Northrop are concepts-by-intuition, or concepts that are more or less immediately perceived by our sensory organs (or their extensions) without recourse to a deductively formulated theory. The color "blue," as perceived by our eyes, would be an example of a concept-by-intuition, whereas "blue" as a wavelength of light would be the corresponding concept-by-postulation.

The distinction he makes between the two follows the logic that concepts-by-intuition are simple concepts, the meaning of which is immediately obvious, while concepts-by-postulation are less obvious concepts that require explicit definitions. Concepts-by-postulation are also called *constructs*. Examples of concepts-by-intuition include judgments, feelings, evaluations, norms, and behaviors. Most of the time, it is quite obvious that a text presents a feeling (x likes y), a norm (people should behave in a certain way), or behavior (x does y). We will return to the classification of these concepts later. Examples of concepts-by-postulation might include "ethnocentrism," different forms of "racism," and "attitudes toward different objects." One item on its own in a survey cannot present an attitude or racism, for example. For such concepts, more items are necessary, and therefore, these concepts need to be defined. This is usually done using a set of items that represent concepts-by-intuition. For example, attitudes were originally defined (Krech et al. 1962) by a combination of cognitive, affective, and action tendency components. In Figure 1.1, an operationalization of the concept-by-postulation "an attitude toward Clinton" is presented in terms of concepts-by-intuition, questions, and assertions representing the possible responses.

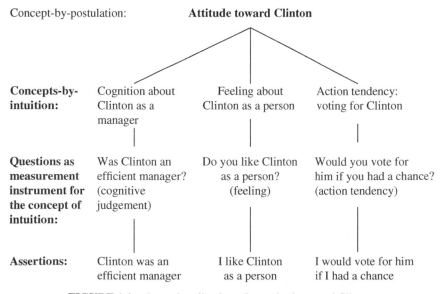

FIGURE 1.1   Operationalization of an attitude toward Clinton.

Three assertions are presented at the bottom of Figure 1.1. There is no doubt that the assertion "Clinton was an efficient manager" represents a cognitive judgment, that the assertion "I like Clinton as a person" represents a feeling, and that the assertion "I would vote for him if I had a chance" represents an action tendency. From this, it follows that the questions connected to such assertions represent measurement instruments for "cognitions," "feelings," and "action tendencies," respectively. Given that there is hardly any doubt about the link between these assertions, questions, and the concepts mentioned, these concepts are called *concepts-by-intuition*. However, the reverse relationship is not necessarily true. There are many different cognitive judgments that can be formulated regarding Clinton, whether as leader of his party or as world leader, for example. On this basis, we can conclude that there are many different possible "cognitions," "feelings," and "action tendencies" with respect to Clinton. But normally, after selecting a specific aspect of the topic, one can formulate a question that reflects the "concept-by-intuition."

In contrast to concepts-by-intuition, concepts-by-postulation are less obvious. In our example in Figure 1.1, the concept-by-postulation "attitude toward Clinton" has been defined according to the attitude concept with the three selected components. However, this choice is debatable. In fact, currently, attitudes are often defined on the basis of "evaluations" (Ajzen and Fishbein 1980) and not on the components mentioned previously. Although these two operationalizations of attitudes differ, both define attitudes on the basis of concepts-by-intuition.

As early as in 1968, Blalock complained about the gap between the language of theory and the language of research (Blalock 1968). More than two decades later, when he raised the same issues again, the gap had not been reduced (Blalock 1990). Although he argues that there is always a gap between theory and observations, he also asserts that not enough attention is given to the proper development of concepts-by-postulation. As an illustration of this, we present measurement instruments for different forms of racism in Table 1.1.

Several researchers have tried to develop instruments for new constructs related to racism. The following constructs are some typical examples: "symbolic racism" (McConahay and Hough 1976; Kinder and Sears 1981), "aversive racism" (Kovel 1971; Gaertner and Dovidio 1986), "laissez-faire racism" (Bobo et al. 1997), "new racism" (Barker 1981), "everyday racism" (Essed 1984), and "subtle racism" (Pettigrew and Meertens 1995). Different combinations of similar statements as well as different interpretations and terms have been employed in all of these instruments. Table 1.1 illustrates this point for the operationalization of symbolic and subtle racism. It demonstrates that five items of the two constructs are the same but that each construct is also connected with some specific items. The reason for including these different statements is unclear; nor is there a theoretical reason given for their operationalization.

The table depicts "subtle racism" as defined by two norms (items 1 and 2), two feelings (items 5 and 6), and four cognitive judgments (items 7a–7d as well as some other items). It is not at all clear why the presented combination of concepts-by-intuition should lead to the concept-by-postulation "subtle racism," nor is the overlap in the items and the difference between items with respect to subtle and symbolic racism (the two

TABLE 1.1   Operationalization of subtle and symbolic racism

Items	Subtle	Symbolic
1. Os living here should not push themselves where they are not wanted.	+	+
2. Many other groups have come here and overcame prejudice and worked their way up. Os should do the same without demanding special favors.	+	+
3. It is just a matter of some people not trying hard enough. If Os would only try harder, they could be as well off as our people.	+	+
4. Os living here teach their children values and skills different from those required to be successful here.	+	
5. How often have you felt sympathy for Os?	+	+
6. How often have you felt admiration for Os?	+	+
7. How different or similar do you think Os living here are to other people like you:		
a. In the values that they teach their children?	+	
b. In religious beliefs and practices?	+	
c. In their sexual values or practices?	+	
d. In the language that they speak?		+
8. Has there been much real change in the position of Os in the past few years?	+	
9. Generations of slavery and discrimination have created conditions that make it difficult for Os to work their way out of the lower class.		+
10. Over the past few years, Os have gotten less than they deserve		+
11. Do Os get much more attention from the government than they deserve?		+
12. Government officials usually pay less attention to a request or complaint from an O person than from "our" people.		+

"O" stands for member(s) of the out-group, which include "visible minorities" or "immigrants."
"+" indicates that this request for an answer has been used for the definition of the concept by postulation mentioned at the top of the column.

concepts-by-postulation), at all clear or accounted for. Even the distinction between the items assigned to "blatant racism" and the items corresponding to the other two constructs has been criticized (Sniderman and Tetlock 1986; Sniderman et al. 1991).

One of the major problems in the operationalization process of constructs related to racism is that the researchers are not, as Blalock suggested, thinking in terms of concepts-by-intuition, but only in terms of questions. They form new constructs without a clear awareness of the basic concepts-by-intuition being represented by the questions. This observation leads us to suggest that it would be useful to first of all study the definition of concepts-by-postulation through concepts-by-intuition and secondly the link between concepts-by-intuition and questions. In this chapter, therefore, we will concentrate on the definition of concepts-by-postulation through concepts-by-intuition. In the next chapters, we will continue with the relationship between concepts-by-intuition and questions.

1.2   DIFFERENT WAYS OF DEFINING CONCEPTS-BY-POSTULATION THROUGH CONCEPTS-BY-INTUITION

The best way to discuss the definition of concepts-by-postulation through concepts-by-intuition might be to give an example. In this case, however, we will not use the example of measuring racism. We will come back to this concept in the exercises of Chapter 2. Here, let us use a simpler example: the measurement of "job satisfaction." We define this concept as the feeling a person has about his/her job. We believe that though this feeling exists in people's minds, it is not possible to observe it directly. We therefore think that an unobserved or *latent variable* exists in the mind, and we denote it as "job satisfaction" or "JS." Note that we do not always expect that for concepts used in the social sciences, a latent variable exists in people's minds. For example, for the concept "the nuclear threat of Iran," there will be no preexisting latent variable for many respondents (Zaller 1992). In such a case, people will make up their minds on the spot when asked about that concept, that is, they will create a latent variable. With respect to job satisfaction, however, we think the case will be different, provided we ask the right question(s).

Many different ways of measuring job satisfaction have been developed. The following is a typical illustration of the confusion that exists around how to measure concepts. A meta-analysis of 120 job satisfaction studies found that the majority use "ad hoc measures never intended for use beyond a particular study or specific population" (Whitman et al. 2010). They found that a mere 5% of studies used a common and directly comparable measure. It will become clear that this can lead to incomparable results across studies (Wanous and Lawler 1972).

At first glance, however, the measurement of job satisfaction may appear straightforward because it can be seen as a concept-by-intuition.

1.2.1   Job Satisfaction as a Concept-by-Intuition

Measuring job satisfaction can appear to be a simple task if one thinks of it as a concept-by-intuition that can be measured with a direct question (see question 1.1):

   1.1   How satisfied or dissatisfied are you with your job?
      1. Very satisfied
      2. Satisfied
      3. Dissatisfied
      4. Very dissatisfied

Indeed, many past studies (Blauner 1966; Robinson et al. 1969; NORC 1972) as well as more recent ones (ESS 2012) have relied on this direct question or a variation of it. Such an operationalization assumes that people can express their job satisfaction in the answer to such a simple question. However, we must accept that errors will be made in the process, whether due to mistakes in respondents' answers or in interviewers' recordings of them.

In Figure 1.2, we present this process through a path model. This model suggests that people express their job satisfaction directly in their response with the exception

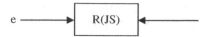

FIGURE 1.2   A measurement model for a direct measure of job satisfaction.

of some errors. The variable of interest is job satisfaction. This *latent* or *unobserved* variable is presented in the circle. The responses to the direct question presented in 1.1 can be observed directly. Such variables are usually presented in squares, while the random errors, inherent in the registration of any response, are normally denoted by an "e." This model suggests that the verbal report of the question is determined by the unobserved variable job satisfaction and errors. As shown in the model, the response to the JS question is denoted as R(JS). We will use this notation throughout the book.

This approach to measuring job satisfaction with a direct question presupposes that the meaning of job satisfaction is obvious to everyone and that people share a common interpretation of it. In other words, it assumes that when asked about their job satisfaction, all respondents are answering the same question.

The approach discussed here, assuming that the concept of interest is a concept-by-intuition that can be measured by a direct question, can be applied to many concepts, such as "political interest," "left–right orientation," "trust in the government," and many other attitudes. However, it has also been criticized as being oversimplistic.

For example, with respect to the direct measure of job satisfaction, some argue that asking people about their degree of job satisfaction is naïve because such a question requires a frank and simple answer with respect to what may be a complex and vague concept (Blauner 1966; Wilensky 1964, 1966). These researchers deny that job satisfaction can be seen as a concept-by-intuition. Others have said that such a direct question leads to too many errors and offers too low reliability (Robinson et al. 1969). Let us therefore look at the alternatives. We will first discuss the complexity problem and then follow with the reliability issue.

1.2.2   Job Satisfaction as a Concept-by-Postulation

As we have seen earlier, some people say that the use of a direct question is far too simple because job satisfaction is a complex concept. For example, Kahn (1972) suggests that people can be satisfied or dissatisfied with different aspects of their job, such as the work itself, the workplace, the working conditions, and economic rewards.

1.2.2.1   Operationalization Using Formative Indicators
Many scholars have suggested that one's feelings about one's job are based on their satisfaction with its different aspects. Clark (1998) mentions that the following aspects are highlighted in the literature: salary and working hours, opportunities for advancement, job security, autonomy in the work, social contacts, and usefulness of the job for society. The simplest operationalization therefore involves defining job satisfaction as the sum or the mean satisfaction with these different aspects of the job.

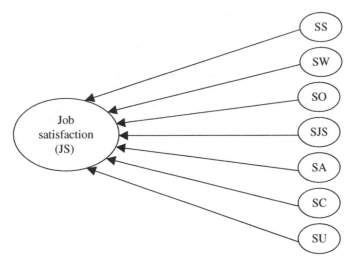

FIGURE 1.3 The operationalization of job satisfaction by a set of formative indicators where SS=satisfaction with the salary, SW=satisfaction with the working hours, SO=satisfaction with opportunities for advancement, SJS=satisfaction with job security, SA=satisfaction with autonomy, SC=satisfaction with contacts, and SU=satisfaction with usefulness of the job.

Note that in this case, we suppose that job satisfaction is based on the combined evaluation of the different aspects (in some way or another). This is different from the situation we depicted in the preceding text. In the previous section, we suggested that an opinion of job satisfaction determines the response, which is the measure for job satisfaction. Here, we are suggesting that it is the level of satisfaction with the different aspects of a job that determines or forms a person's job satisfaction. Therefore, the measures of these aspects are called *formative indicators* for the concept-by-postulation. This leads to a very different picture as shown in Figure 1.3.

So far, we have only defined the concept-by-postulation through other concepts that are causes of job satisfaction. We have done this in order to get from the concept-by-postulation to the concepts-by-intuition. If this theory is correct, then we can ask respondents about their satisfaction with these different aspects and, therefore, obtain information about their job satisfaction. For example, we can ask:

1.2   *How satisfied or dissatisfied are you with the following aspects of your job? Give your judgment in a number from 0 to 10 where 0 means completely dissatisfied and 10 completely satisfied*
Your salary (SS)
Working hours (SW)
Opportunities for advancement (SO)
Job security (SJS)
Autonomy in the work (SA)
Social contacts (SC)
Usefulness of the job for society (SU)

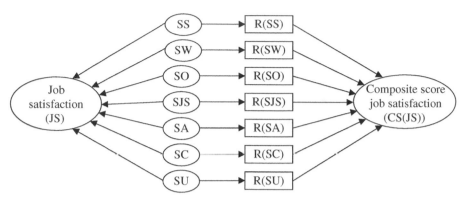

FIGURE 1.4 The complete description of the operationalization of job satisfaction using formative indicators.

The total score for "job satisfaction" can now be obtained as a weighted or an unweighted sum of the scores on all aspects or as a weighted or unweighted mean of the scores over all the aspects. The complete picture of this operationalization is presented in Figure 1.4.

In the case of an unweighted sum, the composite score for job satisfaction (CS(JS)) will have a maximum value of 70, indicating maximal satisfaction, and a minimum value of 0, indicating complete dissatisfaction. The higher the score, the more satisfied a person is, according to this approach. This score is supposed to be a good measure for the latent variable "job satisfaction." However, this is only true if the basic theory is correct and the observed scores for satisfaction with the different aspects do not contain too many errors.

This procedure is used very often for operationalization of complex concepts or, as we call them, concepts-by-postulation. The idea is to determine the different aspects of the concept-by-postulation and ask questions about these aspects. These so-called indicators can be concepts-by-intuition that can be directly converted into questions as shown for job satisfaction. However, it may be that these aspects are still too complex themselves and need to be decomposed even further before one ends up with concepts-by-intuition.

Although this procedure for the operationalization of concepts-by-postulation with formative indicators seems very logical, it also has some very serious limitations:

1. All important aspects need to be included. If not all important issues are present, the measurement is incomplete and therefore will be *invalid*.
2. The importance given to each aspect can vary from person to person. Scherpenzeel and Saris (1996) have shown that this is the case for the different aspects of job satisfaction. It therefore becomes necessary to ask respondents about how they perceive the importance of each aspect and then to use these importance scores as weights in calculating the total evaluation or "composite score."

3. Ignoring these differences would mean that the *researcher* determines what job satisfaction is. This definition might be quite different from the latent variable that exists in the mind of the respondent.

4. It is not necessarily true that all aspects affect the latent variable of interest. It has been found that for some people, the latent variable (job satisfaction) determines the satisfaction with a specific aspect (Scherpenzeel and Saris 1996).

5. This approach requires, at minimum, as many questions as there are aspects evaluated. The number of questions may double, however, if we consider that respondents will evaluate the importance of each aspect differently.

As we can see, this approach of measuring concepts-by-postulation using formative indicators encounters significant problems. The last point, for example, suggests that one needs 7 or 14 questions in order to measure the concept of job satisfaction, while in the first approach, only one question is necessary. In the first approach, it is assumed that people make this evaluation automatically in their minds. Given that the costs of the latter procedure are much higher than in the direct question approach, one needs very strong evidence that the complex approach will lead to much better results before deciding to use it. Otherwise, it is best to rely on the direct question approach. The problems mentioned in points 1–4 at the very least suggest our doubts concerning this approach.

1.2.2.2 Operationalization Using Reflective Indicators

Given these problems in operationalization and the possible unreliability of the direct question, we will present an alternative procedure, once again illustrated with the example of job satisfaction. This third procedure echoes the first procedure in that it assumes that an individual's job satisfaction has an effect on their other opinions. In several studies, Kallenberg (1974, 1975, 1977) has suggested that in addition to using the direct question, an indicator that we shall call "other job" be used. The idea is that an individual who is very satisfied with his/her job will want to continue in the same job, whereas someone who is dissatisfied will prefer the possibility of another job. Another indicator he has suggested is denoted as "recommendation." Here, the assumption is that satisfied people will recommend their jobs to friends, while dissatisfied people will not. A third one is called "choose again," which is based on the idea that someone who is satisfied would choose this job again if he had the opportunity to do so, whereas a dissatisfied person would choose a different job. As we can see, all these cases operate on the assumption that job satisfaction determines opinions regarding other indicators, which is also the case of the direct question approach. In other words, such indicators "reflect" an individual's feeling of job satisfaction. For this reason, we shall call them "reflective indicators." In this case, the concept-by-postulation (job satisfaction) affects the different indicators. This is also illustrated in Figure 1.5.

If the different indicators are seen as concepts-by-intuition, then one can develop direct questions for each of them. This possibility has indeed been used in several

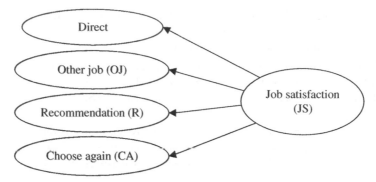

FIGURE 1.5   The measurement model for a concept-by-postulation with reflective indicators.

studies by Kallenberg and others. For example, one could use the following questions to measure the concepts-by-intuition that are used as reflective indictors for job satisfaction:

    1.3   *Would you say that you strongly agree, agree, neither agree nor*
          disagree, disagree, or strongly disagree with the following statements?
          – Overall, I am satisfied with my job (D).
          – I would like to have a different job (OJ).
          – I would recommend my job to a friend (R).
          – I would choose my job again if I had the opportunity (CA).

Note that the responses to these questions are expected to be a consequence of the opinions about these concepts-by-intuition. This leads us to extend the model in Figure 1.5 by including these effects as well as the possibility of errors (Fig. 1.6).

    The reason for which researchers suggest using not only one question, such as the direct question, is that they expect that this question alone will contain too many errors. The idea is therefore that the combination of responses to several questions, which are all observable indicators of the concept of interest, will provide a more reliable measure of that concept. This explains why researchers normally use a weighted or an unweighted sum or mean of the observed scores on the different indicators as the measure for job satisfaction. We present the final model of this measurement process in Figure 1.7.

    It will become clear that the same process can be applied for many other concepts-by-postulation. This is therefore also an illustration of a general approach. One can look for different reflective indicators for a specific concept of interest. If these indicators represent concepts-by-intuition, the concepts can be directly transformed into questions. After collecting responses to these questions, the researcher can combine the scores and obtain a composite score for the concept-by-postulation being measured. Of course, the composite score is only as good as the theory used in the model and the size of the measurement errors in the observed variables.

    Note that the fundamental difference with the previous approach is that here, the indicators are a consequence of the latent variable of interest: they *reflect* this variable

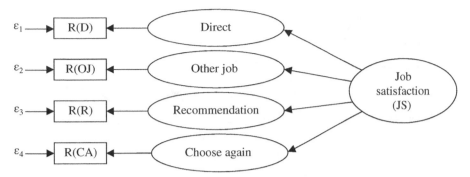

FIGURE 1.6   The measurement model for "job satisfaction" using concepts-by-intuition as reflective indicators.

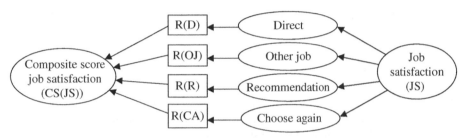

FIGURE 1.7   The complete model for measurement of "job satisfaction" using reflective indicators.

and are not indicators that *form* the latent variable of interest. In these two approaches, the causal direction of the relationships between the indicators and the concept of interest go in the opposite direction as shown in the figures.

The reflective indicators approach seems to make a lot of sense and appears less risky than the formative indicators procedure. The latter example also shows that fewer questions are necessary. In fact, the number of questions needed is determined by the reliability of the questions. The better the questions are, the fewer questions are needed to get a good composite score for the concept of interest. However, this also highlights one of the weak points of this approach. Let us look once more at the suggested questions and ask ourselves if they are really measuring the same concepts.

With respect to the direct question, it is most likely only measuring how a person feels about his job. No other perceptions will influence this response. However, when asked about the attractiveness of having a different job, the respondent will not only consider her level of job satisfaction but also how satisfied she could be in other jobs. Similarly, with respect to the question "recommend to a friend," the respondent will not only reflect on his personal job satisfaction but also on the capacities of his friend as well as his friend's own job satisfaction. This is represented as a causal diagram in Figure 1.8.

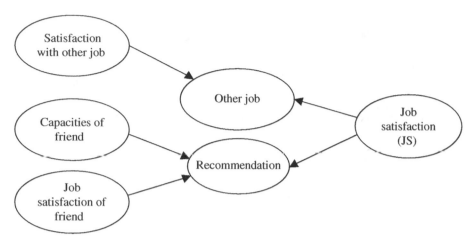

FIGURE 1.8   The systematic and unsystematic effect of other variables in the measurement process.

The effects of other variables are not just random errors that occur due to mistakes. These errors are systematic effects of specific variables. While we wanted to measure job satisfaction, it turns out that some of these reflective indicators are not just influenced by the variable of interest but also by other variables. This leads to systematic errors in the observed responses because they not only represent the variable of interest but also other variables besides random errors. One might expect that when calculating the composite score, the number of random errors decreases because they cancel each other out. But in fact, the systematic errors remain present in the composite score whenever the number of indicators is small, as is the case in survey research.

Saris (1981) has studied this problem and found that the indicators "recommend to a friend" and "choose again" have an overlap of only 70% after correction for random measurement errors. These indicators therefore contain considerable specific components. In psychological tests, this problem is generally less severe because these tests consist of 50 or more questions. In those cases, it may very well be that the systematic effects cancel each other out. In survey research, however, in which only two to four indicators are used, the same cannot be expected. We shall therefore explore an alternative procedure that allows us to avoid this problem.

1.2.2.3   Operationalization Using Reflective Indicators Varying Only by the Method

The logical solution to this problem would be to avoid using different questions and instead use the same direct question repeatedly in order to increase the reliability of the composite scores. While this simple procedure can be used in physics, it leads to problems in the social sciences because of memory effects. Therefore, this solution requires special attention.

To avoid memory effects, researchers should allow for sufficient time between the first presentation of questions and the next. We discuss how much time is needed in Chapter 9, arguing that for a questionnaire with similar questions, a gap of at least 25 minutes is sufficient (Van Meurs and Saris 1990). In order to reduce the time slightly, it is also possible to vary the form of the question so that the content of the question remains the same while ensuring that the respondent cannot rely on the memory of the first answer in responding to the reformulated question. Let us illustrate this point for the direct question, making the plausible assumption that this question measures the concept-by-intuition "job satisfaction." In this, case we could use two different formulations of the same question as follows:

1.4   *Overall, how satisfied or dissatisfied are you with your job?*
    – Completely dissatisfied
    – Quite dissatisfied
    – A bit dissatisfied
    – A bit satisfied
    – Quite satisfied
    – Completely satisfied

1.5   *Overall, how satisfied or dissatisfied are you with your job?*
    Express your opinion by putting a cross on the following scale.
    The more satisfied you are, the more to the right you put the cross.

    Completely   *Completely*
    dissatisfied   *satisfied*
    _____

The advantage of this procedure is that the questions are exactly the same. The only difference lies in the way the respondents have to express their opinions. As such, the two questions are really measuring the same thing. The difference is that one of the formulations may obtain fewer random errors than the other. As we will discuss later, by combining the scores in a composite score, one can even reduce these errors.

1.3   SUMMARY

The first issue we discussed in this chapter was distinguishing between concepts-by-intuition and concepts-by-postulation. We have seen that concepts-by-intuition are easily transformed into questions. Concepts-by-postulation cannot be operationalized directly in survey questions. They are normally defined by some combination of concepts-by-intuition.

Next, we presented different ways of developing measures for concepts of interest. At times, it is possible to see the concept of interest as a concept-by-intuition. When that is the case, a direct question that will measure the concept of interest with a great deal of certainty can be formulated.

However, as an alternative, one can also treat the concept of interest as a concept-by-postulation by specifying its different aspects, asking questions that focus on these aspects, and combining the scores into a composite score attributed to the concept of interest. This procedure relies on evaluations of each of the aspects and is commonly used by researchers, but as we highlight, it is not without its problems. The most important of these has to do with the fact that respondents can interpret the relative importance of the different aspects differently. In this case, researchers must ask additional questions with respect to the relative importance of each aspect, which results in doubling the number of questions. The list of aspects should also be complete; otherwise, the operationalization is incomplete and therefore invalid, since one is not actually measuring the concept that was intended to be measured.

As another alternative, we can think of indicators for the concept of interest that are a consequence of the latent variable. These so-called reflective indicators have the problem that they can also be affected by other variables to such an extent that the observed responses themselves contain unique components that undermine the measurement of the concept of interest.

The solution to this problem is to use reflective indicators that are measuring exactly the same thing, for example, two forms of the direct question. In this case, we have the problem of repeated observations, but by making the time gap between the measurements large enough, this problem can be overcome.

We have illustrated these alternative procedures for measurement in order to show that there is not just one possibility for measurement but many possibilities. In general, all four of the possibilities we have mentioned here can be used for the development of measures for concepts of interest. This, however, raises the question of how to evaluate the quality of the different possible measures for the concepts of interest.

This book sets out to answer this question. Instead of immediately proceeding to analyze the relationships between concepts and their measures, however, we will concentrate first on the link between concepts-by-intuition and their questions. Only once we understand this relationship and can say something about the quality of a single question can we discuss the quality of the measures for concepts-by-postulation. The idea is that in order to speak of the quality of concepts-by-postulation, the elements on which the concepts-by-postulation are built need to be identified. For example, if we realize that the question about "other job" measures not only job satisfaction, we will be more reluctant to use this indicator as an indicator for job satisfaction. Such prudence is necessary to prevent the construction of concepts-by-postulation that are unclear and likely to produce confusing results in data analysis. Therefore, we will first discuss the link between concepts-by-intuition and questions and return to the construction and the evaluation of concepts-by-postulation in Chapter 14 of this book.

EXERCISES

1. Try to formulate questions that represent concepts-by-intuition and concepts-by-postulation with formative and reflective indicators for the following concepts:
   a. Life satisfaction
   b. Happiness
   c. The importance of the value "honesty"

2. In practice, it is seldom clear whether the questions suggested measure what they are supposed to measure. Some examples follow below.

   The following proposal has been made to measure "left–right orientations" in politics. The authors said:

   "The left–right orientation contains two components:
   * *Egalitarianism: a policy of equality of incomes*
   * *Interventionism: a policy of government intervention in the economy by, for example, nationalization"*

   Items 1–3 in the following list are supposed to measure the egalitarian element; the next two, interventionism:

   How strongly do you agree or disagree with the following items?
   Agree completely, agree very much, agree, neither agree nor disagree, disagree, disagree very much, disagree completely
   1. *It is not the government's role to redistribute income from the better off to the worse off.*
   2. *It is the government's responsibility to provide a job for everyone who wants one.*
   3. *Management will always try to get the better of employees, if it gets a chance.*
   4. *Private enterprise is the best way to solve Britain's economic problems.*
   5. *Major public services and industries ought to be under state ownership.*
      a. Check whether these assertions represent the concepts they are supposed to represent.
      b. Try to improve the assertions that seem incorrect.

3. Let us now look at the questionnaire you have developed yourself:
   a. Do the questions measure what they are supposed to measure?
   b. Did you use concepts-by-intuition or concepts-by-postulation?
   c. Is it possible that other variables affect the responses than just the variables you would like to measure?
   d. If you think that some of your questions are wrong, try to improve them.

2

FROM SOCIAL SCIENCE CONCEPTS-BY-INTUITION TO ASSERTIONS

In the first chapter, we discussed the operationalization of concepts-by-postulation through concepts-by-intuition. There, we said that the relationship between the concepts-by-intuition and questions is obvious. However, the transformation of concepts-by-intuition into questions is not so simple, at least if we want to provide a procedure that almost certainly leads to a question that represents the concept-by-intuition. One of the reasons for this is that there exist so many different concepts that one cannot make rules for all of them. Another reason is that there are so many possible ways of formulating questions.

In this chapter, we try to simplify the task by only indicating that the many different concepts that are used in the social sciences can be classified in general classes of basic concepts. For these basic concepts, one can formulate valid questions. In order to make this step simpler, we will first indicate how, for all these concepts, assertions can be formulated that represent them with certainty. In the next chapter, we will then show how these assertions can be transformed into questions.

We start with illustrating the link between many concepts from the European Social Survey (ESS) and basic concepts of the social sciences. Then, we will discuss the basic structures of assertions, after which we will specify which forms of assertions can be used for basic concepts. By following this process, a researcher only has to determine to which class of basic concepts his/her concept-by-intuition belongs and then obtain the corresponding assertions that will measure this concept with certainty.

Design, Evaluation, and Analysis of Questionnaires for Survey Research, Second Edition.
Willem E. Saris and Irmtraud N. Gallhofer.
© 2014 John Wiley & Sons, Inc. Published 2014 by John Wiley & Sons, Inc.

2.1 BASIC CONCEPTS AND CONCEPTS-BY-INTUITION

As was said earlier, there is a nearly endless list of possible concepts in the social sciences. We cannot, of course, specify how to formulate questions for all of these concepts, but if we can reduce the number of concepts by classifying them into a limited number of classes of basic concepts, then this problem may be solved.

In order to illustrate what we mean, we have made a list of the concepts-by-intuition that have been measured in round 1 of the ESS, and we have tried to indicate the classes of basic concepts to which they belong. The results of the process are presented in Table 2.1. In this table, we do not give examples of all possible basic concepts; instead, we show that there are many different concepts-by-intuition that can be seen as specific cases of basic concepts. For example, evaluations of the education system of a country and the evaluation of one's own health have in common that both are evaluations (good or bad), while the subject in this case is very different. If we know how sentences that express an evaluation can be formulated, we can apply this rule to both concepts to formulate questions that will measure what we want to measure with certainty. Other examples can be interpreted in the same way. This is just an illustration.

TABLE 2.1 The classification of a number of concepts-by-intuition from the ESS into classes of basic concepts of the social sciences

Question ID	ESS concepts	Basic concepts
B33	Evaluation of services	Evaluation
C7	Health	Evaluation
A8	Social trust	Feeling
B7	Political trust	Feeling
B27	Satisfaction with…	Feeling
C1	Happiness	Feeling
B1	Political interest	Importance
S	Value benevolence	Importance
B2	Political efficacy	Judgment
B28	Left–right placement	Judgment
C5	Victimization of crimes	Evaluative belief
C16	Discrimination	Evaluative belief
B44	Income equality	Policy
B46	Freedom of lifestyle	Policy
C9	Religious identity	Similarity or association
C10	Religious affiliation	Similarity or association
C14	Church attendance	Behavior
C15	Praying	Behavior
	Political action	Behavior
A1	Media use	Frequency or amount
C2	Social contacts	Frequency or amount
C20	Country of origin	Demographic
F1	Household composition	Demographic
F2	Age	Demographic

It will be clear that the number of concepts-by-intuition is nearly unlimited, but we will show in Section 2.4 that they can be classified in a limited number of basic concepts.

In the following, we define the basic concepts and indicate the form of the assertions that apply to them. However, before we can do so, we first have to say something about the different possible structures of assertions.

2.2   ASSERTIONS AND REQUESTS FOR AN ANSWER

In order to clarify the link between basic concepts-by-intuition and verbal expressions of concepts, the linguistic components of the sentences that represent the different concepts must be discussed first. The starting point of the discussion is the sentence structure. A *sentence* is defined as a group of words that when written down begins with a capital letter and ends with a full stop, a question mark, or an exclamation mark. But, a sentence can also be classified according to its *linguistic meaning* where a distinction is made between *declarative* sentences or *assertions*, *interrogative* sentences or *requests*, *imperative* sentences or *orders*, and *exclamations*. As we will see later in this section, the first three linguistic forms of sentences are used to elicit answers from a respondent and not only in the interrogative form. Therefore, we speak of "requests for answers" and not of questions. The fourth form is not used in survey research.

Most of the items in Table 1.1 were declarative sentences or assertions representing specific concepts-by-intuition. The respondents are asked whether they agree or disagree with these assertions. It is not necessary to use such statements. It is also possible to use normal requests. But we will show how an assertion (example 2.1) can be transformed into a request (2.2). The assertion is:

2.1   *Immigrants living here should not push themselves where they are not wanted.*

To transform this assertion into a request, we only have to add "Do you think that" and then we get:

2.2   *Do you think that immigrants living here should not push themselves where they are not wanted?*

In this or similar ways, any statement can be transformed into a request.

It is also possible to transform any request into an assertion (Harris 1978; Givon 1984). The assertion that corresponds to the aforementioned request has already been given. Another example of a request is item 8 in Table 1.1. The request was as follows:

2.3   *Has there been much real change in the position of black people in the past few years?*

By inverting the term "there" and the auxiliary verb "has," we obtain from this request the following assertion:

2.4   *There has been much real change in the position of black people in the past few years.*

Similar changes can be applied on any request in order to get an assertion.

Instead of requests or assertions, surveys sometimes use instructions or directives that are called "imperatives" in linguistic terminology. These imperatives can also be transformed into assertions. The following example illustrates this:

2.5a   *Tell me if you are in favor of the right to abortion.*

This imperative can be transformed into an assertion as follows:

2.5b   *I am in favor of the right to abortion.*

We have shown in the preceding text that imperatives and interrogatives can be used to elicit answers from the respondents and can also be linguistically transformed into assertions or statements. Although this is true, it should be clear that there are fundamental differences between "requests requiring an answer" and the related assertions. In fact, a request for an answer, whatever the form of the request may be, presents the respondent with a set of possible answers, called the *uncertainty space* by Groenendijk and Stokhof (1997). On the other hand, an assertion is a specific choice from the set. Take example 2.5c, where the request was:

2.5c   *Tell me if you are in favor of the right to abortion.*

This request for an answer allows not only for the assertion 2.5d:

2.5d   *I am in favor of the right to abortion.*

but, equally, for the assertion 2.5e:

2.5e   *I am not in favor of the right to abortion.*

Although this inequality exists between the requests for an answer and the assertions, we prefer to discuss the link between concepts and requests for an answer on the basis of the related assertions (we need to keep in mind that there is an almost unlimited number of forms that requests of an answer can take[1]). The use of assertions therefore simplifies the discussion. In Chapters 3 and 4, we will discuss how these assertions can be transformed into requests for an answer. In order to discuss the link between the basic concepts and their related assertions, the next section introduces the structure of assertions.

2.3   THE BASIC ELEMENTS OF ASSERTIONS

Sentences can be divided into sentence constituents or phrases and their syntactic functional components. In this section, we will discuss the decomposition of assertions into these elements in order to determine how concepts-by-intuition can be formulated into assertions and which parts of assertions can indicate the concept-by-intuition that is represented.

---

[1] In the survey literature, the term "stem" of a question is used (Bartelds et al. 1994; Dillman 2000) in a similar manner to the term assertion, but the term "stem" is used for different meanings. Consequently, we prefer the term "assertion."

In linguistics, a simple assertion is decomposed into two main components: a noun phrase (NP) and a verb phrase (VP). An NP consists of one or more words with a noun or pronoun as the main part. A VP is a phrase that has one or more verbs. But next to the verb, VPs contain all the remaining words in the sentence outside of the NP, which can be complements, objects, or adverbials. The reader should be aware that we use here the definition of VP as employed in transformational generative grammar (Richards et al. 1993: 399). Example 2.6a might illustrate this:

> 2.6a   *Clinton   was a good president.*
>   NP   +   VP

Example 2.6a shows a *simple sentence or clause* where the NP is "Clinton" and "was a good president" is the VP. Although this decomposition into NP and VP is very common, a more detailed decomposition is more useful for our purposes. This decomposition is indicated in 2.6b and in all the following examples. One can always use the distinction between NP and VP but we will concentrate on the specific parts of these components[2]:

> 2.6b   *Clinton   was   a good president.*
>   Subject + predicator + subject complement

As example 2.6b illustrates, "Clinton" functions as the *subject* that indicates what is being discussed in the sentence. The *predicator* or the verb is "was" and connects the subject with the remaining part of the sentence, which is again a noun ("president") with an adjective ("good") as modifier of the noun. This specific remaining part expresses what the subject *is* and is therefore called a *subject complement*. Predicators that indicate what a subject is/was or becomes/became are called *link verbs (LV predicator)*. Other examples of verbs that can function as link verbs (connecting a subject with a subject complement) are "get," "grow," "seem," "look like," "appear," "prove," "turn out," "remain," "continue," "keep," "make," and so on (Koning and van der Voort 1997: 48–49). We suggest that the negations of these verbs are also classified as link verbs, for example, "not look like," "being unlike," and "being different from." According to the linguistic functions of the words, the sentence structure of example 2.6b can be formalized as structure 1:

> Structure 1: subject + LV predicator + subject complement

It can easily be shown that one can make different assertions that refer to different concepts using this structure. As an illustration, we will create different assertions using as subject "my work" and as link verb "is" while the subject complement varies across the examples:

> 2.7a   *My work is useful.*
> 2.7b   *My work is pleasant.*
> 2.7c   *My work is important.*
> 2.7d   *My work is visible.*

---

[2] The linguistic aspect of this section is based on the work of Koning and van der Voort (1997). We would like to thank Dr. Van der Voort for his useful comments on this chapter.

We can see through these examples that in changing the subject complement (which is each time a different adjective), the sentence refers to a different concept-by-intuition. These examples refer to an evaluation, a feeling, an importance judgment, and a neutral cognitive judgment. We will see later that structure 1 is the basic structure for assertions expressing evaluations, feelings, importance, demographic variables, values, and cognitive judgments.

A second relevant linguistic structure is illustrated in example 2.8a:

> 2.8a   *My mother   had washed   the clothes.*
>   Subject   +   predicator   +   direct object

This example has a subject ("my mother"), a predicator ("had washed"), and a direct object ("clothes"). Koning and van der Voort (1997: 52) define a *direct object* as the person, thing, or animal that is "affected" by the action or state expressed by the predicator. The linguistic structure of example 2.8a can thus be summarized as structure 2:

> Structure 2: subject + predicator + direct object

It can easily be shown through examples that in changing the predicator in this structure, we are changing the concept-by-intuition that the assertion refers to. In the examples, we always use "I" as subject and "my clothes" as direct object. By varying the predicator, we formulate different sentences that refer to different concepts-by-intuition:

> 2.8b   *I washed my clothes.*
> 2.8c   *I should wash my clothes.*
> 2.8d   *I shall wash my clothes.*
> 2.8e   *I prefer to wash my clothes myself.*
> 2.8f   *I hate to wash my clothes.*

Although the subject and the direct object remain the same, variation in the verb changes the meaning of the assertion. In the aforementioned sequence of appearance, the sentences refer to a behavior, a norm, a behavioral intention, a preference, and a feeling. Note that example 2.8e even displays a second direct object ("myself").

As we will show, structure 2 can be used to formulate relations, preferences, duties, rights, actions, expectations, feelings, and behavior, to name only a few examples. These will be discussed further in the following sections. Structure 2 has predicators called *lexical verbs* in linguistic terminology. This means that these verbs have specific meanings in contrast with link verbs (structure 1) that are only grammatical tools. Thus, the use of various lexical verbs in predicators explains to a great extent why the concepts change in these assertions. Sometimes, the lexical verb is preceded by an auxiliary verb such as "should" (2.8c) and "shall" (2.8d). Its function in 2.8c is to modify the lexical verb in the predicator into an obligation, and in this way, it contributes to the change of the concept-by-intuition. In example 2.8d, the auxiliary "shall" modifies the lexical verb into the future tense, and this contributes again to a change of the concept-by-intuition.

There is a third linguistic structure relevant to the context of expressing assertions. Example 2.9a illustrates its structure:

2.9a   *The position of the blacks*   *has changed.*
    Subject   +   predicator

Example 2.9a has a subject "the position of the blacks" and a predicator "has changed." In linguistics, these verbs that are not followed by a direct object are called *intransitive*. The basic structure of these assertions can be summarized in structure 3:

Structure 3: subject + predicator

It can be shown that the meaning of the sentences is easily changed by changing the predicator as previously in structure 2. However, the number of possibilities is much more limited because of the reduced number of intransitive verbs. Some examples are provided in the following:

2.9b   *I will go to sleep.*
2.9c   *I slept.*

Here, the subject is "I" and the first sentence (2.9b) indicates a behavioral intention, while the second (2.9c) is a behavior. Here are two more examples:

2.9d   *The position of blacks will change.*
2.9e   *The position of blacks has changed.*

In 2.9d, the subject is "the position of blacks," and the first sentence indicates a future event and the second, 2.9e, a past event. This structure is frequently used to present behavior, behavioral intentions, and past and future events.

So far, we have discussed the basic components of three possible linguistic structures of assertions that can be extended with other components, as will be explained in the next sections.

2.3.1   Indirect Objects as Extensions of Simple Assertions

The first extra components that can be added to the basic structures discussed earlier are indirect objects. An *indirect object* is defined as the person, and sometimes also the thing, that benefits from the action expressed by the predicator and the direct object (Koning and van der Voort 1997: 56). Examples 2.10a and 2.10b are an illustration:

2.10a   *Honesty   is*   *very important*   *to me.*
    Subject + LV predicator + subject complement + indirect object

Example 2.10a has structure 1 but an indirect object "to me" is added to it. Example 2.10b illustrates the same extension for structure 2:

2.10b   *He*   *bought*   *an apartment*   *for his mother.*
    Subject + predicator + direct object + indirect object

In this example, the subject "he" is connected by the predicator "bought" and followed by a direct object "apartment" and then an *indirect object* "for his mother." The general structure of this assertion is the same as structure 2 with the addition of an indirect object.

2.3.2   Adverbials as Extensions of Simple Assertions

Another component that can be added to the basic structure is an adverbial. An *adverbial* gives information about when, where, why, how, and under what circumstances or to what degree something takes place, took place, or will take place. Adverbials can occur in all three structures and can have quite different forms (Koning and van der Voort 1997: 59). Examples 2.11, 2.12, and 2.13 illustrate this:

2.11   *Clinton   was   president   from 1992 to 2000.*
       Subject + predicator + subject complement + adverbial

This is an extension of structure 1 with an adverbial indicating *when* it happened.

2.12   *My mother   had washed   the clothes   in the washing machine.*
       Subject + predicator + direct object + adverbial

This is an extension of structure 2 with an adverbial indicating *the way* it was done.

2.13   *He   worked   a lot.*
       Subject + predicator + adverbial

This is an extension of structure 3 with an adverbial indicating *a degree* of working.

2.3.3   Modifiers as Extensions of Simple Assertions

Another very common component attached to nouns is a modifier. A *modifier* specifies a noun. The modifiers can be placed before and after the noun and can be related to the subject but also to the object. Examples 2.14, 2.15, and 2.16 illustrate the use of modifiers for the three basic structures:

2.14   *The popular Clinton   was   president.*
       Subject (modifier + noun) + predicator + subject complement

This is an extension of structure 1 with a modifier for the subject "Clinton".

2.15   *My mother   had washed   the dirty   clothes.*
       Subject + predicator + direct object   (modifier + noun)

This is an extension of structure 2 with a modifier of the noun in the direct object.

2.16   *The son of my brother   died.*
       Subject (noun + modifier) + predicator

This is an extension of structure 3 with a modifier (of my brother) attached to the subject. The NP as a whole including the modifier is seen as the subject not just the main word in the phrase. For that reason, we have put the modifier and the noun in brackets because together they form the phrase mentioned before. In this way, the basic structure is immediately evident.

2.3.4   Object Complements as Extensions of Simple Assertions

Koning and van der Voort (1997: 54) define the object complement as a noun, adjective, or prepositional phrase that follows the direct object and expresses what the direct object is or becomes. Please see examples 2.17 and 2.18:

2.17   They   are driving   me   crazy.
        Subject + predicator + direct object + object complement

2.18   I   consider   him   as a friend.
        Subject + predicator + direct object + object complement

These structures of 2.17 and 2.18 are the same as structure 2 with an additional object complement "crazy" or "as a friend." Although this kind of expression occurs seldom in survey research, for the sake of completeness, it has been presented here.

2.3.5   Some Notation Rules

So far, we have described three distinct forms of assertions that are relevant for concepts-by-intuition in the social sciences.

Structure 1 of an assertion connects the grammatical subject (x) by means of a link verb (I) in the predicator to a subject complement (sc). The form of this assertion is denoted simply by (xIsc). In principle, the "sc" could be anything, but the most frequently occurring sc's are denoted as follows:

c   denotes a neutral judgment like "large/small," "active/passive," and "obvious"
ca   denotes a relation such as "(to be) the cause/reason/source of"
d   denotes a demographic variable like "age," "profession," and "date of birth/marriage"
e   denotes an evaluation like "good/bad," "valuable," and "advantageous/
         disadvantageous"
f   denotes a feeling or affective evaluation such as "nice/awful," "pleasant/
         unpleasant," and "happy/unhappy"
i   denotes "important," "interesting"
pr   denotes a preference such as "for/against" and "in favor/in disfavor"
ri   denotes a right like "permitted/allowed/justified/accepted"
s   denotes "similarity" or "dissimilarity" such as "alike/unlike" and "similar/dissimilar"

The subject (x) can also be represented by anything, but we use specific symbols for frequently occurring subjects for coding purposes:

g   stands for government or politicians
o   denotes anyone or everybody
r   denotes the respondent himself
v   denotes a value

Structure 2 is denoted by (xPy), where the grammatical subject (x) is connected by the lexical verb (P) to the predicator "y," which contains a direct object in the simplest form. Also the same subjects as mentioned previously can be applicable. In this structure, the predicators play a major role. Since there are some very frequently employed lexical verbs for predicators that relate to the intuitional concepts of social science, we will denote them with specific symbols:

C	indicates relationships where the subject causes or influences the object
D	indicates deeds such as "does," "is doing," "did," or "has done"
E	indicates predicators specifying expectations such as "expects" or "anticipates"
F	specifies feelings as links such as "like/dislike," "feel,"[3] and "worry about"
FD	indicates a predicator referring to future deeds such as "will," "intends," and "wishes"
(H+I)	specifies a predicator that contains words like "has to" or "should" and "is necessary," followed by an infinitive
HR	specifies predicators like "has the right to" or "is allowed to"
J	specifies a judgment connector such as "consider," "believe," and "think"
PR	indicates predicators referring to preferences such as "preferred to"
S	indicates relationships where a similarity (closeness) or difference (distance) between the subject and the object is indicated

Structure 3 for assertions will be denoted by (xP). Here, the predicator (P) and a subject (x) are present without a direct object. An adverbial can follow the predicator. The same choices can be made for the subject and the predicator as enumerated previously.

Having discussed the basic structures of simple assertions in general, the next section will discuss the characteristics of the typical assertions for the most commonly used concepts-by-intuition in survey research.

2.4 BASIC CONCEPTS-BY-INTUITION

In this section, we will describe how assertions that are characteristic of the basic concepts-by-intuition, employed in survey research, can be generated. Most researchers dealing with survey research (Oppenheim 1966; Sudman and Bradburn 1983; Smith 1987; Bradburn and Sudman 1988) make a distinction between factual or demographic requests, requests of "opinion" or "attitudes" and, where they arise, requests of knowledge and behavior. The terms *opinion* and *attitude* are often used in these studies for any type of subjective variables. "Attitude" is not discussed here because we consider attitudes as concepts-by-postulation. Since we want to make a distinction between different kinds of opinions, the term "opinion" itself is also not used in this book.

---

[3] Note that verbs such as "like," "feel," and "resemble" are linguistically mostly considered as linking verbs followed by a subject complement. However, we prefer to classify them according to their semantic meaning as feelings and similarity like lexical verbs. But the part that follows should grammatically always be considered as a subject complement.

In the sections that follow, the structure of the connected assertions is introduced for different concepts. We start with the so-called subjective variables.

2.4.1   Subjective Variables

By subjective variables, as stated, we understand variables for which the information can only be obtained from a respondent because the information exists only in his/her mind. The following basic concepts-by-intuition are discussed: evaluations, importance judgments, feelings, cognitive judgments, perceived relationships between the x and y variables, evaluative beliefs, preferences, norms, policies, rights, action tendencies, and expectations of future events. We begin with evaluations.

Evaluations are seen by most researchers as concepts-by-intuition of attitudes (Fishbein and Ajzen 1975; Bradburn and Sudman 1988; Van der Pligt and de Vries 1995; Tesser and Martin 1996). Their structure (xIe) generates assertions that certainly are expressions of "evaluations" (a_e). Typical for such assertions is that the subject complement is evaluative. Examples of evaluative words are good/bad, positive/ negative, perfect/imperfect, excellent/poor, superior/inferior, favorable/unfavorable, satisfactory/unsatisfactory, sufficient/insufficient, advantageous/disadvantageous, useful/useless, profitable/unprofitable, and lucrative/unlucrative. Examples 2.19 and 2.20 are typical examples of assertions indicating an evaluation:

    2.19   Clinton was a good president.

It is very clear that this assertion indicates an evaluation: the (x) is "Clinton," the evaluative subject complement (e) is "a good president," and the link verb predictor (I) is "was."

    2.20   Their work was perfect.

This is also clearly an evaluative assertion where the subject is "their work," the linking verb is "was," and the subject complement is "perfect." Using structure 1 combined with an evaluative subject complement ensures that the assertion created is an evaluation of the chosen subject.

Importance is the next concept to discuss. The structure of an "importance" assertion (a_i) is (xIi), which means "x is important." This assertion has the same form as the assertions indicating evaluations. The only difference is that the subject complement is in this case an expression of "importance." Example 2.21 illustrates this:

    2.21   My work is important.

"My work" is the subject (x) and "important" represents the subject complement (i), while "is" is the link verb (I). Values are often used as subjects. A value (v) can be defined as a basic goal or state for which individuals strive such as "honesty," "security," "justice," and "happiness" (Rokeach 1973; Schwartz and Bardi 2001). A typical example is:

    2.22   Honesty is very important to me.

In example 2.22, the (x) is the value "honesty," the predicator (I) is "is," and "very important" is the subject complement of "honesty," while "to me" is an indirect object. There is no doubt that assertions generated with structure 1 and an importance subject complement represent importance judgments.

Feelings or affective evaluations have in the past been considered as belonging to evaluations (Bradburn and Sudman 1988; Van der Pligt and de Vries 1995). However, more recently, a distinction has been made between cognitive evaluations and affective evaluations or feelings (Abelson et al. 1982; Zanna and Rempel 1988; Bagozzi 1989; Ajzen 1991). Three basic assertions can be formulated to express feelings. First, (a_f) can be in the form of (xIf) as example 2.23 illustrates:

2.23   *My work is nice.*

Example 2.23 reads as follows: (x) is "my work," (I) is the link verb predicator "is," and (f) is the *affective* subject complement "nice." It will become clear that other feeling words can be used as a subject complement, which will be discussed. However, structure 1 combined with a feeling subject complement generates an assertion that certainly expresses a feeling.

The second structure that can be used to express feelings is (xFy), which is an example of structure 2 discussed previously. An example is assertion 2.24:

2.24   *I like my work.*

In the case of 2.24, "I" is (x), the verb in the predicator "like" is a feeling (F), and "my work" is grammatically a subject complement (see note 3). There is no doubt that this assertion expresses a feeling toward "my work." It is also quite clear that other feelings can be expressed by using a different feeling verb like "hate" or any other feeling word, as in 2.25. Therefore, structure 2 with a predicator as a verb that expresses a feeling generates an assertion that represents a feeling.

The third possible structure is (xPy_f) as shown by example 2.25:

2.25   *Politicians make me angry.*

Example 2.25 reads as follows: (x) is "politicians," (P) stands for the verb form "make," while "me" is the direct object and "angry," expressing a feeling (f), is the object complement. This is one of the few examples of this structure in survey research. Nevertheless, combining structure 2 with a feeling object complement will generate an assertion that will also express a feeling.

Thus, (f) or (F) stands for feelings (fear, disgust, anger, sadness, contempt, shame, humility, hope, desire, happiness, surprise, etc. (Cornelius 1996)) that could be grammatically either *lexical verbs* (frighten, fear, scare, terrify, disgust, offend, repulse, enrage, infuriate, despise, disdain, reject, surprise, amaze, astonish, etc.) or *subject or object complements* (afraid, distressed, ashamed, angry, disappointed, happy, lucky, crazy, etc.).

With "f," the subject or object complement form is denoted and with "F," the lexical verb in the predicator is indicated. The use of "f" or "F" makes a difference in the structure of the assertion but not in the concept presented.

Cognitions have been discussed in the psychology literature as one of the basic components of an attitude (Krech and Crutchfield 1948; Bradburn and Sudman 1988;

Ajzen 1989; Eagly and Chaiken 1993; Van der Pligt and de Vries 1995). Two kinds of cognition have been mentioned in the literature. The first is a *cognitive judgment*. The structure of an assertion representing a cognitive judgment (a_j) is (xIc), which denotes that x has characteristic c. We use c to indicate that a specific type of subject complement must be used. Subject complements of cognitive judgments pertain to neutral connotations such as active/passive, small/big, limited/unlimited, aware/unaware, reasonable/unreasonable, usual/unusual, regular/irregular, ordinary/extraordinary, conservative/progressive, direct/indirect, big/small, slow/quick, left/right, planned/unplanned, practical/impractical, flexible/inflexible, heavy/light, and predictable/unpredictable. It is important to note that the main requirement is that the subject complements do not represent "evaluations," "feelings," and "importance." Example 2.26 displays a typical assertion of a cognitive judgment:

    2.26   Our family was large.

In 2.26, the subject complement (sc) is the neutral term "large." This example shows that structure 1 combined with a neutral subject complement will generate assertions that express cognitive judgments.

    The second concept in the class of cognitions is *a relationship* between a subject x and an object y. However, we need to make a distinction between two relationships: *causal relationships* and *similarity or dissimilarity and connectedness relationships*.

    Causal relationships are, for example, studied in attribution theory (Kelley and Michela 1980). There are two structures for causal relationships (a_c): structure 1 and structure 2. Structure 1 can be used if the subject complement indicates a cause (xCsc). Example 2.27 illustrates this possibility:

    2.27   New laws were the cause of the change of the position of black people.

There is no doubt that example 2.27 represents a causal relationship where "new laws" (x) is the subject, "were" (I) is the link verb, and "the cause of the change of the position of black people" is the subject complement (sc) with several modifiers.

    Structure 2 combined with *a causal or influence predicator* is also typical for assertions indicating a causal relationship. The formal structure can be represented by (xCy), which means (x has a causal relationship with y). Examples of cause or influence indicating lexical verbs are produce, bring about, provoke, create, replace, remove, alter, affect, accomplish, achieve, attain, or lead to. All are used in the sense of being the outcome or consequence of something. Note that relations are expressed by lexical verbs and not adjectives. Example 2.28 is an assertion that indicates a causal relationship:

    2.28   New laws have changed the position of black people.

Example 2.28 indicates a causal relationship where the (x) "new laws" have changed (C) "the position of black people" (y). This example demonstrates that structure 2 assertions with a causal predicator will always indicate a causal relationship.

    Other types of relationships frequently studied in social science refer to the *similarity/dissimilarity* or *distance/closeness* between objects (e.g., Stokes 1963; Rabinowitz et al.1991) or *connectedness between subjects* (Harary 1971; Helmers et al. 1975; Knoke and Kuklinski 1982; Ferligoj and Hlebec 1999). Examples include being attached to, resembling, being similar, identical/different, being like/unlike, and being close. To express such similarity relations through assertions (a_s), structure 1 can be used with a similarity or dissimilarity expressed in the subject complement (xIs) or structure 2 with a similarity or dissimilarity-expressing predicator (xSy). We start by illustrating the use of structure 1. An example of the relationship in the sense of membership is found in 2.29:

2.29 *He is strongly attached to the labor party.*

In example 2.29, the (x) is "He," the link verb predicator (I) is "is," and the subject complement is "strongly attached" followed by an indirect object "to the labor party." To indicate dissimilarity, one can use a negation of the assertion in 2.29 "do not resemble" or "are different from." Example 2.30 is an example of a dissimilarity assertion:

2.30 *Republicans are different from Democrats.*

In example 2.30, the (x) is "Republicans," the link verb predicator (I) is "are," and then follows the subject complement "different" with the indirect object "from Democrats," expressing the negation of similarity.

So far, we have shown that structure 1 can be used to express similarity relations. However, structure 2 can also be used for the same purpose as the following three examples illustrate. A first example is given in 2.31:

2.31 *European Liberals resemble American Conservatives.*

Here, the subject (x) "European Liberals" is said to "resemble," the predicator (S), and the (y) is "American Conservatives." The reader should be aware as stated previously (note 3) that we consider "resemble" as a lexical verb but the "y" (American Conservatives) that follows is grammatically a subject complement. A second example is presented in 2.32:

2.32 *Republicans differ from Democrats.*

In this example, "Republicans" are again the subject (x), the predicator indicating dissimilarity (S) is "differ from," and the direct object (y) is "Democrats."

Example 2.33 expresses the same concept-by-intuition by employing structure 3:

2.33 *Their opinions varied.*

In this example, "Their opinions" is the subject and the dissimilarity predicator is "varied."

Assertions about relationships indicate the views that respondents hold about the relationship between a subject and an object and not just about one subject. In this respect, relational assertions provide a different type of information than cognitive judgments, although both have been called *cognitions* in the academic literature as long as the assertions indicate neutral judgments.

Preferences are frequently asked in consumer research, in election studies, and in studies of policies where items from the most preferred to the least preferred are compared (Torgerson 1958; Von Winterfeld and Edwards 1986). The structure of a preference assertion (a_{pr}) is embedded in structure 2 with a lexical verb in the predicator, indicating preference, and denoted as xPRyz..., which means (x prefers y above z...) as in the example 2.34:

    2.34   I prefer the Socialist Party above the Conservative and Liberal Parties.

Here, "I" indicates (x), "prefer" is the preference verb (PR), the direct object (y) is "the Socialist Party," and the text "above the Conservative and Liberal Parties" indicates an object complement (z). As 2.34 demonstrates, several items are compared, and one is preferred to the others. Often, no explicit comparison is made but the assertion is based on an implicit comparison. Example 2.35 displays this form:

    2.35   I favor a direct election of the president.

In example 2.35, "I" indicates again (x), "favor" is the preference verb (P), and (y) contains only a direct object with a modifier "a direct election of the president." This assertion thus indicates explicitly the preference of a direct election of the president. Implicit in this assertion is the comparison with the opposite of direct elections, which are indirect elections.

    Another frequently occurring type of assertion indicating a preference in survey research pertains to structure 1 and is indicated by (xIpr). Examples 2.36 and 2.37 illustrate this:

    2.36   I am for abortion.
    2.37   I am against abortion.

In these examples, "I" indicates the subject, in this case, the respondent (r); the link verb predicator is "am"; while "for abortion" (2.36) and "against abortion" (2.37) are preference subject complements (p). In these cases, the explicit preference is expressed in the subject complement.

    Norms are also central to social research (Sorokin 1928; Parsons 1951; Homans 1965). Coleman (1990: 242) defines them as specifications of "what actions are regarded by a set of persons (o) as proper or correct." Structure 2 with an obligation indicating word (H) in the predictor followed by an infinitive (I) can be used to express a norm $(a_n) = (o(H + I)y)$, which means that someone should do something to the direct object (y). Example 2.38 illustrates this concept:

    2.38   Immigrants should adjust to the culture in their new country.

In example 2.38, the "immigrants" are the persons (o) for whom this norm holds; "should" stands for the obligation indicating part (H) of the predicator, which also contains the infinitive "adjust to" (H + I); while the direct object (y) with a modifier is "the culture in their new country." For norms, also structure 3 can be used as the following example illustrates:

    2.39   The children should go to sleep.

This assertion also indicates a norm, but does not contain a direct object. In that case, the structure indicates (o) "the children" and the predicator consists of the obligation indicating auxiliary (H) "should" and the infinitive (I) "go to sleep."

Policies are an important topic in political science research. They are used to determine what the public thinks about different measures of the government (Sniderman et al. 1991; Holsti 1996). A policy assertion (a_p) has the structure (g(H + I) y), which means (the government should do something for y). Example 2.40 displays a policy assertion:

2.40 The government should not allow more immigrants.

In example 2.40, "the government" is (g); the predicator is "should not allow," which contains the obligation indicating word "should" and the infinitive "allow"; while the direct object is " more immigrants."

Structure 3 can also be used with policies as example 2.41 illustrates:

2.41 The government has to resign.

In example 2.41, there is no direct object; therefore, structure 3 is applicable and the form is (g (H + I)). The only difference between norms and policies is that there is another subject. Norms are used for explaining the behavior of people (o), while policies indicate obligations for the government (g).

Rights, specifically requests for an answer dealing with civil right issues, are often queried in political science research (Sniderman et al. 1991). These perceived rights can be expressed using structure 1 where the subject is the matter at stake (e.g., abortion) and as subject complement (ri) an expression of permission such as "accepted," "allowed," or justified," which we will denote by (xIri). An example of this type of concept is the following:

2.42 Abortion is permitted.

However, rights can also be expressed using structure 2. Then, the assertion (a_{ir}) must contain a combination (oHRy), which means (someone has the right y). Example 2.43 illustrates our point:

2.43 Immigrants also have the right to social security.

The "immigrants" indicate (o) and "have the right to something" indicates the typical combination of the verb "have" and the direct object "the right to something" (HR). The "to something," in this case "social security," is a modifier within the direct object.

Action tendencies are often considered as the third component of an attitude (Ajzen and Fishbein 1980; Sudman and Bradburn 1983; Bradburn and Sudman 1988; Eagly and Chaiken 1993). An action tendency is what one intends to do in the future. The concept action tendency (a_t) can be represented in structure 2 or 3 where the predicator indicates a future deed of the respondent or (rFDy), which means r will do y. An example could be the following:

2.44 I want to go to the doctor.

Example 2.44 is a structure 3, where "I" is (x); "want to go" is the predicator (FD) indicating a future deed; and "to the doctor" is an adverbial. Structure 2 is also possible if the verb requires a direct object:

2.45 I will do my homework soon.

Example 2.45 uses structure 2 because there is a direct object (y) "my homework." It is followed by the adverbial "soon." In both cases (2.44 and 2.45), the most typical is the predicator that expresses a future deed of the respondent.

Expectations of future events (Graesser et al. 1994) are anticipations of events in which the respondent is not involved. The structure for an expectation (a_{ex}) is the same as in the case of action tendencies. The only difference is that the subject is not the respondent (r) but another grammatical subject (x). This means that the structure is xFD or xFDy. Examples are 2.46 and 2.47:

    2.46   The storm will come.

    2.47   The storm will destroy many houses.

So far, all assertions have been clear about the concepts that they are supposed to represent. There are, however, also assertions used for which the meaning is not so clear. This is sometimes done intentionally but more often than not by mistake. One of such types of assertions will be discussed subsequently under the heading "evaluative beliefs."

Evaluative beliefs (a_{eb}) can be represented by many different types of assertions. Typically, they have a *positive or negative connotation* (Oskamp 1991). Assertions presenting causal relationships are often used in this context. But because of their evaluative connotation, they indicate not only a causal relationship but also an evaluation of it. Therefore, they are called "evaluative beliefs." These assertions are indicated by a_{eb}. In case of a causal relationship, one structure is represented by (xCy_e). Example 2.48 illustrates this:

    2.48   The budget reform has led to prosperity in the United States.

The "budget reform" is (x), "prosperity in the United States" is (y), and "has led to" is the causal predicator (C.) The noun "prosperity" referring to object (y) is clearly a word with a positive connotation (e), and therefore, one can say that this statement also expresses an evaluation, besides the fact that it expresses a relationship, which is typical for evaluative beliefs (y_e). A slightly different form of an evaluative belief is that the relationship predicator (C) contains a positive or negative connotation that is indicated by $(xC_e y)$:

    2.49   The war destroyed a lot of buildings.

In example 2.49, the subject (x) is "the war," which "destroyed" (C_e) "a lot of buildings" (y).

Behavioral assertions, which will be discussed in more detail in the paragraph on objective variables, can also become evaluative beliefs. Example 2.50 illustrates this:

    2.50   The Netherlands prospered in the seventeenth century.

In this example, the predicator "prospered" expresses a past deed with a positive connotation (D_e), which makes it an evaluative belief with the form $(xD_e y)$.

These previous examples demonstrate that structures that do not contain explicitly evaluative terms can nevertheless indicate evaluative beliefs. In such a case, the assertion has to contain words with an evaluative connotation such as to prosper, prosperity, succeed, success, flourish, fail, failure, miss, loss, destroy, spoil, and kill.

Assertions indicating the concept "evaluative belief" can thus have the structure of several different assertions. Here, we have mentioned only causal relations and behavior. What makes these assertions indicate an evaluative belief is the evaluative connotation of some words. Without this evaluative connotation, the assertions cannot be seen as indicating "evaluative beliefs." Assertions, representing evaluative beliefs, have sometimes been used purposely by researchers to avoid socially desirable answers.

With this, we conclude our introduction to the concepts-by-intuition that fall under the subjective variables category. These assertions are based on information that can be obtained only from respondents, whose views cannot be verified because they are personal views that represent subjective variables.

2.4.2 Objective Variables

By *objective variables*, we mean variables for which, in principle, information can also be obtained from a source other than the respondent. One could think of administrations of towns, hospitals, schools, and so on. Commonly, these variables concern factual information such as behavior, events, time, place, quantities, procedures, demographic variables, and knowledge.

Behavior concerns present and past actions or activities of the respondent himself/herself (Sudman and Bradburn 1983; Smith 1987). Structures 2 and 3 with an activity-indicating predicator (D) can be used to specify the behavioral assertion (a_b). The typical form for structure 2 is (rDy), which means that the subject or respondent does or did y; with structure 3, it is (rD). It will be clear that the structure of this assertion is the same as the structure for an action tendency. However, its content differs fundamentally from the latter. Action tendencies deal with subjective matters (likely future behavior), while behavior is factual and in principle controllable. Examples 2.51 and 2.52 show this structure:

2.51   *I am studying English.*

2.52   *I was cleaning.*

In example 2.51, "I" stands for (r), "studying" is the action indicating predicator (D), and "English" is the direct object (y). In example 2.52, the subject "I" is again the respondent, while the action that indicates the predicator is "was cleaning." In this case, there is no direct object. Therefore, it is an example of structure 3, while example 2.51 employs structure 2.

The facts mentioned in these assertions can, in principle, be checked by observation as opposed to subjective variables such as a behavioral intention ("a person is planning to go to the hospital").

An *event* represents another example of an objective variable. It pertains to other people's actions that are presently ongoing or have occurred in the past. The structure of this assertion (a_{ev}) is the same as the previous one, except that the subject is not the respondent and therefore it is (xDy) or (xD). Examples of assertions characteristic to this concept are 2.53 through 2.55:

2.53   *My brother is studying.*

2.54   *My mother had washed the clothes.*

2.55   *The shopping center has been burglarized twice.*

In example 2.53, (x) is "my brother," "is studying" stands for the action predicator (D), and there is no direct object that makes it an example of structure 3. Example 2.54 belongs to structure 2. It has "my mother" as (x), the action predicator (D) is "had washed," and (y) is "the clothes." Example 2.55 belongs again to structure 3 with an adverbial as extension: (x) is "the shopping center," and "has been burglarized" represents the action predicator (D), which is followed by the adverbial "twice."

Demographic variables are used in nearly all surveys and are mentioned in all attempted classifications of data (Oppenheim 1966; Sudman and Bradburn 1983; Converse and Schuman 1984; Smith 1987; Bradburn and Sudman 1988). We represent demographic variables by the assertion (a_d). Structure 1 should be used for judgments (xId). The subject (x) is frequently the respondent or another person in his/her environment, but it differs from a judgment by the fact that the subject complement is limited to certain factual topics such as the respondent's gender, age, or occupation, summarized by (d). Examples 2.56 and 2.57 illustrate these assertions:

2.56   *I am 27 years old.*

2.57   *I am married.*

It will be clear that the structure of these assertions is the same as the one of an evaluation or a judgment. The only difference is the type of subject complement specified.

There are also assertions that relate to *knowledge* (a_k). They could ask, for example, who the 35th president of the United States was or which Russian leader had sent nuclear missiles to Cuba. The assertion to answer the first request would be structure 1, and the second would be structure 2. Examples 2.58 and 2.59 are examples of this type:

2.58   *Kennedy was the 35th president of the United States.*

2.59   *The Russian leader Khrushchev had sent nuclear missiles to Cuba.*

The structure of these assertions requires historical or political knowledge of the respondent. These knowledge assertions can have any structure for objective variables. Our first example reads as follows: "Kennedy" is (x), "was" stands for the link verb predicator (I), and "the 35th president of the United States" is the subject complement (sc). Therefore, the structure can be modeled as $a_k = (\text{xIsc})$.

The second example has the structure of an event: (x) is "the Russian leader Khrushchev," the action predicator (D) is "had sent," and (y) is "nuclear missiles," while "to Cuba" is an adverbial.

Often, information is requested in surveys about *time and place* of behavior or events. In an assertion, this information is presented by adverbials indicating time-/place-specific components. Examples 2.60 and 2.61 illustrate this:

2.60   *I worked yesterday.*

2.61   *I stayed in a hospital in Chicago.*

Thus, the focus shifts in these two examples from the act to the specification of the time (2.60) or the place (2.61).

The first assertion is a time assertion $a_{ti} = (rDti)$. It reads as follows: "I" is (r), "worked" is the behavioral connector (D), and "yesterday" is the time adverbial. The second example is a place assertion $a_{pl} = (rDpl)$, where "I" is (r), (D) is "stayed," and "in a hospital in Chicago" constitutes two place adverbials, indicated in the structure of the assertion by (pl). The reader may note that it is structure 3 that applies to time and place assertions.

Quantities can also be specified by structure 2. The assertion that can be formulated for quantities has the form $(a_{qu} = rDqu)$. Example 2.62 illustrates this:

2.62   *I bought 2 packs of coffee.*

In example 2.62, "I" stands for (r), "bought" is (D), and "2 packs of coffee" is (y), the direct object. "2 packs" indicates the quantitative information (qu) and the modifier "of coffee" specifies the substance.

Assertions concerning *procedures* can be formulated similarly using structure 3 as $(a_{pro} = (rDpl, pro)$. An example is 2.63:

2.63   *I go to my work by public transport.*

"I" is (r), "go" is (D), "to my work" is a place adverbial (pl), and "by public transport" is an adverbial that indicates the procedure (pro).

2.4.3   In Summary

In this review, we have described most concepts-by-intuition used in the survey literature. In these sections, we have tried to make the structure of these assertions explicit. Table 2.2 summarizes them.

We are aware that these concepts can also be expressed in different ways; however, the purpose of this exercise was *to suggest structures where there is no doubt that the generated assertions indicate the desired concepts-by-intuition*. Table 2.2 shows that some concepts can be presented in assertions with different structures. Further research is required to determine whether there is a difference in the responses for different types of linguistic structures.

The table can also be used to detect which kind of concept has been used in assertions applied in practice. This is a more difficult task because there are different extensions of these simple sentences. Some of these extensions or variations in formulations will be discussed in the following sections. These extensions will make the coding of requests for answers more difficult than the production of proper assertions for the presented concepts.

2.5   ALTERNATIVE FORMULATIONS FOR THE SAME CONCEPT

Grammar provides a variety of different ways of expressing the same proposition; this is what some linguists call "allosentences," which are found in particular syntactic constructions and certain choices between related words (Lambrecht 1995: 5). We can select a form that is appropriate according to where we want to place the

TABLE 2.2 The basic structures of simple assertions

Basic concepts		Structure 1 $xIsc$	Structure 2 xPy	Structure 3 xP
Subjective variables				
Evaluation	a_e	xIe	—	—
Importance	a_i	xIi		
Values	a_i	vIi	—	—
Feelings	a_f	xIf	xFy or xPf	—
Cognitive judgment	a_j	xIc	—	—
Causal relationship	a_c	$xIca$	xCy	—
Similarity relationship	a_s	xIs	xSy	—
Preference	a_{pr}	$xIpr$	$xPRy \, (z\ldots)$	—
Norms	a_n	—	$o\,(H+I)\,y$	$o\,(H+I)$
Policies	a_p	—	$g\,(H+I)\,y$	$g\,(H+I)$
Rights	a_{ir}	$xIri$	$xHRy$	—
Action tendencies	a_t	—	$rFDy$	rFD
Expectations of future events	a_{ex}	—	$xFDy$	xFD
Evaluative belief	a_{eb}	—	$xP_e y$ or xPy_e	xP_e
Objective variables				
Behavior	a_b	—	rDy	rD
Events	a_{ev}	—	xDy	xD
Demographics	a_d	xId	—	—
Knowledge	a_k	$xIsc$	xPy	xP
Time	a_{ti}	—	—	$xDti$
Place	a_{pl}	—	—	$xDpl$
Quantities	a_{qu}	—	$xDqu$	—
Procedures	a_{pro}	—	—	$xDpl, pro$

emphasis. Emphasis is placed mostly on new information in a sentence, but it also might be desirable to place it on parts that are assumed to be known or otherwise known as background information (Givon 1984: 251; Lambrecht 1995: 51). Some grammatical constructions that are syntactically different have the same content (Givon 1984; Huddleston 1997; Lambrecht 1995), but they add emphasis to different parts of the sentence. The constructions studied in this section occur frequently in survey requests and are called *active/passive* and *existential*.[4] We begin with an example of the *active voice*:

2.64   *New laws have changed the position of black people.*

---

[4] Linguists also mention the "cleft construction." This means that a single sentence is divided in two parts (cleft), each with its own predicator, while one is highlighted. To illustrate this, we give an example:

"It was new laws that changed the position of the black people"; or "it was the position of the black people that changed new laws." According to our experience, such constructions do not occur frequently in requests for answers; therefore, we discuss them only briefly in Chapter 3.

This assertion (2.64) means that the subject "new laws" is the so-called agent and the direct object is "the position of black people." If one reads this sentence, the emphasis seems to be on "new laws." If we change this assertion into the passive voice, we obtain example 2.65:

2.65   *The position of black people was changed by new laws.*

In the passive voice (2.65), the emphasis is on "the position of black people" that becomes the grammatical subject, while the agent becomes the adverbial "by new laws." To transform the passive assertion of example 2.65 into an existential construction, we need to put the word "there" at the beginning of the sentence and we obtain example 2.66:

2.66   *There has been a change in the position of black people due to
       new laws.*

In example 2.66, the subject "the position of black people" is substituted by "there," and the word "change" is highlighted.

The different formulations in examples 2.64 through 2.66 express the same concept, which is a relationship. But they emphasize different parts in the sentence. However, it is not clear how respondents react when they are confronted with these different forms. It can be that they pay attention only to the concept. On the other hand, they also can answer differently to the various grammatical forms. This is an issue that requires further empirical studies.

2.6   EXTENSIONS OF SIMPLE SENTENCES

Until now, we have focused on the basic structure of assertions; however, in reality, assertions have a lot of variation. They are expressed in sentences much longer than have been studied so far. Often, indirect objects, modifiers, or adverbials are added to the simple sentences. In this section, we will address this issue.

2.6.1   Adding Indirect Objects

An additional component that can be added to the simple sentences without changing the concept represented in the sentence is an indirect object. Examples 2.67 and 2.68, given previously, illustrate this:

2.67   *Honesty is very important to me.*
2.68   *He bought an apartment for his mother.*

These examples show that adding the indirect object component "to me" or "for his mother" does not change the concept the assertion refers to. The same holds true when modifiers are added to a sentence.

2.6.2   Adding Modifiers

As we stated previously, a *modifier* gives a specification to a noun. The modifiers can be placed before and after the noun and be related to the subject and to the object. Previously, some examples of this type were given as significant (2.14, 2.15, and 2.16). These examples demonstrated that normally, modifiers are not complications for the assertions. Whether we say "Clinton" or "the popular Clinton" or "dirty clothes" instead of just "clothes" will rarely lead to serious interpretation problems for most respondents. However, the modifiers can be longer; for example, "the most famous president of the United States" can be written instead of just "president." If both the subject and the object have a modifier, the meaning of the sentence can become quite complicated. Therefore, they should be used with moderation; they can be helpful but they can also lead to rather complex sentences.

2.6.3   Adding Adverbials

In contrast to the previous additions to sentences discussed, adding an adverbial will *change the concept* most of the time. The reason is that adding such an adverbial implies providing specific information that becomes the focus of the attention (Givon 1990: 712). Structure 3 sentences often contain adverbial components or just an adverb. For example:

  2.69   He worked full-time.

In this sentence, the emphasis is not on whether he did or did not work, but on the fact that he worked "full-time" and implicitly not "part-time." So, this assertion expresses something about his work and is still an assertion expressing demographic information. But in the following example (2.70), a change in concept takes place:

  2.70   He worked hard.

In adding the adverb "hard," the attention shifts from working or not working to "hard" or "lazy working," which expresses a cognitive judgment of one person about another. Take note that the concept has shifted from an objective variable to a subjective one. Examples 2.71 and 2.72 display concept shifts from objective to subjective variables, where the adverb has an evaluative (2.71), followed by an emotive (2.72) connotation:

  2.71   He worked very well.
  2.72   He worked with pleasure.

These sentences express an evaluation (2.71) and a feeling (2.72).

  In Section 2.4.2, we gave other examples of assertions for which the concept of intuition changed by adding adverbials with respect to time, place, quantity, or procedure.

2.7   USE OF COMPLEX SENTENCES

So far, we have discussed only simple sentences or clauses, with only one subject and predicator. In contrast, complex sentences consist of more subjects and predicators because of additional clauses. Examples 2.73a–2.73d illustrate assertions with complex clauses (where subj. = subject and pred. = predicator):

2.73a	*Immigrants*	*who come from Turkey*	*are generally friendly.*
	Subj.1	Subj.2 Pred.2	Pred.1
2.73b	*Abortion is*	*permitted if a*	*woman is raped.*
	Subj.1	Pred.1	Subj.2 Pred.2
2.73c	*While driving home,*	*he had*	*an accident.*
	Pred.1	Subj.2	Pred.2
2.73d	*The Social Democrats*	*performed better than*	*the Conservatives.*
	Subj.1	Pred.1	Subj.2

Examples 2.73a and 2.73b display two subjects and two predicators, as the definition requires. The reader may note that example 2.73a displays a complex clause where the second clause "who came from Turkey" is embedded, or nested, in the first one. In the other examples, the second clauses follow the first clause (2.73b–273d). There are thus two ways of joining sentences: linearly or by embedding them.

In example 2.73c, the first subject is missing but implied, since it is the same as in the main clause "he." Example 2.73d omits the second predicator, and it seems to be implied since it has the same meaning as the first. The sentence would read correctly with "than the Conservatives *did perform.*"

Complex sentences can be built from coordinate clauses by linking them with coordinating conjunctions such as "and," "but," or "neither," in which case they are considered the "same" level and called *main clauses*. Coordinate clauses can become rather problematic in survey research, as we will discuss in the following chapter, but from a linguistic perspective, this type of complex sentence is clear and therefore, we will concentrate on subordinate clauses in the next sections.

Examples 2.73a–2.73d expressed complex clauses consisting of a main clause and a subordinate clause. If the *subclauses* that are linked to the rest of the sentence by subordinating conjunctions ("who" 2.73a, "if" 2.73b, "when" 2.73c, "than" 2.73d) are omitted, then the remaining part is the main clause: "Immigrants are generally friendly" (2.73a), "Abortion is permitted" (2.73b), "He had an accident" (2.73c), or "The Social Democrats performed better in the elections" (2.73d).

At the beginning of this chapter, we discussed the grammatical elements of simple clauses, which were the subject predicator, direct object, indirect object, object complement, and adverbial. All these parts of sentences *except the predicator* can also be expressed by a subordinate clause in complex sentences (Koning and Van der Voort 1997: 84–90).We will illustrate this by an example:

2.74a   *Problems in Turkey caused emigration to Europe.*

2.74b   *Problems in Turkey caused Turkish people to emigrate to Europe.*

Example 2.74a is a simple clause of structure 2 (subject + adverbial + predicator + direct object + adverbial). In example 2.74b, the direct object + adverbial "emigration to Europe" are substituted by a subordinate clause "Turkish people to emigrate to Europe." It is thus characteristic of complex sentences that a component of a simple sentence is substituted by a subclause.

Having provided the necessary linguistic background to understand complex assertions, we will study them in more detail in the next sections.

2.7.1 Complex Sentences with No Shift in Concept

The simple expression "emigration to Europe" (2.74a) has been substituted by the more elaborate subclause "Turkish people to emigrate to Europe" (2.74b). This example illustrates that the meaning of the two assertions is similar but that the second formulation (2.74b) is much longer than the first. The subject (x) of assertion 2.74b is "problems in Turkey," which is followed by the causal predicator (C) "caused" and where the (y) is mentioned consisting of another assertion (a behavioral one (a_b)) that reads as follows: "Turkish people (x) emigrated (D) to Europe (y)." This interpretation of the assertion can be verified by asking: "what did the problems in Turkey cause?" Example 2.74 illustrates that the object in the previous assertion is substituted by another one. This complex assertion can be written more formally as (xRa_b). In this case, both assertions, the simple one and the complex one, represent the same concept (a relationship), but the second assertion (2.74b) is much more complex than the first (2.74a). Whether complexity of assertions makes a difference for the respondent is still an empirical question.

2.7.2 Complex Sentences with a Shift in Concept

Substitutions of the sentence components y or x that represent different concepts can be employed for nearly all assertions discussed previously. Previously, we gave an example where the complex and the simple assertion represented the same concept (2.74a,b). There are, however, cases where the two concepts present in the complex assertion are different. In the following, we provide several examples. A common example is the judgment of a relation. The relational assertion used (2.75) is one we have seen before:

2.75   *Problems in Turkey caused emigration to Europe.*

A judgment of this relation, a_r, is formulated in examples 2.76a and 2.76b:

2.76a   *That problems in Turkey caused emigration to Europe is quite certain.*
2.76b   *It is quite certain that problems in Turkey caused emigration to Europe.*

The equivalent meaning of the two linguistic variants of example 2.76 consists of the main sentence "(it) is quite certain" and the subordinate clause "that problems in Turkey caused emigration to Europe." However, the structure of these assertions (2.76a,b) is not (xIc) but $(a_r Ic)$. Therefore, the assertion (a_r) "problems in Turkey caused emigration to Europe" takes the place of the subject x, the predicator is "is," and the subject complement is "quite certain." By asking oneself "what is quite certain?" (2.76b), we can conclude that the subject "it" can be substituted by "that problems in Turkey caused emigration to Europe." The phrasing of example 2.76a is a clearer example of this type of assertion, but 2.76b can be classified in the same category. Krosnick and Abelson (1991) discuss the use of such complex assertions, in particular the certainty about an opinion as a measure of *opinion strength.*

Evaluations can also be formulated with respect to assertions. Example 2.77 illustrates this point:

    2.77   It is bad that the problems in Turkey caused emigration to Europe.

In 2.77, the structure is (a_cIe) and therefore, this is an evaluation of an assertion.

In the same way, importance judgments can be formulated (2.78):

    2.78   It is important to me that the Conservative Party continues to be strong.

While "that the Conservative Party continues to be strong" is an assertion on its own (a_e), in this statement, an assertion concerning importance is formulated (a_jIi). Krosnick and Abelson (1991) discuss the requests using this type of complex assertion also as measures of "opinion strength."

Feelings can be formulated in the same way. Example 2.79 begins with the judgment (a_j):

    2.79   Most immigrants are hardworking.

For this assertion (2.79), we can formulate an assertion for a feeling (2.80):

    2.80   I am glad that most immigrants are hardworking.

In Example 2.80, the subject complement "glad" is extended by the subclause "that most immigrants are hardworking," which functions as an adverbial within the subject complement and could be paraphrased by "about the hardworking immigrants." The structure of 2.80 is ($sIf\ a_j$).

As a last example, we show how a right is formulated on the basis of an evaluative belief in order to demonstrate the general approach. The evaluative belief $a_{eb} = (xD_ey)$ is illustrated by example 2.81:

    2.81   Immigrants exploit our social security system.

The assertion of a right (a_{eb} I ri) can then be formulated in example 2.82:

    2.82   It is unacceptable that immigrants exploit our social security system.

These examples have shown how this approach is used in general. Please keep in mind the complexity that can result from this approach. It is especially true when subject x and subject complement y are both substituted by assertions. Therefore, we do not recommend them for survey research, even though there is evidence that they are quite common in research practice. We did not include complex assertions in Table 2.2; however, the reader should be aware that any x and y mentioned in Table 2.2 can be replaced by a complete assertion. We did not include this option in the table because the main clause will still indicate the same concept whatever the concept in the subclause may be.

2.7.3   Adding Conditions to Complex Sentences

Another commonly used extension of an assertion is the *use of conditionals*. They express the circumstances under which something stated in the main clause can occur. They can express real or unreal things (Yule 1998: 123–152). In survey requests, both types of conditionals are used. Examples 2.83 and 2.84 show assertions with a real conditional:

> 2.83   *Abortion is permitted if a woman is raped.*
>
> 2.84   *If immigrants work harder, they will be as well-off as our people.*

Example 2.83 clearly expresses a woman's right to abortion if she has been raped. Formally, it can be summarized as (xHRy |con) where "|con" indicates the condition. Example 2.84 indicates a right depending on the prior occurrence of the "if" clause: ((xFDy) |con)

Also, sometimes, unreal events are expressed in complex sentences. This is shown by examples 2.85 and 2.86:

> 2.85   *If immigrants worked harder, they could be as well-off as our people.*

or

> 2.86   *If immigrants had worked harder, they could have been as well-off as our people.*

Clearly, the evaluative state ("they could be as well-off") in example 2.85 is unlikely because the "if" clause, describing the willingness of the immigrants to work harder, is in the past tense. In example 2.86, the evaluative state in the main clause ("they could have been as well-off as our people") is impossible only because the "if" clause expressed in the past perfect implies that the condition was not fulfilled.

It is difficult to understand what concept is represented by these assertions (2.85 and 2.86). Our best guess is that they represent two concepts: a relationship suggesting that hardworking immigrants will be as well-off as our people and the cognition that immigrants did not work hard, suggesting it is their own fault that they are in a worse situation. If researchers have difficulty in understanding what is being asserted by such assertions, it is very likely that the respondents will also be confused, which can lead to rather low reliability and validity scores. Nevertheless, assertions like this are not uncommon in survey research, as demonstrated in Table 1.1, item 3 (Chapter 1).

2.8   SUMMARY

This chapter has discussed three basic assertion structures that can be used to represent most concepts-by-intuition from the social sciences. We also have indicated how the most commonly applied basic concepts-by-intuition in survey research can be expressed with certainty in assertions specifying these concepts. These rules are summarized in Table 2.2. The knowledge summarized in Table 2.2 can be used in two ways.

The table can be used to specify an assertion for a certain type of concept according to the criteria specified there. For example, if we want to specify an evaluation about immigrants, we know that the structure of the sentence should be (xIe). Therefore, we can formulate a statement such as "immigrants are good people." If we want a feeling (xIf), we can write "immigrants are in general friendly." If we want a cognitive judgment (xIc), the statement is "immigrants are in general hardworking." If we want to formulate a cognition concerning the reasons why immigrants come here, the structure is (xRy), and a possible assertion would be "Problems in their own country cause emigration to Europe." In the same way, assertions can be formulated for any other concept.

Table 2.2 can also be used to detect which kind of concept has been used in assertions applied in practice. The elementary structures of the assertions refer in a simple way to the concepts mentioned. However, we have to say that the assertions can be made rather lengthy by use of complex sentences, subordinate clauses, time and place statements, and conditions. The use of such complicating possibilities can cause that the meaning of the assertions becomes much less intuitively clear than in the simple assertions. It is an interesting topic of further research to study what kinds of complications are possible without shifting the meaning of the request or assertion for the respondent.

Now that we have been able to specify standard assertions for the basic concepts of the social sciences, the task of the researcher would be to determine what type of basic concept his/her specific concept-by-intuition is. If that is done, sentences that represent with certainty this concept can be formulated. In the next chapter, we will indicate how these assertions can be transformed into questions or, as we call it, requests for answers.

EXERCISES

1. Formulate assertions concerning the al-Qaida network in terms of:
   a. A cognition
   b. An evaluation
   c. A feeling
   d. A relationship
   e. A norm
   f. A policy
   g. A behavioral intention
   h. A future event
   i. A behavior

2. Guttman (1981, 1986) suggested the use of facets designs to create measurement instruments. The facet design presented in Table 2.3 has been developed in discussions between the members of the International Research Group on Methodological and Comparative Survey Research (Saris 1996). The purpose of

TABLE 2.3   Facet design for ethnocentrism

Aspects of life	Judgment	Evaluative belief in/out group	Evaluation	Norms	Policies
Way of life					
Religion					
Economic					
Political					
Personal					

this table is to show that one can systematically formulate statements for different concepts-by-intuition mentioned above the columns. This can be done for the different aspects of life indicated in the rows of the table.

a. Can you specify an assertion for each cell of the table using our procedure?

b. Can the items in the rows be used to measure a concept-by-postulation?

c. Can the items in the columns be used to measure a concept-by-postulation?

3. Measurement instruments are not always carefully developed in research. Examples are the measurement instruments presented in Table 2.4.

TABLE 2.4   Operationalization of subtle and symbolic racism

Items

1. Os living here should not push themselves where they are not wanted.
2. Many other groups have come here and overcame prejudice and worked their way up. Os should do the same without demanding special favors.
3. It is just a matter of some people not trying hard enough. If Os would only try harder, they could be as well-off as our people.
4. Os living here teach their children values and skills different from those required to be successful here.
5. How often have you felt sympathy for Os?
6. How often have you felt admiration for Os?
7. How different or similar do you think Os living here are to other people like you?
   a. In the values that they teach their children
   b. In the religious beliefs and practices
   c. In their sexual values or practices
   d. In the language that they speak
8. Has there been much real change in the position of Os in the past few years?
9. Generations of slavery and discrimination have created conditions that make it difficult for Os to work their way out of the lower class.
10. Over the past few years, Os have received less than they deserve.
11. Do Os get much more attention from the government than they deserve?
12. Government officials usually pay less attention to a request or complaint from an O person than from our people.

O stands for member(s) of the out-group, which includes any minority group member(s).

    a. Indicate where the different items of Table 2.4 fit in the facet design presented in Exercise 2.

    b. Can these items in Table 2.4 be used to form a concept-by-postulation?

4. For the ESS pilot study, a proposal was made by Shalom Schwartz to measure basic human values. The suggestion for one of the items was as follows:

Here, we briefly describe some people. Please read each description and think about how much each person is or is not like you. Put an X in the box to the right that shows how much the person in the description is like you.

HOW MUCH LIKE YOU IS THIS PERSON?						
	Very much	Much	Somewhat	A little	Very little	Not at all
Thinking up new ideas and being creative is important to her. She likes to do things her own original way.						

    a. Specify the concepts that are present in this survey item.

    b. Check if these assertions represent the concepts they are supposed to represent.

    c. If needed, try to improve the survey item.

5. Check over your own questionnaire from Chapter 1 exercises to see:

    a. What the basic concepts-by-intuition behind your requests are

    b. If your assertions indeed reflect these concepts-by-intuition

3

THE FORMULATION OF REQUESTS FOR AN ANSWER

So far, we have discussed the distinction between concepts-by-postulation and concepts-by-intuition (Chapter 1). We also studied the way basic concepts-by-intuition used in survey research can be expressed in assertions (Chapter 2). In this chapter, we will continue with the discussion of how assertions can be transformed into requests for an answer.

While the choice of the topic of requests and the selection of concepts are determined by the research goal of the study, the formulation of questions or requests for an answer, as we call them, provides much more freedom of choice for the designer of a questionnaire. A great deal of research has been done on the effect of different ways in which requests are formulated (Schuman and Presser 1981; Billiet et al. 1986; Molenaar 1986). Also, a considerable part of the literature is devoted to devise rules of thumb for the wording of survey items (Converse and Presser 1986; Dillman 2000). On the other hand, relatively little attention is given to the linguistic procedures for the formulation of requests for answers in the survey literature.

Therefore, in this chapter, we will discuss different procedures to transform the assertions, discussed in the last chapter, into requests for an answer. In doing so, we make use of a large body of research in linguistics, especially Harris (1978), Givon (1984), Weber (1993), Graesser et al. (1994), Huddleston (1994), Ginzburg (1996), and Groenendijk and Stokhof (1997). The rules for the formulation of requests for an answer in English will be presented in the text, but in general, these formulation rules also apply in other languages such as German, Dutch, French, and Spanish. If they are different in one of the languages just mentioned, it will be indicated in the appropriate section by a note.

Design, Evaluation, and Analysis of Questionnaires for Survey Research, Second Edition.
Willem E. Saris and Irmtraud N. Gallhofer.
© 2014 John Wiley & Sons, Inc. Published 2014 by John Wiley & Sons, Inc.

3.1 FROM CONCEPTS TO REQUESTS FOR AN ANSWER

The term "request for an answer" is employed, because the social science research practice and the linguistic literature (Harris 1978; Givon 1990; Weber 1993; Graesser et al. 1994; Huddleston 1994; Ginzburg 1996; Groenendijk and Stokhof 1997; Tourangeau et al. 2000) indicate that requests for an answer are formulated not only as requests (interrogative form) but also as orders or instructions (imperative form), as well as assertions (declarative form) that require an answer. Even in the case where no request is made and an instruction is given or a statement is made, the text implies that the respondent is expected to give an answer. Thus, the common feature of the formulation is not that a request is made but that an answer is expected.

If an assertion is specified for a concept, the simplest way to transform it into a request for an answer is to add a prerequest in front of the assertion. This procedure can be applied to any concept and assertion. Imagine that we want to know the degree of importance that the respondents place on the value "honesty," as in examples 3.1a–3.1d:

3.1a *Honesty is very important.*
3.1b *Honesty is important.*
3.1c *Honesty is unimportant.*
3.1d *Honesty is very unimportant.*

To make a request from these assertions, prerequests can be added in front of them, as for example:

3.2a *Do you think that honesty is very important?*
3.2b *Do you think that honesty is important?*
3.2c *Do you think that honesty is unimportant?*
3.2d *Do you think that honesty is very unimportant?*

Using such a prerequest followed by the conjunction "that" and the original assertion creates a request called an *indirect request*. The choice of one of these possible requests for a questionnaire seems rather arbitrary or even incorrect as this specific choice of the request can lead the respondent in that direction. Therefore, a more balanced approach has been suggested:

3.2e *Do you think that honesty is very important, important, unimportant,*
       or very unimportant?

In order to avoid such an awkward sentence with too many adjectives, it is advisable to substitute them with a so-called WH word like "how," as in example 3.2f:

3.2f *Can you specify how important honesty is?*

This is also an indirect request with a prerequest and a subclause that starts with a WH word and allows for all the assertions specified earlier (3.1a–3.1d) as an answer and other variations thereof.

Instead of indirect requests, *direct requests* can also be used; the most common form is an interrogative sentence. In this case, the request can be created from an assertion by the inversion of the (auxiliary) verb with the subject component. The construction of direct requests by the inversion of the verb and subject component is quite common in many languages but also other forms can be used.[1]

Let us illustrate this with another example "vote intention," which is a behavioral intention. It can be formulated in an assertion of structure 2 (Chapter 2) with an auxiliary verb indicating that the action will be in the future. This leads to the following possible assertions:

> *3.3a   I am going to vote for the Social Democrats.*
> *3.3b   I am going to vote for the Republicans.*

One can transform these assertions into direct requests by inverting the auxiliary verb and the subject, while a simultaneous change from the first to the second person for the subject is also necessary. It leads to examples 3.4a and 3.4b:

> *3.4a   Are you going to vote for the Social Democrats?*
> *3.4b   Are you going to vote for the Republicans?*

Here, the requests can be seen as "leading" or "unbalanced" because they have only one possible answer option. It could be expected that a high percentage of respondents would choose this option for this reason. Therefore, the requests can be reformulated as follows:

> *3.5   Are you going to vote for the Social Democrats or the Republicans?*

A different way to formulate a direct request is also possible. We have seen that the point of interest is the party preference. Therefore, one can also omit the names of the parties in the request and place a "WH word" in front of the request. In this case, one is interested in the party preference that people intend to vote for. Hence, the words "Social Democrats" and/or "Republicans" are omitted and the WH word "what" followed by the more general term "party" is placed in front of the request, which leads to the following request for an answer:

> *3.6   What party are you going to vote for?*

The advantage of this format is that it is not biased to a political party, by mentioning only one possibility or giving first place to a party in the sentence word order.

This overview shows that two basic choices have to be made for formulating a request for an answer: the use of direct or indirect requests and whether to use WH words. The combination of these two choices leads to four different types of requests, which we will describe in the following; however, before doing so, we will discuss one other distinction.

Besides the interrogative form, two other grammatical forms of a request for an answer are also possible, the second of which is the imperative form. In its basic form, the request consists only of an instruction to the respondent, as for example:

---

[1] In French, it is also possible to place the question formula "Est-ce que" in front of a declarative sentence to indicate the interrogative form. Spanish, for instance, constitutes an exception since one does not have to use the inversion, as rising intonation of the declarative form is already enough. Interrogatives are indicated by two question marks, one in front of the clause (¿) and the other at the end of the clause (?).

> *3.7 Indicate the party you are going to vote for:*
>   *1. Republicans*
>   *2. Social Democrats*
>   *3. Independents*
>   *4. Don't know*

Example 3.7 illustrates that requests for an answer can also have another grammatical form than an interrogative one. Example 3.7 is colloquially known as an instruction or in grammatical terms, it is referred to as an "imperative." This is another example of a direct request for an answer.

The third grammatical form, a declarative request, is not possible as a direct request for an answer but only as an indirect request. Illustrations are examples 3.8 and 3.9. Both examples have a declarative prerequest, followed by a WH word and an embedded interrogative query:

> *3.8 I would like to ask you what party you are going to vote for.*
> *3.9 Next, we ask you whether you are going to vote for the Republicans or the Democrats.*

Although these are statements from a grammatical perspective, it is commonly understood that an answer to the embedded interrogative part of the sentence is required.

This overview shows that many different requests for an answer can be formulated to measure concepts like "the importance of the value of honesty" or "vote intention." However, it is important to note that whatever the request form used, there is no doubt that all these requests measure what they are supposed to measure. Therefore, there is no real difficulty with making an appropriate request for a concept if the assertions represent the concept of interest well. It only points further to the importance of the previous chapter in the whole process of designing requests.

3.2 DIFFERENT TYPES OF REQUESTS FOR AN ANSWER

After the introduction of the basic forms of requests for an answer, we will now examine how the different requests can be formulated in more detail. Table 3.1 summarizes the different types of requests for answers occurring in survey interviews according to their grammatical form and use in survey research. The table shows that not all theoretically possible combinations can be formulated; direct instructions with WH words are impossible because they automatically become indirect requests. Indirect requests with embedded declarative statements are also only possible without WH words because these subclauses have to begin with the conjunction "that" to be considered as declarative. We will discuss and illustrate the remaining options starting with direct requests.

3.2.1 Direct Request

We have already given several examples of direct requests. Therefore, we will be relatively brief about this type of request. We start with the direct instructions.

TABLE 3.1   Different types of requests for answers

WH words	Direct request	Indirect request	
WH word not present	Direct instruction	Imperative + interrogative	—
	Direct request	Interrogative + interrogative	Interrogative + declarative
	—	Declarative + interrogative	—
WH word present	—	Imperative + interrogative	—
	Direct request	Interrogative + interrogative	—
	—	Declarative + interrogative	—

3.2.1.1   The Direct Instruction

As was mentioned earlier, the direct instruction consists of a sentence in the imperative mood. This form is not so common in colloquial language but is quite common in written questionnaires or other written formats that one has to fill out for the government and other agencies. In this case, no request is asked but just an instruction is given as to what one has to do. Very common examples in very short form appear on application forms starting with:

> 3.10   *First name: ....*
> 3.11   *Family name: ....*
> 3.12   *Date of birth: ....*

Most people understand these instructions and write the requested information on the dots. In the case where less elementary information is asked for, more instruction has to be given and full sentences are used. Very common examples include the following:

> 3.13   *Select from the specified possibilities the one which fits your family situation best:*
> 1. *Single*
> 2. *Single with child(ren)*
> 3. *Married without child(ren)*
> 4. *Married with child(ren)*

Another example taken from social research is:

> 3.14   *Indicate your opinion about the presidency of Clinton by a number between 0 and 10, where 0 means very bad and 10 very good.*

Another example from mail questionnaires is also:

> 3.15   *Indicate your opinion about the presidency of Clinton by a cross on the line in the following text:*

Very bad   *Very good*

I_____I_____I_____I_____I_____I_____I_____I

In interviewer-administered questionnaires, these instructions are very uncommon and a more polite request form is preferred.

3.2.1.2 The Direct Request

As was mentioned before, a direct request contains an inversion of the (auxiliary) verb and the subject. The word order thus is changed. To illustrate, we start with the following assertions:

3.16   *The position of blacks has improved recently.*

3.17   *I prefer the Republican Party above the Democratic Party.*

From the first example, a direct request can be formed by putting the auxiliary verb "to have" in front of the subject:

3.18   *Has the position of blacks improved recently?*

This is called an *inversion* because the positions of the auxiliary verb and the subject are reverted. Example 3.17 in request form could be:

3.19   *Do you prefer the Republican Party above the Democratic Party?*

Here, the auxiliary is not the verb "to be" but "to do" and the subject is "you."

Direct requests can also be formulated as we have seen using WH words. So let us look at this approach a bit more carefully here. The linguistic literature (Lehnert 1977; Harris 1978; Chisholm et al. 1984; Givon 1990; Huddleston 1994) treats requests introduced by words like "who," "whose," "whom," "why," "where," "when," "which," "what," and "how" as a specific type of request. In English, they are also called *WH-interrogatives*. According to Givon (1990: 739), a specific request is used when the researcher and the respondent share some knowledge but a particular element is unknown to the former, and this component is the asked for element. This element is replaced by an interrogative word and constitutes the focus of the request, which means that the researcher wishes to draw special attention to it. In the previous section, we have seen that one reason to use WH words in front of direct requests is to avoid leading or unbalanced requests. The example was:

3.20   *What party are you going to vote for?*

The advantage of this form is that one cannot be blamed for giving an advantage to one of the two parties by mentioning only one or mentioning one party as the first in the request. However, there are other reasons to use WH-fronted requests. Fronting the word "when" realizes the request to ask for the *time* when the change occurred:

3.21   *When did this change occur?*

By asking a "where" request, one can determine the *place* where the change occurred:

3.22   *Where did this change occur?*

Finally, by asking a "why" request, one can determine the *cause* or *motives* of the change:

> *3.23   Why did this change happen?*

These examples show that WH requests are used to ask a specific element. Grammatically, the WH word stands in the beginning and mostly also a switching of the subject auxiliary occurs. The nature of the WH word determines the missing element. This fronting of the request word occurs in many languages with slight variations.[2]

There are more WH words that can be used. We shall return to this issue in the following sections.

3.2.2   Indirect Request

Indirect requests for an answer necessitate further discussion because they can come in many different forms as indicated in Table 3.1. We will discuss the different forms in sequence. Because it is rather natural to use in these indirect requests WH words like "whether" or "what" or "which," we will not completely separate the two types of requests but give examples with and without WH words.

3.2.2.1   Imperative–Interrogative Requests

As stated previously, an indirect request consists at least of two parts. The first part is the main clause and contains mostly a prerequest by which the researcher indicates a desire to obtain information about something in a neutral or polite way. The queried topic is then embedded and frequently presented by the second part, a subordinate clause. When a neutral prerequest is formulated as an order in the imperative mood, words like "tell," "specify," "indicate," show," and "encircle," are characteristic. The researcher signals by the use of these words to the respondent to inform him about something. The topic the researcher wants to know can, for instance, be specified by another main clause that is formulated as direct request. Examples 3.24 and 3.25 serve as an illustration:

> *3.24   Tell me, did you leave school before your final exam?*
> *3.25   Specify, what were your reasons for leaving school before your*
> *final exam?*

In the first example (3.24), the imperative is followed by a direct query characterized by the inversion of the auxiliary verb and the subject "did you." In the second example (3.25), the imperative is followed by a direct specific query initiated by the WH word "what" and the inversion of the auxiliary and subject "are the reasons." Note that in both requests, the requests are main clauses.

Requests can also be formulated using subordinate clauses. Some examples (3.26 and 3.27) are provided as follows:

> *3.26   Tell me if you left school before your final exam.*
> *3.27   Specify your reasons for leaving school before your final exam.*

---

[2] In French, additionally, the interrogative form "est-ce que" might be put after the specific question word.

Both examples show that the requests for answers are formulated as subordinate clauses and there is also no inversion present as is the case of direct requests. Example 3.26 has "if" as the conjunction of the subordinated clause. Since the subordinate clauses after "if, which, what, whether, who, etc.," function as indirect or embedded interrogatives, we call this kind of requests (3.24, 3.25, 3.26, and 3.27) of the form imperative + interrogative.

Since the researcher wants to elicit information from the respondent, the communication requires some politeness in the interaction. In order to make the prerequests in *imperative mood* more polite, researchers frequently add the word "please," as in 3.28 and 3.29:

> 3.28   *Please tell me did you leave school before your final exam.*
> 3.29   *Please specify your reasons for leaving school before your final exam.*

These examples demonstrate that the grammatical form has not changed, only the utterance is in a more polite tone. We have also shown that indirect requests can be formulated with and without WH words as is generally the case. Therefore, we will not emphasize this issue any further but discuss the use of the WH words in a separate section subsequently.

3.2.2.2   Interrogative–Interrogative Requests

Another way to make prerequests more polite is to change the grammatical form, namely, replace the imperative mood with the *interrogative mood* while formulating prerequests. Research has shown that there seems to be a linguistic continuum where prototypes of imperative forms gradually shade into polite interrogative forms (Chisholm et al. 1984; Givon 1990). In the following, we demonstrate how imperative prerequests in survey research can gradually change into more and more deferent prerequests in interrogative form. Examples could be:

> 3.30a   *Tell me whether you are going to vote in the next election.*
> 3.30b   *Please tell me whether you are going to vote in the next election.*
> 3.30c   *Will you tell me whether you are going to vote in the next election?*
> 3.30d   *Can you tell me whether you are going to vote in the next election?*
> 3.30e   *Can you please tell me whether you are going to vote in the next election?*
> 3.30f   *Could you tell me whether you are going to vote in the next election?*
> 3.30g   *Could you please tell me whether you are going to vote in the next election?*
> 3.30h   *Would you tell me whether you are going to vote in the next election?*
> 3.30i   *Would you please tell me whether you are going to vote in the next election?*
> 3.30j   *Would you like to tell me whether you are going to vote in the next election?*
> 3.30k   *Would you mind telling me whether you are going to vote in the next election?*
> 3.30l   *Would you be so kind as to tell me whether you are going to vote in the next election?*

The first two examples, 3.30a and 3.30b, are in the imperative mood. The remaining examples, 3.30c–3.30l, switch to the interrogative mood, characterized by the inversion of the auxiliary verb and the subject. They use different combinations of the modal auxiliaries such as "will," "can," "could," and "would," indicating that they are asking for permission to ask for something. They start with asking permission by "will," which is gradually more polite than the imperative and are followed by the use of "can," which is a bit more hesitant, where the addition of "please" increases the relative politeness of the sentence. Thereafter, the more polite and more distant form of "could" is introduced, which is again combined with "please" to increase politeness mood. Examples 3.30h–3.30l use the form "would," which is even less forward than the previous forms and therefore adds to the polite feeling. These examples show some gradations of politeness within its uses by adding "please" or combining it with "would you like," "mind," or "be so kind."

The reader may have noticed that logically the answer "yes" to any of these polite interrogative requests signifies that people are either willing to or can give the answer since it is formally related to the prerequest. Even in the polite form, respondents in general will suppose that they are asked to appraise the embedded request presented to them and therefore will answer "yes," meaning that they are going to vote in the next election, or "no," meaning that they are not going to vote. If it is anticipated that polite requests lead to confusion, it is better to avoid using them. An unusual variation of the aforementioned two types is illustrated by the following:

3.30m   *Which issue, tell me, will mostly influence your vote?*
3.30n   *Which issue, would you say, will mostly influence your vote?*

These examples show that the prerequests are placed within the clause that would normally be considered as an embedded subclause.

3.2.2.3 *Declarative–Interrogative Requests*
It is also possible to use *polite declarative* prerequests. They are presented in examples 3.31a and 3.31b:

3.31a   *I ask you whether you are going to vote during the next elections.*
3.31b   *I would like to ask you whether you are going to vote during the next elections.*

It is interesting to note that in examples 3.31a and 3.31b, no actual request is presented. Formally, the two texts are statements. As with the case of polite interrogative requests, research practice and conversational custom make it informally understood to listeners that they have to provide an answer to the embedded part in the sentence.

3.2.2.4 *Interrogative–Declarative Requests*
Finally, it frequently happens that in survey research, prerequests are formulated in the interrogative mood, and the embedded request, in the declarative form. Examples 3.32a and 3.32b illustrate this:

3.32a   *Do you think that the Republicans will win the elections?*
3.32b   *Do you believe that abortion should be forbidden?*

These examples show that the request is introduced by the declarative conjunction "that." The most common form of this type of request for an answer is illustrated by the next example:

3.32c   Do you agree or disagree that women *should have the right to abortion?*

This is a popular form because any assertion can be transformed directly into a request for an answer by adding a prerequest (e.g., "Do you agree or disagree" or "How much do you agree or disagree") in front of the statement. Respondents are often provided with whole series of assertions of this type.

3.2.2.5   *More than Two Parts*

We also stated that a request for an answer consists at minimum of two parts, which in practice can mean that more than one prerequest occurs, which can take all kinds of grammatical forms. Semantically, one of them can be either neutral or polite, and the other may convey a concept-by-intuition where the proper request follows. Examples 3.33 and 3.34 illustrate this:

3.33   *Please tell me whether you think that homosexuals should be given the same rights as heterosexuals.*

3.34   *I would like to ask you whether you can tell if you think that homosexuals should be given the same rights as heterosexuals.*

In example 3.33, the prerequest "please tell me" is a polite imperative, while the second prerequest, "whether you think," introduces a cognitive judgment in an embedded interrogative form followed by a declarative mood "that homosexuals should be given the same rights as heterosexuals," conveying a specific policy.

Example 3.34 illustrates a chain of three prerequests. The first, "I would like to ask you," is a polite declarative statement. The second, "whether you can tell," is a neutral prerequest in interrogative form, and the third, "if you think," relates again to an interrogative constituting a cognitive judgment. The main request is initiated by "that" and conveys a policy.

Here, it is important to state that while the sentences are becoming quite long, the main risk is that the proper request will fall to the background. In the next section, we will formulate some hypotheses concerning the possible effects of the consequences of length and complexity of sentences on the response.

3.3   THE MEANING OF REQUESTS FOR AN ANSWER WITH WH REQUEST WORDS

In all forms of requests for answers, WH request words can be used as we have explored in Section 3.2.1.2, which studied direct requests with a specific introductory word. In that section, it was also mentioned that these words refer to a specific aspect of an issue assuming that the basic idea is known. The example given previously was a request referring to the change of the position of black people in the United States:

3.35   *When* did this change occur?

This request for an answer presupposes that the respondent agrees that a change has occurred; otherwise, the request has no meaning:

3.36   Has the position of the blacks in the United States changed?

It is clear that example 3.35 asks for the objective concept-by-intuition "time," while example 3.36 measures the subjective "judgment" of possible change. Here we see that a change in meaning similar to that described in the last section occurs. The difference is that now the change in concept is not due to a prerequest but to a WH request word, and in general, the concept referred to by the WH word is clear, even though some of these words can imply many concepts in a request for an answer. The different meanings of the requests for an answer using the different WH words are the topic of this section, which will be discussed in sequence from simple to complex.

3.3.1   "When," "Where," and "Why" Requests

The most simple WH requests are the requests starting with the word "when," "where," or "why." These requests are simple because the words indicate only one specific concept. It is common knowledge that "when" asks for a time reference, "where" asks for a location, and "why" asks for a reason. We refer to Section 3.2.1.2 for examples.

3.3.2   "Who" Requests

"Who," "whose," and "whom" are used for asking information about a person or several people. "Who" and "whom" are pronouns that substitute for a noun, while "whose" can also be a determiner that occurs together with nouns, like "whose house." "Who" queries the personal subject. Examples of "who" that query the personal subject are:

3.37   Who is the new president of the United States?
3.38   Who is the most powerful person in the European Union?

Using *"whose"* signifies requests asking for ownership:

3.39   Whose house is this?

On the other hand, *"whom"* requests information about a personal object:

3.40   To whom did you sell the house?

3.3.3   "Which" Requests

The request word "which" is used for *preference* requests such as

3.41   Which party do you prefer?

or

3.42   Which car do you like the most?

It can also be used as an alternative for "who." In combination with "which one" in example 3.43, it refers to a definite set of persons (Givon 1990: 794):

3.43   Which one did it?

"Which" also can be used as an alternative for "why," "where," or "when" if it is used in combination with nouns like "reason," "country," or "period,". For example, instead of "why," one can use "which" to ask about *relations*:

    3.44   Which was the reason for the changes?

Instead of "where," one can use "which" to ask about *places*:

    3.45   In which area did the change take place?

Instead of "when," one can use "which" to ask about *time*:

    3.46   In which period did the change take place?

And instead of "how" requests (which will be discussed later), "which" requests can also be used to ask about *procedures*. For example:

    3.47   In which way do you solve your financial problems?

The reader should be aware that instead of "which," one can, in these cases, also use "what." This request word is the topic of the next section.

3.3.4 "What" Requests

"What" can be used in even more requests as it asks for the subject or the object. One very common use of "what" is in *demographic* requests such as:

    3.48   What is your family name?
    3.49   What is your highest education?
    3.50   What is your age?

It is also used in consumer research to ask for a specific aspect of the *behavior* of customers:

    3.51a   What did you buy?
    3.51b   What did you pay?

In time budget research or studies of leisure time, a more open request type of "what" is used to ask for *behavior*:

    3.52   What did you do (after 6 o'clock)?

"What" in combination with verbs like "cause" or nouns like "motives" or "goals" can also indicate a *relation*:

    3.53   What caused the outbreak of World War I?

"What" can also be used to formulate requests about subjective variables. For example:

    3.54a   What do you think of Clinton's quality as a president?

or

    3.54b   What do you think of President Clinton?

Note that example 3.54a asks for an evaluation. However, it is not clear what concept is measured in example 3.54b. This depends on the answer alternatives. If they are preset, they could be formulated in terms of various concepts. If they are not preset, it depends on what comes to the mind of the respondent at the moment of requesting.

3.3.5 "How" Requests

Special attention has to be given to requests using the term "how." This term can be used in different contexts and in still more ways. The following different uses of the request word "how" will be discussed:

> Measure of a procedure
> Measure of a relationship
> Measure of an opinion
> Measure of quantity
> Measure of extremity
> Measure of intensity

We start with the use of "how" when asking about *procedures*. The request word "how" can first be used to ask about *procedures* that people use to accomplish a certain task. Typical examples include:

> *3.55a   How do you go to your work?*

or

> *3.55b   How did you solve your financial problems?*

Examples 3.55a and 3.55b use the word "how" specifically and similar to the way words like "who," "where," and "when" are used in the previous sections.

A second application of the request word "how" is in requests about *relations* such as:

> *3.56   How did it happen that the position of black people changed?*

In this case, the request asks about the *cause* of the event mentioned. This request is rather close to the procedure request, but the former one asks for a "tool," while the latter one asks for a "cause."

The third application of the "how" request is an *open opinion* request such as:

> *3.57   How do you see your future?*

This request is similar to the open request we mentioned before when we discussed the "what" requests. In fact, often one can substitute "what" for "how."

The fourth use of the request word "how" is in requests about *quantities* and *frequencies* such as:

> *3.58a   How often do you go to the church?*

or

> *3.58b   How many glasses of beer did you drink?*

or

> *3.58c   How many hours a day do you watch television?*

We have put this use of the request word "how" in a separate category because the answer is specific, as in our examples the expected answer is a number. The following applications of "how" are similar, but with different answers.

A fifth application of the "how" request form relates to requests that ask about the *extremity* of an opinion. They modify the request word by an adjective or past participle. Typical examples are:

    3.59a   How good is Mr. Bush as president: very good, good, neither good
           nor bad, bad, or very bad?

or

    3.59b   How interested are you in politics: very interested, interested, a bit
           interested, or not at all interested?

In requests 3.59a and 3.59b, respondents are asked to give more details about their *opinion*. The "how" request form can also be indicated by *extremity*, which can be measured by objective variables. An example is:

    3.60   How many kilos do you weigh: under 50 kilograms, between 50
           and 60 kilograms, between 61 and 70 kilograms, or above
           70 kilograms?

We should mention that this can also be done by a direct request with answer categories. For example, we can ask:

    3.61a   Is Bush a very good, good, neither good nor bad, bad, or very bad
           president?

or

    3.61b   Are you very interested, rather interested, a bit interested, or not at
           all interested in politics?

or

    3.61c   Do you weigh under 50 kilograms, between 50 and 60 kilograms,
           between 61 and 70 kilograms, or above 70 kilograms?

It is unknown whether the direct request or the "how" request is better. However, some experiments have shown that requests with labels as responses are preferable if frequencies are asked for (Saris and Gallhofer 2004).

The sixth application of the "how" request asks for the *intensity of an opinion*. This type looks similar to the previous one, but an argument can be made that it represents a different request form (Krosnick and Fabrigar forthcoming). If this is the case, we do not ask how extreme an opinion is but how strongly people agree with an assertion. For example:

    3.62a   How strongly do you agree with the statement that Clinton was a
           good president?

or

    3.62b   How strongly do you believe that you will get a new job
           next year?

In such requests, the gradation is not asked with respect to the quality of the president or the likelihood of an event but with respect to the strength of an opinion. Therefore, it is called the *intensity* of an opinion.

Most of the specific requests have an equivalent translation in other languages. However, the word "how" has different meanings in romance languages like French and Spanish.[3]

3.4 SUMMARY

In this chapter, we focused on different linguistic possibilities to formulate a request for an answer. We called it a "request for an answer" because not only interrogative forms (requests) are used to obtain information from the respondents; imperative and declarative statements are also commonly employed. What the three request types share in common is that they ask the respondent to make a choice from a set of possible answers.

We have discussed several procedures. The first distinction we made was between direct and indirect requests. Direct requests consist of only one sentence, a request or an imperative, while indirect requests consist of a prerequest in the form of an interrogative, imperative, or declarative sentence with an embedded sentence that contains the real request. We also discussed specific requests introduced by particular request words such as "when," "where," "why," "which," "who," "what," and "how." These request words are used when the researcher wants to get specific information from the respondent about, for example, the time, place, or reason(s) of event. These possibilities are summarized in Table 2.1.

The most important result of this linguistic analysis is that one can formulate very different requests for an answer, while the concept, the topic of research, and the set of possible responses are the same. Logically, that would suggest that the requests provide the respondents with the same choice, and therefore, the requests can be seen as equivalent forms. However, we have to warn the reader that the possibility cannot be excluded that differences will nevertheless be found in the responses for the different forms because the difference in politeness of the forms may have an effect on the response. In Chapter 4, we will demonstrate how to formulate requests for answers that are linguistically very similar but measure different concepts.

By specifying all these different forms,[4] we tried to indicate the diversity of the possibilities to formulate requests for answers in survey research. Although linguists

---

[3] In French, for instance, "how" in procedure, relationship, and opinion requests is translated as "comment." For "how" in frequency requests, "combien," or "avec quelle fréquence," or "est-ce souvent que" is used. The extremity and intensity are expressed by "de quelle qualité est," "dans quelle mesure vous etes d'accord," and so on. In Spanish, "how" in procedure, relationship, and opinion requests is translated by "como." For "how" in frequency requests, "cuanto" is used. The extremity and intensity are expressed by "hasta que punto" or "hasta que grado" and "en que medida."

[4] Linguists (Chisholm et al. 1984; Givon 1984; Huddleston 1988, 1994) also discern some other types of questions that are, in our opinion, typical for normal conversation but not for requests for answers in survey research. To these questions, for instance, belong the so-called "echo questions" that repeat what has been said before because the listener is uncertain about having understood the question well. An example could be: "Am I leaving tomorrow?" "Multiple questions" are also used frequently in conversation (Givon 1990: 799) such as "who said what to whom?" In English, interrogative tags also are quite common such as "he left alone, did not he?" It is clear that such constructions are too informal and therefore are preferably avoided in survey research.

suggest that in many cases the meaning of the requests is the same, this does not mean that respondents will perceive these requests as identical and that they will reply in the same way.

Without claiming that all requests for answers fit in the system developed in this chapter, we think that it is useful to keep these possibilities in mind when formulating requests for answers and analyzing requests for answers to clearly grasp the diverse grammatical forms and the potential differences in meaning of the requests.

EXERCISES

1. For the following two concepts-by-intuition, derive assertions representing these concepts and transform these assertions into different requests for an answer:
   a. Trust of the government
   b. The occupation of the respondent

2. Two requests for an answer have been mentioned in the following:
   a. Is it the position of black people that has changed?
   b. Is it the position of black people that has changed by new laws?
      • What are the potential answers to these requests?
      • Do these answers mean the same?
      • Why is there a difference?

3. How would you formulate a request about:
   a. A perception if women have the right of abortion
   b. The norm that women should have this right
   c. The evaluation of this right
   d. An importance judgment of this right

4. Finally, check for your own questionnaire whether the transformation of the concepts-by-intuition in requests for an answer was done in the proper way. Should you change some requests?

PART II

CHOICES INVOLVED IN QUESTIONNAIRE DESIGN

Part I discussed the basic steps needed to formulate requests for an answer in order to operationalize the measurement of the desired concepts. Part II will show that in survey research many more choices have to be made to design a questionnaire. The following issues will be discussed in sequence:

1. The different ways requests for an answer can be formulated (Chapter 4)
2. The choice of the response alternatives (Chapter 5)
3. The structure of open-ended and closed survey items (Chapter 6)
4. The structure of batteries of survey items (Chapter 7)
5. Other choices in survey design such as the order and layout of the questions and the choice of the data collection method (Chapter 8)

Design, Evaluation, and Analysis of Questionnaires for Survey Research, Second Edition.
Willem E. Saris and Irmtraud N. Gallhofer.
© 2014 John Wiley & Sons, Inc. Published 2014 by John Wiley & Sons, Inc.

4

SPECIFIC SURVEY RESEARCH FEATURES OF REQUESTS FOR AN ANSWER

Chapter 3 examined the various linguistic structures of requests for answers. In this chapter, we will discuss features of requests for an answer that are important with respect to their consequences for survey research. Hence, we will first look at the characteristics of requests that cannot be changed by the researcher because they are connected with the research topic. Then, we will discuss some features that the researcher has influence over, such as the choice of the prerequest and the use of batteries of requests for an answer with the same format. So far, we have discussed only single requests, but if batteries are used, the form of the requests changes significantly.

Other issues that social scientists are concerned with include whether the request is balanced in the sense that equal attention is given to positive and negative responses in the request and whether absolute or relative judgments are asked, as well as whether a condition should be specified within the request. Finally, the request for an answer can include "opinions of others," or "stimuli to answer," or emphasize that a "personal opinion" is asked. In the following sections, these characteristics will be discussed in detail.

4.1 SELECT REQUESTS FROM DATABASES

So far, we have suggested the following method to develop a request. First, it is crucial to determine what needs to be studied, for example, "the satisfaction with the work of the present government" or "the amount of hours people work normally." The first concept is a feeling about the government, and the second is a factual request

Design, Evaluation, and Analysis of Questionnaires for Survey Research, Second Edition.
Willem E. Saris and Irmtraud N. Gallhofer.
© 2014 John Wiley & Sons, Inc. Published 2014 by John Wiley & Sons, Inc.

about the work. Next, typical assertions for these concepts (Chapter 2) need to be specified like the following two examples:

> *4.1a   I am (very) (dis)satisfied with the work of the present government.*
> *4.2a   Normally, I work x hours.*

The last step is to transform these assertions in requests for an answer (Chapter 3), for example:

> *4.1b   Are you satisfied or dissatisfied with the work of the present*
> *government?*
> *4.2b   How many hours do you work?*

This process has little margin for error. The requests measure what was planned to be measured. However, there are other ways of obtaining requests.

For many topics, requests already exist in archives such as the one in Cologne (Germany), Essex (United Kingdom), or Ann Arbor (United States) or "question banks" such as the one of CASS in Southampton. Mostly, the requests are ordered in some type of classification. However, beware that the classification has to be very detailed in order to find the proper requests. For example, the following classification can be used as a first step:

1. *National politics*
2. *International politics*
3. *Consumption*
4. *Work*
5. *Leisure*
6. *Family*
7. *Personal relations*
8. *Race*
9. *Living conditions*
10. *Background variables*
11. *Health*
12. *Life in general*
13. *Other subjective variables*

This first step in classification is not detailed enough because a large number of requests concerning national politics (the first topic) and concerning work (the fourth topic) exist. Therefore, be prepared to invest some time in searching the exact measure of the desired concept. The criterion to evaluate whether a request measures what was intended to be measured is the same as was discussed in the first three chapters. If a concept-by-intuition is studied, a direct request is possible, and Chapters 2 and 3 are applicable. If a concept-by-postulation is being studied, first determine what concepts-by-intuition form the basis for this more abstract concept and then find their direct measures as discussed in the previous two chapters. The most important criterion is, of course, that the possible answers represent assertions that are obvious assertions for the chosen concepts-by-intuition. Chapter 2 provides ample suggestions for this type of check.

4.2 OTHER FEATURES CONNECTED WITH THE RESEARCH GOAL

Directly connected with the research goal and consequently with the choice of concept are some other characteristics of the survey items: the *time reference, social desirability*, and *saliency* or *centrality*. We start with the time reference.

Requests for answers regarding the present situation: feelings at the moment or satisfaction with different aspects of life or opinions about policies, norms, or rights. Requests can also be directed to future events or intended behavior. One can ask whether one will buy some goods in the future or will support some activity or expect changes or events, for instance. Finally, survey items can be directed to the past, asking whether one has bought some thing last week or whether one has been to a physician, dentist, and hospital during the last year. It will be clear that the time period mentioned in the request—past, present, or future—is completely determined by the goal of the research, and the designer of the study normally has no possibility to change this time period. Only the requests about the past give a bit more freedom to the researcher. Let us look at this issue a bit more closely.

The time period indicated in requests is called the *reference period*. It will be clear that the longer the reference period is, the more unlikely it is that one can reproduce the requested information from memory. This holds especially for activities that occur very frequently, for example, media use. For that reason, and as an alternative, researchers use requests about yesterday. Hence, instead of the request in example 4.3a, they ask example 4.3b:

4.3a   *How much time did you spend watching programs on politics or actuality last week?*

4.3b   *How much time did you spend watching programs on politics or actuality yesterday?*

But, because requests like 4.3b lead to unusual results for at least some people, one also asks the request of example 4.3c:

4.3c   *How much time do you spend watching programs on politics or actuality on a normal day?*

It is unclear what time period is used in this request. One could say that the respondent is asked for his/her normal behavior at present. Such a shift in time is of course only possible if the research goal allows it.

One more problem should be mentioned concerning requests referring to the past. It is well-known from research that people have a tendency to see events as closer to the date of the interview than is true in reality. This phenomenon is called *telescoping* (Schuman and Presser 1981). A typical request that reflects this problem is shown in example 4.4:

4.4   *Have you experienced robbery or theft during the last year?*

Respondents are inclined to mention many more cases than should be reported. Scherpenzeel (1995) found that the reported number of cases is twice as high using this request (4.4) than when one asks two requests illustrated by examples 4.5a and 4.5b:

4.5a   *Have you experienced robbery or theft during the last 5 years?*

4.5b   *How about the last year?*

It seems that people can better estimate the point in time if first a larger reference period is mentioned (4.5a) than in a one-step procedure like 4.4.

In general, the designer of a questionnaire has little flexibility with respect to the specification of the time period mentioned in the requests. Basically, he/she has only a choice with respect to the reference period that will be mentioned.

A second characteristic that is directly connected with the choice of the concept is the *social desirability* of some responses. As an example, we can mention that using the direct request about political interest, it is socially desirable for some people to answer that they are interested even if they are not. This happens because the respondents want to make a good impression on the interviewer. This means that differences in responses can be expected between surveys using interviewers and studies that do not use inter-viewers. So, for sensitive requests, differences are expected between personal or tele-phone interviews and mail surveys and other self-completion procedures. For requests about criminal and sexual behavior, very large social desirability effects have been found in this way (Aquilino 1993, 1994; Turner et al. 1998). This suggests that in a study where social desirability can play an important role, one should consider using a data collection method that reduces the effect of social desirability as much as possible.

The third characteristic that is directly connected with the choice of a concept is the *centrality* or *saliency* of the necessary information to answer the requests. In the past, the idea was that people have an opinion about many issues stored in memory that they just had to express in one of the presented response alternatives. Nowadays, researchers have a different view on this process, thanks to the important work of Converse (1964), Zaller (1992), and Tourangeau et al. (2000). It is more likely in many situations that people create their answers on the spot when they are asked a request. They will do that on the basis of all kinds of information that they have stored in memory, and it depends on the context of the request, recent events, and their mood which information will be used and therefore what answer will be given. As a consequence, one can expect quite a lot of variation in answers to the same request at different points in time (Converse 1964; Van der Veld and Saris 2003).

However, one should not exaggerate this point of view. There are requests where most people give more or less the same answer all the time, for example, requests about their personal lives, backgrounds, and living conditions. There are also topics about which some people have rather stable opinions and others do not. For example, with respect to political issues, some people who are very interested and follow what is going on have a clear opinion; there are, of course, also people who are not at all interested in politics and are, therefore, more likely to provide different answers if they are forced to answer requests about these issues. This does not mean that this division will always be the same. It may be that the people, who know nothing about politics, know a lot about consumer goods and education where the politically interested respondents do not know much about these issues. So the saliency of opinions depends on the topic asked and the interest people have in the specific domain of the survey items.

4.3   SOME PROBLEMATIC REQUESTS

Besides the problems unavoidably connected with the research topic, there are also problems that can be avoided such as the so-called double-barreled requests and assertions with more than one component. We will also indicate how to correct them in order to improve the comprehension of the respondents. These complications are also mentioned by Daniel (2000) in his request taxonomy.

4.3.1   Double-Barreled Requests

In the literature about survey research, the problem of requests with several concepts has been extensively discussed (Converse and Presser 1986; Fowler and Mangione 1990; Lessler and Forsyth 1996; Graesser et al. 2000a,b). An example of such a so-called double-barreled request could be:

4.6a   *How do you evaluate the work of the European Parliament and the Commission?*

The problem with such a request with two object complements in 4.6a (the work of the European Parliament and the Commission) is that two simultaneously opposing opinions are possible: a positive opinion about the Parliament and a negative opinion about the Commission. This leads to confusion about how to answer the request. Linguistically, this is a complex sentence built up with the coordinate conjunction "and," and as we stated in Chapter 2, in this case with two different subjects, it can become problematic. To avoid this problem, two requests, each containing one of the object complements, are a solution:

4.6b   *How do you evaluate the work of the European Parliament?*
4.6c   *How do you evaluate the work of the European Commission?*

Another example of two concepts in one request is the following:

4.7a   *Do you agree with the statement that the asylum seekers should be allowed into our country but should adjust themselves to our culture?*

Although such a statement is not unusual in colloquial speech, it can create problems for clear answers in surveys. The reason is that the first part of this statement is a right but the second part is a norm. It is again quite possible that a person is opposed to immigration but thinks that immigrants should integrate once they have entered a country. Again, this respondent can be perplexed about what answer to provide to this request. Splitting this statement into two separate requests creates clarity:

> *4.7b   Do you agree with the statement that asylum seekers should be*
> *allowed into our country?*
>
> *4.7c   Do you agree with the statement that if asylum seekers come to our*
> *country, they have to adjust themselves to our culture?*

The previous examples 4.6a–4.7c showed the problem of double-barreled requests. There are also double-barreled requests that work as intended, as a study for some items of the human value scale of Schwartz (1997) demonstrated. The items are formed by a combination of a value and a norm or a feeling. An example is the following request:

> *4.8   How much are you like this person?*
> *Looking after the environment is important to him/her. He/she strongly*
> *believes that people should care for nature.*

In this case, the importance of a value and a norm is combined in a complex assertion of similarity. This is in principle a typical example of a double-barreled request, but if we ask the two assertions separately with the same prerequest, the correlation between the answers (after correcting for random errors) is so high (.95) that one can assume that these two assertions measure the same (Saris and Gallhofer 2004).

The aforesaid is an interesting example showing that double-barreled requests do not always have to be problematic. However, one should be aware that they can cause problems and should be used only after a careful study of the consequences. In general, such requests can be very confusing for respondents.

4.3.2   Requests with Implicit Assumptions

There are also requests for answers that assume a first component that is not literally asked but is implicitly true in order to respond to the second component. An example could be:

> *4.9a   What is the best book you read last year?*

Here, the hidden assumption is that the respondents actually read books. People who do not read books can be unsure about how to answer this request. If the hidden component is made explicit in a separate request, the problem is resolved:

> *4.9b   Did you read books last year?*

If yes:

> *4.9c   What is the best book you read last year?*

Sometimes, the previously discussed hidden assumption in the first component is stated explicitly in the request, but the focus for answering is on the second component (Emans 1990) such as in example 4.10:

> *4.10   Did you read books last year and what is the best book you read?*

Again, respondents who do not read books will be confused about how to answer the request. Again, the remedy is to split these two requests into two separate requests.

4.4 SOME PREREQUESTS CHANGE THE CONCEPT-BY-INTUITION

Although it is possible to transform assertions in many different ways into requests for answers, it is not always risk-free. In the previous chapter, we have discussed prerequests referring to words such as "saying," "telling," "asking," and "stating," which were used to indicate a simple transfer of information. They did not refer to specific concepts-by-intuition as described in Chapter 2, which might differ from the concept used in the request for an answer. Hence, it can be concluded that using these verbs will not change the concept-by-intuition, as this is shown in the following four different assertions in direct request format:

> 4.11a   *Has the position of black people changed in the last 30 years?*
> 4.11b   *Was Clinton a good president?*
> 4.11c   *Should women have the right to abortion?*
> 4.11d   *Did you live with your parents when you were 14 years old?*

In sequence, these requests represent a judgment (4.11a), an evaluation (4.11b), a right (4.11c), and a behavior (4.11d). However, if at the beginning of the request "tell me," "may I ask," or any other prerequest is combined with any of the abovementioned neutral verbs, the concept measured will not change.

However, be careful with prerequests of survey items such as "think," "believe," "remember," "consider," "find," "judge," "agree," "accept," "understand," and "object," which refer to a cognitive judgment. Linguists like Quirk et al. (1985: 1180–1183) independently classified these verbs in a similar way. One would think that using such verbs in the prerequests would change the concept measured, but it does not always happen, as can be seen in the next three examples:

> 4.12a   *Do you think that the position of black people has changed in the*
>   *last 30 years?*
> 4.12b   *Do you think that Clinton was a good president?*
> 4.12c   *Do you think that women should have the right to abortion?*

There are also verbs that measure feelings such as "like" and "enjoy." If such verbs are used in prerequests in the same way, the concept may change to a feeling about a concept. Examples 4.13a–4.13c illustrate this:

> 4.13a   *Do you like that the position of black people has changed in the*
>   *last 30 years?*
> 4.13b   *Do you like that Clinton was a good president?*
> 4.13c   *Do you like that women should have the right to abortion?*

The structure of the requests is exactly the same, only the meaning of the verb is changed from "think" to "like" (4.12–4.13).

The same effect occurs with adjectives that refer to other concepts like "importance" or "certainty." In the following examples, we see that the concepts asked in the indirect requests are different from the concepts in the direct requests mentioned so far.

> *4.14a   Is it important for you that the position of black people has changed*
> *in the last 30 years?*
> *4.14b   Is it important for you that Clinton was a good president?*
> *4.14c   Is it important for you that women should have the right to abortion?*

These examples clearly indicate that one has to be careful with a change from a direct request to an indirect request for substantive reasons. By selecting an indirect request, the concept-by-intuition measured in the request can change in agreement with the concept expressed in the verb or adjective of the prerequest. That is not the case with the neutral terms that we have used in the previous sections, but this occurs with less neutral terms and not always as we saw in the changed verb examples (4.12a–4.12c) "think" and " believe" and similar ones that measure judgments. This is still an area where further research is needed to investigate when the concept measured changes and when it does not.

In Chapter 2, we mentioned that terms added to an assertion can change the concept. Thus, using a prerequest that is introducing a different concept-by-intuition than the concept connected to the embedded query is referred to as a *complex assertion*. As was stated before, complex concepts seem to confuse people leading to lower reliability of responses (Saris and Gallhofer 2004) and should be avoided if possible.

4.5 BATTERIES OF REQUESTS FOR ANSWERS

In survey research, many requests are asked, one after the other in series. If they are in similar form or can be made similar, then the whole process can be simplified by the use of *batteries of requests*. In batteries, the entire request and answer categories including the introduction, the request in the broadest sense, and the eventual components after the request such as instructions are mentioned before the first stimulus or statement. Subsequently, one stimulus or statement after the other follows without repeating the request and the answer categories, since it is assumed that the respondent already knows them. Written questionnaires present stimuli and statements often in table format where the stimuli or statements are presented in rows and the answer categories or rating scales, in columns. We will call this kind of structure a "battery of requests for answers." The difference between stimuli and statements is that statements are complete sentences, while stimuli do not consist of complete sentences. They can contain a noun, a combination of nouns, or another part of a sentence or a subordinate clause.

From the aforesaid, one can conclude that requests for answers with stimuli or statements are quite different from the requests for answers studied in Chapter 3 because they occur in series. The consequences of this approach, which is typical for survey research, will be discussed in later chapters. Here, we want to present the structure of batteries and to discuss some of the choices that have to be made to construct batteries. We start with the use of stimuli.

4.5.1 The Use of Batteries of Stimuli

Example 4.15 presents a possible formulation of a battery of stimuli:

> 4.15 *There are different ways of attempting to bring about improvements or counteract deterioration in society. During the last 12 months, have you done any of the following? Please mark either "yes" or "no."*

	Yes 1	No 2
A. *Contacted a politician*	☐	☐
B. *Contacted an association or organization*	☐	☐
C. *Contacted a national, regional, or local civil servant*	☐	☐
D. *Worked in a political party*	☐	☐
E. *Worked in a political action group*	☐	☐
F. *Worked in another organization or association*	☐	☐
G. *Worn or displayed campaign badge/sticker*	☐	☐

In this example, "any of the following" stands for the so-called stimulus, which could be a single action, such as "contacted a politician" or "taken part in a strike." Such stimuli batteries can also consist of nouns or combinations of nouns. Example 4.16 illustrates this:

> 4.16 *How satisfied are you with the following aspects of life:*
> 1. *Your income*
> 2. *Your house*
> 3. *Your social contacts*

Another possibility is that a stimulus consists of a part of a verb phrase such as in example 4.17:

> 4.17 *Did you do any of the following?*
> *Shopping*
> *Cleaning*
> *Washing*

The reader should be aware that stimuli also could occur in all kinds of combinations of requests for answers such as example 4.18 illustrates:

> 4.18 *Please tell me whether or not you are satisfied with the following aspects of life:*

One reason to use batteries of stimuli is that the requests and the response categories do not have to be repeated each time. This is very efficient for the questionnaire designer and the printing of the questionnaires and the interviewer, since they have less to write, print, and read. So far, we have not seen any convincing evidence that this approach has a negative effect on the answers of the respondents, although one can expect that they will not answer the requests independently of each other. It is more likely that they make use of their previous answer to judge the next stimulus in case of evaluations on scales. This would lead to correlated errors between the responses; however, Saris and Aalberts (2003) did not find strong evidence for this in their research.

4.5.2   The Use of Batteries of Statements

Very popular in survey research is the indirect request with an interrogative prere-quest using the verb "agree" followed by assertions discussed in Chapter 2, often called "statements." A typical example[1] of such a battery of agree/disagree requests is given below. Example 4.19 is taken from a study of Vetter (1997), but the concept "political efficacy," which is measured here, has already been questioned in a similar way in 1960 in *the American Voter* (Campbell et al. 1960).

Typical for such a battery of statements are the following characteristics:

1. The request for an answer is formulated only once before the first statement.
2. The response categories are mentioned only one time.
3. The formulation of the request for an answer is rather abstract through use of the term "statement" at the place where, normally, the statement itself follows.

> 4.19   *How far do you agree or disagree with the following statements?*
> 1. *Disagree very strongly*
> 2. *Disagree*
> 3. *Neither agree nor disagree*
> 4. *Agree*
> 5. *Strongly agree*

	Possible responses				
Statement	*1*	*2*	*3*	*4*	*5*
A *I think I can take an active role in a group that is focused on political issues.*					
B *I understand and judge important political issues very well.*					
C *Sometimes politics and government seem so complicated that a person like me cannot really understand what is going on.*					

If we abide by the rules we have seen in Chapter 3, the following formulations could also be an alternative:

> 4.20a   *How far do you agree or disagree that you can take an active role in a group that focused on political issues: (1) disagree strongly, (2) disagree, (3) neither agree nor disagree, (4) agree, or (5) strongly agree?*
>
> 4.20b   *How far do you agree or disagree that you understand and judge important political issues very well: (1) disagree strongly, (2) disagree, (3) neither agree nor disagree, (4) agree, or (5) strongly agree?*

[1] These requests for an answer were originally formulated in German. The authors of this text have translated them into English. These requests are not given as examples of very good requests for this section.

> 4.20c   *How far do you agree or disagree that sometimes politics and*
> *government seem so complicated that a person like you cannot really*
> *understand what is going on: (1) disagree strongly, (2) disagree,*
> *(3) neither agree nor disagree, (4) agree, or (5) strongly agree?*

This transformation to a standard indirect request with an interrogative agree/disagree prerequest and a different embedded declarative assertion of each request makes it clear that the battery form is far more efficient.

Krosnick and Fabrigar (forthcoming) make a comparison with direct requests for an answer. They suggest that the popularity of the use of agree/disagree batteries lies in the fact that it reduces the amount of work as we have mentioned earlier and maybe, even more importantly, this approach can be applied to nearly all possible assertions in the same way.

If direct requests for an answer are more desirable, a different form for each assertion is needed, as is illustrated for the same assertions in examples 4.21a–4.21c. The transformation of the battery mentioned earlier to three direct requests leads to the following result:

> 4.21a   *Could you take a very active, quite active, limited role, or no role at*
> *all in a group that is focused on political action?*
> 4.21b   *Can you understand and judge important political issues very well,*
> *well, neither good nor bad, bad, or very bad?*
> 4.21c   *How often does it seem to you that politics and government*
> *are so complicated that a person like you cannot really understand*
> *what is going on: very often, quite often, sometimes, seldom,*
> *or never?*

This transformation again indicates the efficiency of the battery format for the researcher and the interviewers. They do not have to specify and read a different response scale for each separate assertion. Whether the efficiency for the researcher and the interviewer goes together with efficiency for the respondent and with better data is another matter. Saris et al. (2009a) have the following opinion on the matter:

> The goal of agree/disagree requests is usually to place respondents on a continuum. For example, an assertion saying "I am usually happy" is intended to gauge how happy the respondent usually is, on a dimension from "never" to "always." An assertion saying "I like hot dogs a lot" is intended to gauge how much the respondent likes hot dogs, on a dimension from "dislike a lot" to "like a lot." And a statement saying "Ronald Reagan was a superb President" is intended to gauge respondents' evaluations of Reagan's performance, on a dimension ranging from "superb" to "awful."
>
> To answer requests with such statements requires four cognitive steps of respondents (Trabasso et al. 1971; Carpenter and Just 1975; Clark and Clark 1977). First, they must read the statement and understand its literal meaning. Then, they must look deeper into the statement to discern the underlying dimension of interest to the researcher. This is presumably done by identifying the variable quantity in the statement. In the first example above, the variable is identified by the word "usually" it is frequency of happiness. In the second example above, the variable is quantity,

identified by the phrase "a lot." And in the third example, the variable is quality, identified by the word "superb." Having identified their dimension, respondents must then place themselves on the dimension of interest. For example, the statement, "I am usually happy," asks respondents first to decide how happy a person they are. Then, they must translate this judgment into the agree/disagree response options appropriately, depending upon the valence of the stem. Obviously, it would be simpler to skip this latter step altogether and simply ask respondents directly for their judgments of how happy they are.

It is self-evident here that answering batteries of statements is not a simple task for the respondent. Moreover, hundreds of papers have been written about the issue that respondents may have a tendency to simplify their task and to answer all requests in a battery in a same way. This phenomenon is called *response set* or *acquiescence*. The response set will increase the correlation between the answers in the batteries, but this extra correlation is a method effect and has nothing to do with the substance of the requests. Krosnick and Fabrigar (forthcoming) and Billiet and McClendon (2000) have discussed this problem extensively. It is also one of the possible reasons why method effects are found in multitrait–multimethod studies (Andrews 1984; Költringer 1995; Scherpenzeel and Saris 1997; Saris and Aalberts 2003).

Finally, Krosnick and Fabrigar (forthcoming) have made the argument, mentioned in Chapter 2, that the request asking "In how far do you agree" does not estimate the extremity of an opinion but the intensity, which is a different aspect of measurement. The latter aims at the strength of the agreement with the statement, and this is not the same as the extremity of an opinion in the former. If one says "I like ice cream very much," that is not the same as "I very strongly agree with the statement: I like ice cream."

We would like to mention one more complication for this method. As was mentioned earlier, the respondents have to place themselves in the dimension of interest. After careful examination of statement 4.19c, it was suggested that the purpose of the item was to evaluate how often people had the impression that politics and government were too complicated. This was formulated in example 4.21c, which is repeated here in example 4.22:

> 4.22   *How often does it seem to you that politics and government are so complicated that a person like you cannot really understand what is going on: very often, quite often, sometimes, seldom, or never?*

It is very clear what a choice of one of the answer categories means; however, this does not mean that no errors will be made (Hippler and Schwarz 1987) or that people have a clear opinion in their mind of what they should say (Tourangeau et al. 2000). However, several alternatives are available if an agree/disagree format is used, as we show in questions 4.22a–4.22e:

> 4.22a   *How far do you agree or disagree that politics and government <u>very often</u> seem so complicated that a person like me cannot really understand what is going on: (1) disagree very strongly, (2) disagree, (3) neither agree nor disagree, (4) agree, or (5) strongly agree?*

> 4.22b   How far do you agree or disagree that politics and government <u>quite</u>
> <u>often</u> seem so complicated that a person like me cannot really
> understand what is going on: (1) disagree very strongly, (2) disagree,
> (3) neither agree nor disagree, (4) agree, or (5) strongly agree?
>
> 4.22c   How far do you agree or disagree that politics and government
> <u>sometimes</u> seem so complicated that a person like me cannot really
> understand what is going on: (1) disagree very strongly, (2) disagree,
> (3) neither agree nor disagree, (4) agree, or (5) strongly agree?
>
> 4.22d   How far do you agree or disagree that politics and government
> <u>seldom</u> seem so complicated that a person like me cannot really
> understand what is going on: (1) disagree very strongly, (2) disagree,
> (3) neither agree nor disagree, (4) agree, or (5) strongly agree?
>
> 4.22e   How far do you agree or disagree that politics and government <u>never</u>
> seem so complicated that a person like me cannot really understand
> what is going on: (1) disagree very strongly, (2) disagree, (3) neither
> agree nor disagree, (4) agree, or (5) strongly agree?

These statements differ only by the word indicating the frequency of the occurrence of the event of interest. Logically, all these possibilities (and many others) can be employed, and there is seemingly no reason to prefer one over another. But are there practical reasons to prefer one request above the other? In order to check this, let us perform a small thought experiment.

Imagine that you have the idea *very often* that politics is too complicated for you. Now an interviewer comes with the request 4.22a and if you have the idea *very often*, then your answer is simple: strongly agree. Imagine now that you have the idea *often* and you are confronted with the same request 4.22a: both agree and disagree could be chosen. Formally, disagree is better, but with a bit of flexibility, you could as a respondent also say agree. Suppose now that you have the idea *sometimes* and you are confronted with the same request: most likely you would choose disagree.

Now, imagine again that you have the idea *very often* but the request is asked if you have this idea <u>sometimes</u> as in 4.22c. You may be confused as to what to answer because you can say disagree since you have these ideas often but you can also agree as you have them more than sometimes. Suppose now that you *never* have these ideas and the interviewer uses request 4.22c with the term <u>sometimes</u>. You could say "disagree" since you never have these ideas or you can agree depending on your perception of whether "sometimes" is rather close to never.

Our thought experiment shows that the statements in the middle of the scale encounter the problem that people at both sides of the spectrum can give the same answer, which makes further analysis rather problematic. Extreme statements have a lesser issue with this particular problem, but these statements have the problem that people with a different opinion than stated in the request can all choose the same response of disagree. This effect will be even stronger when the extreme statement is very extreme.

The conclusion on the basis of our practical analysis is that if one really wants to use statements, one should choose a statement that represents an extreme position but that is not too far from the opinions of the people; otherwise, no variation will be obtained. This analysis also shows that the choice of the formulation of item 3 in the political efficacy request is definitely incorrect.

Given all the complications of batteries with statements, it is very questionable why this type of formulation is so popular. Further research is required, but we recommend avoiding this approach and using direct requests. It is more work for the researcher and the interviewer, but it simplifies the task of the respondents and probably increases the quality of the answers.

4.6   OTHER FEATURES OF SURVEY REQUESTS

The possible consequences of other features of requests are discussed in the next sections.

4.6.1   The Formulation of Comparative or Absolute Requests for Answers

We move now to a quite different aspect of the formulation of requests for an answer, namely, the use of comparative or absolute judgments. Comparative requests for answers ask about the similarity or dissimilarity of two objects, and they also ask for degrees of similarity. Examples of this type include:

4.23a   *Are you more satisfied with your new job than with the old one?*
4.23b   *Do you earn less money in your new job?*
4.23c   *How much better is your daughter in languages than your son is?*
4.23d   *How much do you prefer languages above science?*
4.23e   *Which political party do you prefer?*

As the first two examples 4.23a and 4.23b illustrate, the inequality can be expressed by "more...than" or "less...than" where the comparison "than" can be implicit as in the second example. But it also can be expressed by comparative adjectives or adverbs such as "much better than" (4.23c) or by words that indicate a preference as shown in the last two examples.

Requests for an answer that ask for an absolute judgment, in contrast, do not express a comparison in terms of more or less than from a reference object. Absolute judgments are very frequently used in survey research. Examples are as follows:

4.24a   *Are you satisfied with your job?*
4.24b   *How satisfied are you with your job?*
4.24c   *How good are you at mathematics?*

Although absolute judgments are very popular in survey research, it is questionable whether people are very good in making such judgments. In psychophysics, this phenomenon has also been observed by Poulton (1968). Similar results have been found by Saris (1988b) in survey research. A famous experiment by Schwarz and Hippler (1987) showed the same results. They asked for the amount of time people spent watching TV and showed that even in such cases many people gave relative judgments, relative to watching patterns of other people, suggested by the specified response categories, and not absolute judgments. We will come back to this example in the next chapter.

4.6.2 Conditional Clauses Specified in Requests for Answers

Sometimes, in requests for answers, clauses are included that refer to something that must happen first so that something else can happen. This is called a "condition" in the narrowest sense, or an event is mentioned that is qualified as uncertain. Such clauses are called *conditional* (Swan 1995: 245, 252), and they restrict the content of the request to this specific condition or event. The following examples can illustrate conditional clauses:

4.25a   *Do you think it is acceptable that a woman has an abortion if she has been raped?*

4.25b   *If the government is reelected, do you believe that they will realize what they had promised before the elections?*

4.25c   *Should refugees be allowed to work in our country, provided they take simple jobs?*

4.25d   *Should Muslims be allowed to build mosques in our country as long as they are not subsidized by the government?*

4.25e   *If you finish your studies in some years, are you planning to work in the field of study?*

4.25f   *Suppose the government increases the income tax next year, would you have to change your lifestyle?*

4.25g   *Imagine you were the president: which of the following measures to change our country would you take first?*

Examples 4.25a–4.25d illustrate conditions in the narrowest sense. The first two are specified by an "if" clause, while the third and the fourth use the expressions "provided" and "as long as," which means that the event mentioned in this clause should occur first before the main clause can be appraised. Examples 4.25e–4.25g refer to uncertain or hypothetical events. Request 4.25e is again formulated with the word "if" and expresses just an uncertain event in the future. Often, reality is too complex to be asked without condition, like requests about abortion.

Request 4.25f uses the word "supposes" and indicates in this example again an uncertain event in the future, while the last example expressed by "imagine" refers—because of the use of the past tense—to a very unlikely event in the future. Respondents may have never thought about these specific hypothetical situations. In that case, they have not premeditated their answer, and it is questionable if these responses have any stability (Tourangeau et al. 2000).

4.6.3 Balanced or Unbalanced Requests for Answers

A *balanced request* for an answer means that it is made formally explicit that both negative and affirmative answers are possible (Schuman and Presser 1981: 180; Billiet et al. 1986: 129). If only one answer direction is provided, the request for an answer is called *unbalanced*. An example of a balanced request could be:

> 4.26   To what extent do you favor or oppose euthanasia?

This request is balanced as it explicitly specifies both answer directions: in favor of and in opposition to. Sometimes, this seems to be a bit exaggerated. For example, one could also have asked:

> 4.27   Do you strongly favor, favor, neither favor nor oppose, oppose, or strongly oppose euthanasia?

Such requests are formulated because the researcher tries to prevent more attention being given to one side of the scale than to the other. In general, it is supposed that a bias in the response will occur in the answer direction that is indicated in the request even though there is no research evidence supporting this assumption. The reason that no errors have been found may be that people are very much familiar with one-sided formulations and are very well able to fill in the missing alternatives themselves (Gallhofer and Saris 1995).

The following example is balanced although the request indicates none of the answer directions:

> 4.28   What do you think about euthanasia?

A request that does not specify the different sides is also considered as balanced in our research, although this is a rather arbitrary decision. Examples of unbalanced requests for an answer could be:

> 4.29a   To what extent do you favor euthanasia?
> 4.29b   To what extent do you oppose euthanasia?
> 4.29c   Some people think that euthanasia should be legalized. In principle, what is your opinion about euthanasia?

Example 4.29a only mentions the positive answer direction, while the negative one should be guessed by the respondent. In example 4.29b, only the negative direction is indicated, and in example 4.29c, only a favorable opinion is mentioned in the survey item.

In the case where the response possibilities go from 0 to positive or from 0 to negative (unipolar scales, Chapter 5), the notion of balance is not applicable because there exists only one direction. An example might illustrate this:

> 4.30   How often do you go to church?

Here, "often" is mentioned in the request, but the request is nevertheless unbiased because this is a unipolar request, as there is only one side. The following request for an answer, however, is more complicated:

> 4.31   To what extent do you favor euthanasia?

This question is unbalanced because only one side of the scale is indicated. However, the unbalanced question can only be unbiased if it is posed to respondents in favor of euthanasia. Otherwise, this request is a "leading" request and that is an extreme form of bias.

4.7   SPECIAL COMPONENTS WITHIN THE REQUEST

Sometimes other components, not necessarily belonging to the request for an answer, are placed in the request. We shall discuss two different components: remarks to stimulate the respondent to answer and remarks that emphasize that the subjective opinions of the respondents are requested. We start with the remarks that are intended to stimulate the response.

4.7.1   Requests for Answers with Stimulation for an Answer

A special stimulation to elicit an answer from the respondent can be included in the requests for answers. They can be in either imperative or interrogative prerequests with all kinds of gradation of politeness as already mentioned in Chapter 3 in connection with procedures to formulate requests for answers. Some examples of a stimulation to answer within requests for answers could be:

> 4.32a   *Tell me, are you going to vote?*
> 4.32b   *Would you be so kind as to tell us what you did before studying at the university?*
> 4.32c   *Could you tell us who is the president of the EU?*

Sometimes, a stimulation for an answer also occurs in other parts of survey items such as introductions or motivations of the researchers, which are discussed in Chapter 6.

The presence or absence of a stimulation to answer requires attention because their presence might make a difference in the readiness of the respondent to comply. If a stimulation is formulated very politely, it might be that the respondent is more inclined to answer, even if this person has no specific opinion and might just give a random opinion because of the extra encouragement to give an answer.

4.7.2   Emphasizing the Subjective Opinion of the Respondent

Like stimulation for an answer, a stimulus for the respondent to give his/her own opinion can occur within requests and encourage the subjects to give an opinion even if he/she hardly thought about the issue. However, this procedure has an effect and will be studied later. Some examples of emphasizing to express his/her own opinion might be:

> 4.33a   *According to you, what is the most important issue in this election?*
> 4.33b   *In your opinion, who is responsible for the economic recession in our country?*
> 4.33c   *What do you think is the main reason for the economic recession?*
> 4.33d   *We would like to know whether you personally think that the death penalty should be implemented.*

The first two examples relate to specific direct requests where the expressions "according to you" or "in your opinion" stress that a personal appraisal is desired. In the third example, the interrogative clause "do you think" emphasizes the subjective opinion, and in the fourth example, the clause "whether you personally think…" functions in a similar way. Emphasis on the subjective opinion also can occur in other parts of the survey item such as in the introduction (see Chapter 6).

4.8 SUMMARY

In this chapter, several decisions in developing a request for an answer are discussed again but this time from the perspective of a survey researcher. The choice of the research topic brings with it some unavoidable consequences. For example, given the research goal, the decision of whether the requests are directed to the past, present, or future is predetermined. The research goal also determines the social desirability of the possible response alternatives and the salience of the topic. However, the format of the question can be chosen, while double-barreled requests, requests with an implicit assumption, and prerequests that change the concept can be easily avoided.

In this chapter, we suggest why the use of batteries is so popular in survey research. The reason is mainly the efficiency of the formulation because the request and the answer categories have to be mentioned only once. To our knowledge, batteries with stimuli do not create problems, but batteries with statements have been criticized heavily by different authors. One reason is the possibility of response set or acquiescence that can generate correlations that are due to the method (use of a battery) and have no substantive meaning. Another problem is that the choice of the statements is rather arbitrary, but the choice will certainly have an effect on the response distributions and most likely also on the correlations with other variables.

Furthermore, several characteristics of requests for an answer have been discussed, which may play a role in the quality of an item. First, the choice between absolute and comparative judgments has been discussed, followed with considerations for the choice between balanced and unbalanced requests. Whether we can say that one characteristic is indeed better than another requires further research. But as far as we know, the balancing of the requests by survey researchers does not seem to be based on empirical evidence while at the same time balancing the requests makes the formulations much more complex.

Finally, it was mentioned that sometimes researchers include requests in the texts to stimulate respondents to give answers or to give their own opinions. These choices also require further research to determine whether adding them to texts has a positive effect on the results.

EXERCISES

1. Look at the following request of an answer:
   Do you have the feeling that homosexuals should have the same rights with respect to marriage and raising children?
   a. What do you think that the researcher wants to measure?
   b. What went wrong in the formulation of this question?

2. Formulate a battery for human values using the following value stimuli: honesty, love, safety, and power.

3. Formulate a battery for human values using the same values mentioned in Exercise 2, but this time, make a statement for each of these values.

4. Several alternative statements can be formulated; indicate how these different statements can be created for the value "honesty."

5. Which of the statements you created in Exercise 3 is the best?

6. How would a human value request be formulated for the value "honesty" using direct requests?

7. Is it possible to formulate this request in an absolute and in a comparative way?

8. Is your request balanced? If so, when could it be considered unbalanced? If not, how could it be balanced?

9. Can you also add texts to the last request to stimulate a response and to emphasize that a personal opinion is asked?

10. With respect to your own questionnaire, discuss whether you have made the best choices while considering the abovementioned options? If so, why?

5

RESPONSE ALTERNATIVES

So far, we have been discussing requests for answers. As was indicated in Chapter 3, the requests can have many different forms, which in turn can create the same response alternatives for the respondent. However, the fact that the same response possibilities are present does not mean that the requests for an answer measure the same thing. Along the same line, it is not immediately clear whether requests for an answer that are identical but differ in the set of possible responses measure different variables. This is an empirical question that has to be answered for different measures. Saris (1981) showed that at least some sets of response scales, although different, will give responses that are identical, except for a linear transformation suggesting that roughly speaking, these measures are indeed identical.

Another issue studied by many people is whether it makes sense to present the respondents with more than only a few categories. Most textbooks suggest, in reference to Miller (1956), that people cannot use more than approximately seven categories. Cox (1980) has argued that Miller's rule does not apply at all to this problem. He suggests that more information can be obtained if more categories are used. This opinion is shared by a few more researchers (Saris et al. 1977; Andrews 1984; Költringer 1995; Alwin 1997).

Finally, there are people, Krosnick and Fabrigar (forthcoming), who suggest that it would be advisable for certain problems, or in general in qualitative research, not to use explicit response alternatives. They suggest that requests with open answer categories are the best because they do not force the respondents in the frame of reference of the researcher.

Design, Evaluation, and Analysis of Questionnaires for Survey Research, Second Edition.
Willem E. Saris and Irmtraud N. Gallhofer.
© 2014 John Wiley & Sons, Inc. Published 2014 by John Wiley & Sons, Inc.

All these options will be discussed in this chapter. The pro and con arguments will be mentioned, and an empirical evaluation of the effects on data quality of the different possibilities will be given in Part III of this book.

5.1 OPEN REQUESTS FOR AN ANSWER

As has been mentioned earlier, some people argue that requests with open answer categories are better than requests with closed categories because people can follow their own thoughts and are not forced in the frame of reference of the researcher. A request that is exemplar for this dilemma and that has been studied frequently is as follows:

> 5.1   *What is the most important problem that our country is confronted*
> *with nowadays?*

This request for an answer can be made as an open request as indicated in the preceding text or with possible responses, chosen on the basis of prior research based on the open request. A comparison between these two requests has been studied several times by Schuman and his colleagues. Schuman and Presser (1981) reported that the results from the two requests are very different. The open request seems to be influenced by events that were recently discussed in the media, while the request with response categories provides a frame of reference indicating what is expected from the respondent. The option of "other" category along with a set of responses can be introduced, but it turns out that this option is not chosen as frequently as expected. Hence, the authors concluded that the given response categories of a request guide respondents in their answer choices.

Subsequent research by Krosnick and Schuman (1988) suggests that there is more consistency across the open and closed request results if the coding of the answers of the open request is more in line with the categories used by less-educated people. This brought Krosnick and Fabrigar (1997) to conclude that open requests are preferable because the effect of the researcher on the result is avoided.

The last statement may be correct for the aforementioned type of request, where a choice out of a multitude of nominal categories is requested; however, the findings need to be investigated further to determine whether they are also true for other open requests for an answer. Therefore, let us explore some other possibilities.

Krosnick and Fabrigar (forthcoming) indicate in another chapter of their book that not all open requests can be trusted at face value. They discuss the open "WHY request and the validity of introspection." In psychology, introspection has been discussed at length by the different schools of thought where some scholars think that only people can know why they do things, and therefore, they should be asked. Other scholars argue that answers based on introspection cannot be trusted. One of the reasons provided is quick memory loss of thoughts concerning the choices made. Therefore, a "think aloud" procedure is suggested, but if one asks for arguments before or while people are making choices, this in itself can influence the process (Ericsson and Simon 1984; Wilson and Dunn 1986), and most of the time, rationalizations of the answer choice are provided. This is not only the view of the behaviorists like Skinner (1953) but also of scholars with a less extreme point of view (Nisbett and Wilson 1977).

Krosnick and Fabrigar (1997), while applying this bulk of research on survey research, comment, "…if results based on introspection requests seem sensible on their surface, we would all be inclined to view them as valid. And yet, as Nisbett and Wilson (1977) and Wilson and Dunn (1986) have made clear, this apparent sensibility may well be the result of people's desire to appear rational, rather than the result of actual validity of introspection." Therefore, Krosnick and Fabrigar (1997) clearly indicate their reservations with the use of introspection procedure with the open request for an answer method. One should, however, also remark that formulating alternative procedures for introspection is not very easily done.

Wouters (2001), in her research, has specified open requests for all kinds of combinations of concepts and request forms that have been mentioned in Chapters 2–4. For example, one could ask:

    5.2   How would you evaluate the presidency of Clinton?

It is clear that an evaluation is asked for but the possible responses are not specified. So, this is an open request, and the respondent can give an answer in many different ways. In a similar way, Wouters (2001) was able to transform nearly all possible closed requests into open-ended requests for answers. Hence, the pertinent question is which of the two forms is better. To answer this question, a lot of research is still needed. Presently, we can say only that closed requests are more efficient than open requests because the former do not require an extra coding phase.

The analysis of Wouters (2001) also showed that it is not always simple to formulate a closed form for all open requests. We will demonstrate our point with the following example:

    5.3a   What do you think about the presidency of Clinton?

Example 5.3a is an open-ended request; however, what is special about this request is that it does not measure a specific concept because respondents can answer with an evaluation (good or bad) but also with a cognition (that Clinton's government was the first to balance the budget) or a relationship (that Clinton's presidency led to an impeachment procedure) as just a few examples of possible answers. Not only is the answer open-ended, but also the concept itself to be measured is not specified. Our hypothesis is that such requests are used to determine what aspect of the object the respondents consider the most important from which are derived further requests about this aspect. If that is true, an alternative in closed form to the open-ended request could be:

    5.3b   What is for you the most important aspect of the presidency of Clinton?
        1. His foreign policy
        2. His national policy
        3. His economic policies
        4. His personal conduct
        5. Others

Another type of open request that is hard to formulate in closed form concerns the enumeration of different events of actions. An example is:

5.4   Can you describe the different events that took place before the demonstration changed into a violent action?

Here, the respondent has to provide a series of events that have occurred in sequence.

From example 5.4, it can be inferred that asking an equivalent request in closed format would require a very different and complex series of requests.

Another type of request for an answer that requires special attention is a request for a frequency or an amount. Examples are found in the following text:

5.5a   For how many hours did you watch TV last night?

5.5b   How much did you pay for your car?

These requests are, in some sense, the opposite of the open requests we have discussed earlier, because now it is very clear how the respondents have to answer. The first request asks for a number that indicates the number of hours they have watched TV, and the second asks for a monetary amount. So people know quite well how they should answer, but nevertheless, the answer is open because no response options have been provided to them (Tourangeau et al. 2000). For these requests that ask numeric answers, closed alternatives have been formulated. They will be discussed in further detail in the section on vague quantifiers.

It might depend on what request type we are about to use or whether we choose an open or closed form. For some requests, alternatives in closed form exist; for others, alternative closed requests are difficult to formulate. For those requests that can be asked in a variety of ways, different aspects should be considered. First, it is important to consider whether more information is obtained through using the open request format. If that is not the case, then it is better to choose the closed form because the processing of the information is much easier. A second issue is whether open and closed requests lead to different response distributions and relationships with other variables. If that is the case, one has to consider which request form is better. Evaluation of the effects on the data quality will be discussed later. If the same results are obtained or the quality is not clearly better for the open requests, then the closed requests should be preferred because of the efficiency in information processing. It will be clear that in our opinion, the conclusion of Krosnick and Fabrigar (1997) is still premature and we think that further research is required before a conclusion about the choice between open and closed requests can be stated with certainty. We speculate that the request choice will depend on the type of issue the request is aiming at as was the case with our example.

5.2   CLOSED CATEGORICAL REQUESTS

The first of the requirements regarding closed response answer categories is that they should be *complete*. In practice, however, sometimes the answer alternatives are not complete, which can result in nonresponse. Such an example is given in the following text:

> 5.6a *What is the composition of your household?*
>   1. *One single adult*
>   2. *Two adults*
>   3. *Two adults with one child*
>   4. *Two adults with two children*
>   5. *Two adults with three children*
>   6. *One adult with one child*

After scanning the answer options for 5.6a, it becomes clear that the answer categories are not exhaustive since there are several variations of adults and children possible and one for commune is missing. Hence, 5.6b is a more complete version:

> 5.6b *What is the composition of your household?*
>   1. *Number of adults*
>   2. *Number of children*

The second requirement is that the answer categories are *exclusive*, or in other words, they should not overlap. An example of overlapping answer categories is found in request 5.7a:

> 5.7a *What is the most important reason why you are against nuclear*
>   *energy?*
>   1. *Too expensive*
>   2. *Too dangerous*
>   3. *Causes environmental problems*
>   4. *Others*

In request 5.7a, the second and third categories are not exclusive because environmental problems can cause dangers, and dangers, like radioactive waste, can cause environmental problems. Therefore, a respondent may be confused about which choice to make. The remedy is to reformulate these two categories in order to make them exclusive:

> 5.7b *What is the most important reason why you are against nuclear*
>   *energy?*
>   1. *Too expensive*
>   2. *The probability of an accident is too high*
>   3. *Too much radioactive waste*
>   4. *Others*

Here, the second category focuses on accidents, and the third, on radioactive waste, which are now distinct and no longer overlap.

A third requirement is that answer categories match with the information provided in the request or statement asked (Lessler and Forsyth 1996; Graesser et al. 2000a,b):

> 5.8a *How far do you agree or disagree with the statement that*
>   *governmental decisions are always carried out?*
>   1. *Completely agree*
>   2. *Agree*

    3. Neither agree nor disagree
    4. Disagree

In the example, the statement refers to an objective concept (a behavior), while the answer categories relate to subjective concepts. The appropriate answer categories would be "true/false." The request could be reformulated in the following manner:

    5.8b Do you think that the following statement is true or false?
        Governmental decisions are always carried out.
        1. True
        2. Neither true nor false
        3. False

Finally, a requirement is that all the response categories represent *the same concept.* Sometimes, a mismatch of answer categories occurs because they concern different concepts and then it is difficult for the respondent to choose a category. Example 5.9 illustrates a case where this is not correct:

    5.9 What is your opinion about a ban on driving a car in downtown area?
        1. Inconvenience
        2. Acceptable

The first category refers to a feeling, while the second is a right. In order to be consistent, it is possible to provide either a feeling (unpleasant/pleasant) or a right (acceptable/unacceptable) as options of the uncertainty space. All requests for an answer with closed answer categories should satisfy the aforementioned requirements.

In the following sections, we want to illustrate the different types of response categories that are available to the survey designer. The first type uses nominal categories without any ordering, while the second type provides ordinal response categories, and the third consists of what is called vague quantifiers.

5.2.1 Nominal Categories

Requests for an answer using unordered response categories are an alternative for the open requests asking for one option out of a set. An example is:

    5.10 What is the most important problem that our country faces at the
        moment?
        1. Terrorism
        2. Unemployment
        3. Racism
        4. Criminality
        5. Others, please specify

Similar requests can be made for the most important aspect of the work and many other topics. There is no ordering in the different response possibilities even though they can be numbered in the questionnaire and certainly in the database, but, the numbers cannot suggest an ordering on any dimension because that dimension does not exist. Response scales that are not ordered are called *nominal* scales.

A special nominal scale is a scale for *dichotomous* responses where only two answers are possible, for example:

> 5.11   *Did you vote in the last elections?*
>> 1. *No*
>> 2. *Yes*

In this case, the scale is officially nominal, indicating no ordering. However, it is possible to use the scale in the ordinal sense and apply analyses that at minimum require ordinal data, and it is arbitrary if the coding by the researcher is completed as 0–1 or 1–0 for the dichotomous scale.

5.2.2   Ordinal Scales

Ordinal response categories require that there is an ordering of the response categories. Such sets of response alternatives are very common in subjective judgments. For example:

> 5.12   *How good do you think Clinton was as president?*
>> 1. *Very bad*
>> 2. *Bad*
>> 3. *Neither good nor bad*
>> 4. *Good*
>> 5. *Very good*

In this case, there is an ordering in the response categories, and one can say that the numbers in front of the categories suggest an ordered scale where 1 is the lowest and 5 is the highest category. Similar scales can be made with any predicate with "high" and "low," "friendly" and "unfriendly," and "active" and "passive," to name only a few examples.

Although such an ordinal scale is called a 5-point scale—a scale with five possible answers—a person with a positive evaluation of Clinton has only two possibilities: good or very good. If it is desirable to have a more precise answer, it can be specified as a 7-point scale such as the one in the following text:

>> 1. *Very bad*
>> 2. *Rather bad*
>> 3. *Bad*
>> 4. *Neither good nor bad*
>> 5. *Good*
>> 6. *Rather good*
>> 7. *Very good*

Along the same lines, one can also construct a 9- or 11-point scale. Keep in mind that there is a limit to the possibilities of labels for the different categories and that it is also possible to specify ordinal scales with labels for only a limited number of categories. Common examples are the following:

> 5.13a *How good do you think Clinton was as president? Express your opinion in a number between 0 and 10 where 0 = very bad and 10 = very good.*

or

> 5.13b *How good do you think Clinton was as president? Express your opinion by placing an x at the point of the scale that best expresses your opinion.*

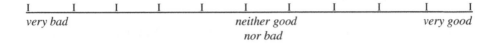

very bad   *neither good*   *very good*
                                  nor bad

Examples 5.13a and 5.13b are both 11-point scales; the distinction is that the former has only two labeled categories, while the latter has three labeled categories and that the first request uses numbers, while the second is a typical example of what is called a "rating scale."

Many alternative presentations can be developed with ordinal response scales. What is important is that the categories are ordered in some way from low to high. It can also be done by pictures of faces that are more or less happy or ladders where each step indicates a different level of satisfaction (Andrews and Wittey 1974) or a thermometer where the increasing grades indicate the warmth of the feelings of respondents toward parties and party leaders. The United States' National Election Studies are exemplar for this type of creative ordinal response scale grading.

When developing ordinal scales, a range of decisions is at the researchers' disposal. We will discuss some of these choices with their alternatives. First, we have seen that either all or some of the possible responses can be labeled. Therefore, the responses can be *completely labeled or partly labeled*.

In examples 5.13a and 5.13b, the numbers in front of the categories were ordered in the same way from low to high as the labels, and they started with the lowest or most negative category. It can also happen that there is no *correspondence* between the category labels and the numbers or that the scale does not go from low or negative to positive but vice versa.

All the scales presented so far are *symmetric* around the middle of the scale, which means that there are as many categories at the positive as at the negative side.

In general, it is advisable to use symmetric scales; the reason can be demonstrated by the example:

> 5.14 1. *Very unhappy*
> 2. *Unhappy*
> 3. *Neither unhappy nor happy*
> 4. *Happy*
> 5. *Rather happy*
> 6. *Very happy*

This example demonstrates that it appears awkward to be using an asymmetric scale in this case. However, if we know that all respondents' answers are on the happy side of the scale, it is not very efficient to use a 5-point scale from "very unhappy" to "very happy" because the distribution of happiness in the population is reduced to a 2-point scale. Therefore, an asymmetric 5-point scale is more appropriate and precise:

> 5.15   1. *Not happy*
>   2. *A bit happy*
>   3. *Happy*
>   4. *Rather happy*
>   5. *Very happy*

Example 5.15 has a 5-point scale that favors the positive side, while the "not happy" side of the scale is represented by only one response category. Such a scale presupposes knowledge about the happiness of a survey population; otherwise, such an asymmetric scale is biased.

So far, except for the last example, all sets of response scales were also *bipolar*, which means that there are two opposite sides of the scales: positive to negative or active to passive. The last scale of happiness was made one-sided or *unipolar*, but happiness itself is in principle a bipolar concept, going from unhappy to happy. Therefore, we also said that the unipolar scale presupposed knowledge of the distribution of feelings within the population. There are, however, also concepts that are typically unipolar. For example, "attachment to a party" goes from "no attachment" to "strong attachment" because it is impossible to imagine a negative side of the scale of attachment.

The discussion in the preceding text has served to demonstrate that both the provided scale for responses and the concept can be in *agreement* with each other (both bipolar or both unipolar) or in disagreement if the concept is bipolar, but the responses are only unipolar, as in example 5.15.

So far, we have used a *neutral category or a middle category*, but it is not always necessary to do so. If it is necessary to force people to make a choice in a specific direction, then the middle category can be omitted. Schuman and Presser (1981) have shown that this has no effect on the distribution of the respondents over the positive and negative categories. However, it might have the effect that fewer people are willing to answer the request because, according to them, their response is not provided and consequently they choose for a "don't know" or "refusal" (Klingemann 1997).

The *"don't know"* category has been the subject of serious investigation. Research has centered around the question of whether it should be offered, and if so, in what form. One can ask, for instance, before the request itself is asked whether people have an opinion or not about the topic in question. This is the most explicit "don't know" check. The second possibility is to provide "don't know" explicitly as one of the response options. The third possibility is that "don't know" is not mentioned but that it is an admissible response alternative that can be found on the questionnaire of the interviewer but is not mentioned as a possibility to the respondent. Finally, there is the possibility of omitting it altogether.

Providing the "don't know" option explicitly creates several obstacles. The most important issue is that respondents can choose this option for several reasons that have nothing to do with their own opinion. Krosnick and Fabrigar (forthcoming) mention that this option is chosen because respondents do not want more requests or because they do not want to think about the request and therefore an acceptable option "don't know" is easily available. The authors call this "satisficing behavior of a respondent."

Schuman and Presser (1981) argue that people who normally would say that they "don't know" would make a difference in the relationships between variables under investigation. They report on a study where without respondents using the "don't know" category, the correlation between two variables was close to 0, while with them, it went up to .6.

Another problem with people choosing "don't know" is that fewer representatives of the population are left for the analysis. If the option is available for several requests, the number of people with complete data on a larger set of variables can decrease, and it becomes questionable whether the respondents who are left in the sample are on the whole representative for the population. These three arguments have led researchers to allow for the "don't know" option but only if the respondent explicitly asks for it. However, whether this is the most scientific course of action, we will evaluate later.

So far, the focus of our discussion has been the specification of response categories for subjective variables. However, ordinal response categories are also used for objective variables such as the frequency of activities or categories of income and prices. An example could be:

5.16a   *How often do you watch TV during a week?*
   1. *Very often*
   2. *Often*
   3. *Regularly*
   4. *Seldom*
   5. *Never*

If we had omitted the response alternatives, this could have been an open-ended request, but researchers often add response categories to such requests, and the issue is that respondents can differ in their interpretation of the different labels: what is "often" for one person means "seldom" for another. It all depends on the reference point of the respondent. Therefore, these ordinal scales are called *vague quantifiers*. We could have also asked the following:

5.16b   *How often do you watch TV during a week?*
   1. *Every day*
   2. *Five or six times*
   3. *Three or four times*
   4. *One or two times*
   5. *Never*

This request is more precise and less prone to different interpretations. Even so, 5.16b is an ordinal scale because it is not clear what numeric values the categories 2–4 represent.

TABLE 5.1   The results of Hippler and Schwartz with respect to TV watching

Categories at the low side		Categories at the high side	
Categories	Percentage of respondents	Categories	Percentage of respondents
<.5 hour	11.5		
.5–1.5 hours	53.8		
1.5–2.5 hours	34.7	<2.5 hours	70.6
>2.5 hours	0	>2.5 hours	29.4
Total	*100*		*100*

Similar scales can be used for income and prices with the option of using vague quantifiers or more precise category labels. Hippler and Schwarz (1987) made a remarkable observation when they varied the category labels in an experiment about the amount of time people watch TV. In it, they did not use vague quantifiers like those of example 5.16a but two different and separate categorizations for the number of TV viewing hours. Their results are presented in Table 5.1. The table shows that the different categories had a considerable effect on the responses. Their explanation was that respondents do not have an answer readily available for this type of request. Instead, they use the response scale as their frame of reference. Respondents estimate their TV watching time on whether they view themselves as more or less TV watching than other persons. Therefore, if they consider that they watch more TV than others, they will choose the high end of the scale, and vice versa. This experiment shows that even for objective variables, the answers do not represent absolute judgments but relative judgments. It has been suggested that people always make relative judgments. If that is so, it is better to adjust the approach of asking requests to the human judgment factor. We will investigate this problem in the next section in more depth.

5.2.3   Continuous Scales

Another form for response possibilities is to give respondents instructions to express their opinions in numbers or lines. This approach was developed with the idea that it would result in more precise information than would the other methods discussed in previous sections.

Originally, such approaches were used in psychophysics. For an overview, we refer to Stevens (1975). Presently, these measurement devices have been introduced in the social sciences by Hamblin (1974), Saris et al. (1977), Wegener (1982), and Lodge (1981). When these approaches were introduced, special procedures, called *magnitude estimation* and *line production*, were used. The basic idea is rather simple and will be illustrated by several examples of procedures used in practice. Originally, a request for an answer was formulated as follows:

   5.17a   Occupations differ with respect to status. We would like to ask you to
           estimate the status of a series of occupations. If we give the status of

*the occupation of a schoolteacher a score of 100, how would you
evaluate the other occupations? If an occupation has a status that is
twice as high as that of a schoolteacher, give a twice larger number
or 200. If the status of the occupation is half that of a schoolteacher,
divide by 2, which gives 50.*

What is the status of a physician?

*Of a carpenter?
And so on.*

People are asked to match the ratios of status judgments with the ratios of numbers.
This could also be done using "line production," as has been shown in the following
instruction:

5.17b   *Occupations differ with respect to status. We would like to ask you to
estimate the status of a series of occupations. We express the status of
the occupation of a schoolteacher by a standard line as follows:*

———————————————————

*If an occupation has a status that is twice as high as that of a school-
teacher, draw a line which is twice as long. If the status of the occupation
is half of that of a schoolteacher, draw a line that is half the size of the
standard line.*

What is the status of a physician?

*Of a carpenter?
And so on.*

With these procedures, a striking precision of responses was obtained (Hamblin
1974; Saris et al. 1977). However, in their embryonic stage, these approaches were
used only for evaluation of stimuli, as we have previously indicated. Currently, other
concepts are also measured in this way. For example, we could reformulate the
frequently asked satisfaction request using continuous scales as follows:

5.18a   *How satisfied are you with your house? Express your opinion with
a number between 0 and 100, where 0 is completely dissatisfied and
100 is completely satisfied.*

This request differs in several points from the original instruction. The first point is
that the ratio estimation is no longer mentioned. The reason is that the results are not
very different whether one gives this instruction explicitly, while at the same time,
omitting this instruction makes the formulation much simpler. The second point is
that two *reference points* have been mentioned instead of just one. This is due to
research showing that people use different scales to answer these requests if only one
reference point is provided, while using two reference points, it is less of a concern

(Saris 1988b). A condition for this conclusion is that *fixed reference points* are used. With fixed reference points, we mean that there is no doubt about the position of the reference point on the subjective scale in the mind of the respondent. For example, "completely dissatisfied" and "completely satisfied" must be the endpoints of the opinion scale of the respondent. If we would use "dissatisfied" and "satisfied" as reference points, then respondents may vary in their interpretation of these terms because some of them see them as endpoints of the scales, while others do not.

The disadvantage of using numbers is that people tend to use numbers that can be divided by 5 (Tourangeau et al. 2000). This leads to rather peaked distributions of the results. This can be largely avoided by the use of line length instead of requesting a numerical evaluation. For request 5.18a, the instruction, using line length as response mode, would be as follows:

> 5.18b   *How satisfied are you with your house? Express your opinion in length of lines, where completely dissatisfied is expressed by the following line:*
>
> –
>
> *and completely satisfied by the following line:*
>
> _____
>
> *Now express your opinion by drawing a line representative of your opinion:*

The disadvantage of the line production is, of course, that later the lines need to be measured. This is a challenge if paper-and-pencil procedures for data collection are used, but with computer-assisted interviewing (CAI), the programs can measure the length of lines routinely.

Although these methods gained some popularity around the 1980s, they are still not frequently employed. One reason is that researchers want to continue with existing measurement procedures and do not want to risk a change in their time series due to a method change. Another reason is that several researchers have argued that the lines do not increase precision a lot. The most outspoken author is Miethe (1985). Some other people (Alwin 1997; Andrews 1984; Költringer 1995) do not agree with Miethe's argument, and they have shown that better data is indeed obtained if more categories are used. In the next section, we will argue why we think that it is better to use more than 7-point category scales and why we prefer line drawing scales as a standard procedure.

Before moving to the next section, we should clarify a point about the measurement level of continuous scales. So far, we have discussed nominal scales and ordinal scales; however, it is interesting to know what kind of measurement level is obtained using the continuous scales discussed here. One may think that the scales discussed represent ratio scales given the ratio instructions originally requested. However, Saris (1988a) has found that the line and number responses are nearly perfectly linearly related (after correction for measurement error and logarithmic transformation), and he concludes that on the basis of these results, the measurement level of these continuous scales is log-interval (Stevens 1975). This means that the data obtained with the suggested response procedure, after logarithmic transformation, can be analyzed using interval-level statistics. From this, it follows that continuous scales have a higher measurement level than do the previously discussed category scale procedures.

5.3 HOW MANY CATEGORIES ARE OPTIMAL?

Most researchers are in agreement that it is better to use more than two categories if it is possible, and they are even inclined to accept that 7-point scales are even better. For example, Krosnick and Fabrigar (1997) make this recommendation very explicitly and conclude not to use more categories. Several studies share this opinion and they have tried to indicate that people cannot provide more information than suggested by a 7-point category scale.

However, we are of the opinion that respondents are capable of sharing more information. This can be shown by asking people the same judgment three times: once expressed on a category scale and once expressed in numbers and once expressed in lines. If people did not have more information than can be expressed in the number of categories of the scale, the correlation between line and number judgments of stimuli placed in the same category of the category scale would be 0. This is, however, not the case. The correlation between the line and number responses of stimuli that all received the same categorical scale score can go as high as .8. This reveals that people have indeed more information than they can express in the verbal labels of the standard category scales (Saris 1998a).

Why this extra information normally is not detected has to do with the problem that the respondents may use different scales in answering requests even from one occasion to the next. Saris (1988b) calls this type of phenomenon "variation in the response function." He suggests that respondents answer very precisely but in their own manner. Figure 5.1 illustrates this phenomenon.

In this figure, respondent 1 expresses herself in rather extreme words compared to the others: if she has an opinion that is close to 0, she also gives responses close to 0, and if she has an opinion close to 100, she also gives responses close to 100. The other two respondents give much more moderate responses even though they have the same opinions. Of course, this is just a fictional illustration of the problem. For empirical illustrations, we refer to Saris (1988b). In this illustration, we have assumed that all respondents will give the response 50 if they have an opinion of 50 about the evaluated stimuli. In practice, this is only necessarily so if one reference stimulus is provided with a standard response of 50; otherwise, this point will also vary across respondents.

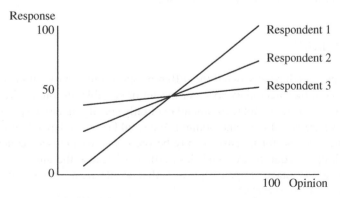

FIGURE 5.1   Variations in the response function.

Let us now look at what happens when only one stimulus is provided for which all respondents have an opinion of 100. In accordance with Figure 5.1, we see that the respondents will give a different response even though they have the same opinion. This means that the varying responses cannot be explained by substantive variables. They are a consequence of the differences in response function and could be mistakenly interpreted as measurement error. This is a problem for researchers because this kind of variation will occur while the respondents may have very precise, reliable responses if you look at their individual data.

The variation in responses due to variation in response function is larger at the extreme ends of the scale and closer at the middle. This phenomenon can explain that extension of scales with more categories, for example, above 7, will increase what is seen as measurement error, and it is for this reason that many researchers believe that they do not gain more information by increasing the length of the scales.

On the basis of our research, with respect to the amount of information that people can provide and the problem of variation in response functions, we would like to suggest that people often have more information than they can express in the labels of the standard 7-point category scales, but increasing the number of categories also increases the problem that respondents will start to use their own scale. The latter problem can be reduced by the use of more than one fixed reference point. If two fixed reference points are given on the response scale, then the endpoints of the opinion and response scale are the same for all people, and if a linear response function is used, the responses will be comparable. It has been shown that in that case, the variation in response functions is indeed smaller. In this way, it is possible to obtain more information from respondents using response scales with 7-point scales (Saris and de Rooy 1988).

That such procedures are not so difficult to formulate has been illustrated earlier because the last examples of continuous scales (examples 5.18a and 5.18b) provided in the previous section satisfied the aforementioned criteria. It was also mentioned there that the line production is the better procedure because the respondents will not round off their answers when using the line method. In Part III of this book, where we discuss the empirical evidence for the effects of the different choices that we discuss here, we will come back to this issue.

5.4   SUMMARY

In this chapter, we have discussed the different options that exist with respect to the specification of the uncertainty space or the set of possible responses. We have seen that some researchers do not recommend explicitly specifying any response options. However, we are not of the same opinion. We would say that depending on the context, an open request for an answer may be preferable to a closed request. On the other hand, open requests are much less efficient because the answers have to be coded, but the advantage of open requests is that people are not forced into the frame of reference of the researcher.

One type of open request, the "WHY requests," was given special attention in this chapter because it is commonly used. However, we share Krosnick and Fabrigar's (forthcoming) view in not recommending this type of request because respondents may be led into rationalizations and may not give their true reasons for the answer. It was also shown through a research review that introspection is not a very scientifically valid procedure.

Furthermore, we have seen that there are some requests that are difficult to translate into closed request form, such as open requests about sequences of events and open requests that are open with respect to the concept measured. In those specified cases, open requests are probably the preferred method. Therefore, it depends on the topic, context, and researcher's intent, whether open or closed requests should be selected for a request for an answer.

With respect to closed requests, a distinction was made between nominal and ordinal categorical response scales and continuous response scales. There are many forms of categorical scales, especially ordinal scales. Several examples were discussed. In doing so, we introduced choices that are connected with the development of such scales such as:

- Correspondence between the labels and the numbers of the categories
- Symmetry of the labels
- Bipolar and unipolar scales and agreement between the concept and the scale
- The use of neutral or middle category
- The use of "don't know" options
- The use of vague quantifiers or numeric categories
- The use of reference points
- The use of fixed reference points
- The measurement level

Additionally, we introduced the advantages and disadvantages of choosing the number of possible responses. Our logical argument is that more information can be obtained than is possible in the standard 7-point category scales if we allow respondents to provide more information. However, in order to obtain responses that are comparable across respondents, at least two fixed reference points need to be specified in the response procedures that are connected to the same responses across all respondents. In this context, we suggested that line production scales provide better results than magnitude estimation; since respondents have a tendency to prefer numbers that can be divided by 5, this leads to peaked response distributions and this does not happen with line production scales.

It should not be concluded that the line production scales should be used for all topics and at all times. If researchers do not need more information than "yes" or "no," it does not make sense to force the respondents to use a continuous scale. Also the continuity in survey research often requires the use of the standard category scales. The continuous scales may have a future when CAI becomes more popular.

EXERCISES

1. In the following text is an example of a request for an answer:
    All in all, nowadays, are you feeling very happy, quite happy, not so happy,
    or not at all happy?
    1. *Very happy*
    2. *Quite happy*
    3. *Not so happy*
    4. *Not at all happy*

    What can you say about this response scale with respect to:
    a. The correspondence between the labels and the numbers of the categories?
    b. The symmetry of the labels?
    c. The bipolar and unipolar scales and agreement between the concept and the
        scale?
    d. The use of a neutral or middle category?
    e. The "don't know" option?
    f. The use of vague quantifiers or numeric categories?
    g. The use of reference points?
    h. The use of fixed reference points?
    i. The measurement level?

2. Could you reformulate the request in order to improve the quality of the request in
    the light of the evaluation on the different characteristics mentioned in Exercise 1?

3. Is it also possible to formulate this request in an open request form? If so, how?

4. Is it also possible to formulate this request using continuous scales? If so, how?

5. Which of the three scales would be the most attractive one and why?

6. One could also have asked: *How are you these days?*
    a. Do you see a problem with this request?
    b. Is it possible to reformulate this request in a closed form?

7. Now look at your proposal for a questionnaire. Do you think that you have
    chosen the best response categories? If not, make improvements and indicate
    why you have made these improvements.

6

THE STRUCTURE OF OPEN-ENDED AND CLOSED SURVEY ITEMS

So far, we have discussed the basic form of requests for an answer, but often, they are placed in a larger textual unit called a "survey item," which consists of an entire text that requires one answer from a respondent (Saris and de Pijper 1986). Andrews (1984) defined a survey item as consisting of three different parts of text or components, namely, an introduction, one or more requests for an answer, and a response scale. Molenaar (1986) uses quite similar components. In this chapter, we propose distinguishing even more components of a survey item. First, we will describe the components and thereafter, we will present different structures of survey items for open and closed requests. The structure of batteries of requests for an answer, such as those using stimuli or statements, will be the topic of Chapter 7. We close this chapter with a discussion of the advantages and disadvantages of the different forms of open-ended and closed survey items.

6.1 DESCRIPTION OF THE COMPONENTS OF SURVEY ITEMS

Figure 6.1 shows the basic components of a survey item. The reader should notice that we make a distinction between parts embedded in the request for an answer as discussed before and parts that can be juxtaposed before or after the request for an answer.

In our opinion, the following parts can be added: an introduction, a motivation, information regarding the content, information regarding a definition, an instruction

Design, Evaluation, and Analysis of Questionnaires for Survey Research, Second Edition.
Willem E. Saris and Irmtraud N. Gallhofer.
© 2014 John Wiley & Sons, Inc. Published 2014 by John Wiley & Sons, Inc.

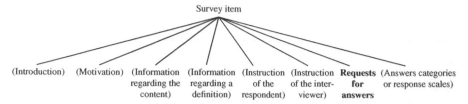

Survey item

(Introduction)   (Motivation)   (Information regarding the content)   (Information regarding a definition)   (Instruction of the respondent)   (Instruction of the inter-viewer)   **Requests for answers**   (Answers categories or response scales)

FIGURE 6.1   Decomposition of a survey item into its components.

of the respondent, an instruction for the interviewer, the request for an answer, and response categories or scales, as shown in Figure 6.1.

The components indicated within parentheses in Figure 6.1 are optional. This implies that the request for an answer is the core unit of a survey item, and it also means that the simplest form of a survey item is just an open request for an answer and nothing more. The figure also demonstrates that a survey item can consist of many more components. How many and which ones will be discussed further. But, first, we begin with a description and illustration of the different components.

Introductions (INTRO) are meant mainly to indicate the topic of the request for an answer to the respondent. In general, they consist of one or more sentences. Examples are as follows:

6.1   *Now, a couple of questions follow about your health.*
6.2   *The next question is on the subject of work.*

Sometimes, two requests for an answer are formulated, and the first request functions just as an introduction because no answer is asked for it. The second request for an answer is the one to be answered that is indicated by the answer categories. Examples 6.3 and 6.4 are an illustration:

6.3   *Would you mind telling me your race or ethnic origin (INTRO)?*
      Are you white, black, Hispanic American, American Indian or Alaskan native, or Asian or Pacific Islander?
      1. White but not Hispanic
      2. Black but not Hispanic
      3. Hispanic
      4. American Indian or Alaskan native
      5. Asian or Pacific Islander

6.4   *What is your opinion on each of the following proposals (INTRO)?*
      Could you tell me if you are for or against it? There should not be death penalty anymore.
      1. For it
      2. Against it

The next component of a survey item we introduce is called *motivation (MOTIV)*. This part of text explains the broader purpose of the research to stimulate the respondent to answer the question(s). It consists of one or more sentences and contains keywords like "purpose," "research," and "representative." Examples 6.5 and 6.6 demonstrate our point:

> 6.5 *We are doing research to find out the best way to ask questions.*
>
> 6.6 *For the statistical processing of a survey, it is important that the research is representative of the entire population. In order to obtain this, we need to know the general range of incomes of all people whom we interview.*

Information regarding the content (INFOC) clarifies or explains something about the content of the survey item. It is included in a survey item because many people do not have an opinion about many issues (Converse 1964). Linguistically, it consists of one or more sentences. Examples 6.7–6.9 illustrate this concept:

> 6.7 *The European Political Union may include a common defense arrangement involving the member states of the European Community. Successive Irish governments have accepted that moves toward the European Political Community could mean scraping Ireland's policy of military neutrality.*
>
> 6.8 *There are different ways by which people can show their disagreement with the measures employed by the government.*

Frequently, the explanation or the clarification contains arguments for and/or against a point of view. Kay (1998) has used this approach to test the stability or strength of opinions. Example 6.9 is an illustration:

> 6.9 *Since the crime rate among young people has been drastically increasing over the last few years, some citizens and political parties think that the government has to take strong action against crime.*

However, example 6.9 provides only one-sided information. Using such information, one can get very different results depending on the information given. If arguments are given, they should include arguments for both points of view (Sniderman and Theriault 2004), as is done in example 6.10:

> 6.10 *As you probably know, some politicians fear that within a few years they will have to deal with an energy shortage. They therefore propose building more nuclear reactors. Other politicians warn about the dangers of nuclear energy and therefore suggest that no new reactors should be built.*

Saris et al. (1984) have developed a choice questionnaire using this approach in order to solicit *well-considered opinions* from respondents. For an elaborate discussion of this approach and its evaluation, we refer the reader to Neijens (1987).

We also defined a component about *information regarding a definition (INFOD)*. This part of text defines some "concept" used in the survey item like "abortion" or "euthanasia" or some scales. It can consist of one or more sentences but frequently, it is shorter than a sentence, implying that it is embedded in another component, which is often an instruction to the respondent or a request for an answer. Illustrations of this component type might look like examples 6.11–6.13:

6.11   *By abortion, we understand the deliberate termination of a pregnancy by a physician.*

6.12   *The net income is the amount you receive after tax deduction.*

6.13   *You get a ladder with steps that go from 0 at the bottom to 10 at the top; 5 is the middle step at the ladder. At the top of the ladder are the very best feelings you might expect to have and at the bottom of the ladder are the worst feelings.*

The next two components relate to *instructions*. Researchers can give instructions to the respondents or the interviewers. Linguistically, they are characterized by sentences in the imperative mood or polite variations of it. Here, we discuss only instructions that are used outside the request for an answer. Instructions can also be used within the request for an answer that has already been discussed in Chapter 3. Examples 6.14 and 6.15 illustrate *instructions to the respondents* (INSTRR):

6.14   *What do you think of President Bush? Express your opinion in a number between 1 and 0, where 0 is very bad and 10 very good.*

6.15   *Look at the card and tell me which answer seems to fit your situation.*

In the first example, the survey item begins with a request for an answer and continues with an instruction for the respondent (INSTRR). In the second example, the positions are reversed. Examples of *instructions of interviewers* (INSTRI) are as follows:

6.16   *Hand over the show card. Only one answer is possible.*

6.17   *Read out the following text. If unclear, repeat instructions.*

The next component of a survey item indicated in Figure 6.1 is the *request for an answer (REQ)*. We will not repeat our discussion of this form because it has already been detailed in Chapter 3.

The last component of a survey item presented in Figure 6.1 relates to *answer categories or response scales (ANSWERS)*. They are optional, as open requests for an answer do not require them and respondents have to give their own answers. Since Chapter 5 is entirely devoted to this topic, we only wish to alert you to the presence of this component at this time.

6.2   DIFFERENT STRUCTURES OF SURVEY ITEMS

In this section, we will discuss some different structures of survey items occurring in questionnaires as a consequence of researchers' choices. We will also indicate the position of the components in the item as far as possible. First, we present the structures encountered in a number of selected questionnaires. For this purpose, we used a sample of 518 Dutch survey items selected on the basis of a random procedure from a larger sample of 1527 survey items by Molenaar (1986: 34–44). Since this sample contains only requests for an answer with closed answer categories, we added a sample of 103 open-ended Dutch requests for an answer from a database of the Steinmetz archive collected by Wouters (2001). A convenience sample of factual requests for an answer is studied on the basis of a collection of questionnaires from the Dutch Gallup institution NIPO, the Telepanel, and the Dutch Bureau of Statistics of the late 1980s and early 1990s.

In order to compare these Dutch survey items with structures of English survey items, we collected 200 closed and open-ended requests for an answer that actually do not constitute a representative sample, from the Institute of Social Research (ISR, Ann Arbor, MI) questionnaires from the period 1979 to 1981, the Eurobarometer (1997), and survey items from Sniderman et al. (1991). Factual requests for an answer were also collected from Sudman and Bradburn (1983). A similar collection of 250 German survey items coming from surveys from the IFES in Austria was also used for comparative purposes.

The aforementioned databases of survey items serve as an overview of different structures that occur in practice. At the end of the chapter, we present a quantitative estimate of the frequency of occurrence of the structures for subjective variables on the basis of the random sample of survey items collected by Molenaar (1986).

In this chapter, we will separately discuss two groups of survey items: open-ended requests for an answer and closed ones. This distinction is made because we expected a considerable difference between them.

6.2.1 Open-Ended Requests for an Answer

First, we illustrate the structure of an open-ended survey item that consists only of a request for an answer. There are no answer categories or rating scales mentioned since the request for an answer is open-ended. Examples 6.18–6.20 illustrate this type of structure:

6.18   *What is, in your opinion, the most important problem with which our country is confronted nowadays (REQ)?*

6.19   *Please give me the reasons why you changed your job last year (REQ).*

6.20   *How many hours a day do you watch television (REQ)?*

It will be obvious that the first two examples (6.18 and 6.19) are open-ended subjective requests for an answer where respondents are free to give their answers. Example 6.20 is a factual request for an answer, where the respondent provides the appropriate answer in terms of a number, which we also consider as open-ended since no answer categories are provided (Tourangeau et al. 2000). The following structures of open-ended requests for an answer contain two components. The first one that we illustrate consists of an introduction and a request for an answer.

6.21   *Now, we would like to ask a question about your job (INTRO). What do you think of your present job (REQ)?*

In this example, the first sentence is a typical introduction, while the second sentence is an open request. Sometimes, the introduction is also formulated as a question:

6.22   *What do you think about the political situation in Europe (INTRO)? Do you think that the developments go in the right direction (REQ)?*

The first request in this example must be seen as an introduction because no answer is expected from it. The second request in example 6.22 indicates that an evaluation is requested, but as it is an open-ended request for an answer, answer categories are not provided.

Another structure of an open-ended survey item used is in combination with a closed request for an answer. This structure consists of three components, namely, a preceding closed request for an answer with answer categories either embedded or specified separately where an open-ended request for an answer follows. In this case, an answer to both requests is expected. We mention this type of question because the closed question is normally used only as an introduction, while the actual open request is not complete as its content relates to the closed question. This can be shown by the following open-ended examples such as "could you explain your answer?" or "could you tell me why?," which are not sufficient alone. An example of this combination is as follows:

6.23   *Do you think that in our country many or few people (ANSWER categories) make easy money (closed REQ)?*
   1. Many people
   2. Few people (ANSWER CATEGORIES)

   How did you get this idea (REQ open-ended)?

It is also obvious that this open-ended request for an answer relates to the specific answer given and asks for more detailed information.

6.2.2   Closed Survey Items

First, we will discuss the structure of a closed survey item that consists only of a request for an answer with explicitly mentioned answer categories.

Examples are as follows:

6.24   *Do you think that we should have a world government for all countries (REQ)?*
   1. Yes
   2. No (ANSWER CATEGORIES)
6.25   *Please tell me if your parents were members of a church (REQ).*
   1. Yes
   2. No (ANSWER CATEGORIES)

These structures are rather normal in mail surveys or other self-completion surveys. In such surveys, the response categories are presented after the request for an answer and are not embedded in the request for an answer. In oral surveys presented by an interviewer, such forms are rather unusual, except in the case of very simple requests for an answer with clear "yes" or "no" answer categories. In surveys presented by an interviewer, the answer categories are also often presented on a card given to the respondent (show card) or embedded in the request for an answer. This latter structure is the next case to be discussed.

In interviewer-administered surveys, closed survey items consist of a request for an answer in which answer categories are embedded. They are mentioned in the interview form again after the request. The interviewer does not repeat them again. These second set of answer categories are presented for administrative purposes. Therefore, in these cases, we indicate their presence in the printed form of the questionnaire by enclosing them in parentheses. Examples 6.26 and 6.27 illustrate this structure:

6.26   *Do you own or rent your home (REQ + ANSWERS)?*
   1. (I rent the home
   2. I own my home) (ANSWER CATEGORIES)
6.27   *Please tell me whether you consider our president a good or bad leader or neither good nor bad (REQ + ANSWERS).*
   1. (Good leader
   2. Neither good nor bad
   3. Bad leader) (ANSWER CATEGORIES)

After some discussion of the simplest structures of closed survey items, we will show how structures of survey items can be expanded by adding components between the request for an answer and the answer categories.

The structures earlier mentioned can be expanded by inserting, for instance, an instruction for the respondent between the request for an answer and the answer categories:

6.28   *Are you working in the household or do you have a paid employment (REQ + ANSWER categories)?*
   Please mark only one answer category. If you engage in both activities, choose the one you consider the most important (INSTRR).
   1. Works in household
   2. Has paid employment (ANSWER CATEGORIES)

The inserted component can also be an instruction to the interviewer that leads to the following example:

6.29   *What is the main reason why you might not go and vote at the next European elections (REQ)?*
   Interviewer: show card, one answer only (INSTRI).
   1. I am not interested in politics.
   2. I am not interested in the European elections.
   3. I am against Europe.
   4. I am not well enough informed to vote.
   5. Other reasons (ANSWER CATEGORIES).

The added component can also be information regarding a definition, for example:

6.30   *Are you a member of a public library (REQ)?*
   By "public library," we understand a library other than at a school or university (INFOD).
   1. Yes
   2. No (ANSWER CATEGORIES)

The examples just mentioned demonstrated that survey items can be expanded by inserting an instruction for the respondent, an instruction for the interviewer, or information regarding a definition after the request for an answer before the answer categories.

The next examples will present extensions of survey items by inserting components such as an introduction, information regarding the content, an instruction for the interviewer, or a motivation from the researcher before the request for an answer. Typical examples are as follows:

> 6.31 *The next question concerns the upcoming elections (INTRO).*
> *Please tell me if there were elections tomorrow would you go to vote (REQ)?*
> *1. Yes*
> *2. No (ANSWER CATEGORIES)*

> 6.32 *People look for different things in their job. Some people like to earn a lot of money. Others prefer interesting work (INFOC).*
> *Which of the following five items do you most prefer at your job (REQ)?*
> *1. Work that pays well*
> *2. Work that gives a feeling of accomplishment*
> *3. Work where you make most decisions yourself*
> *4. Work where other people are nice to work with*
> *5. Work that is steady with little chance of being laid off (ANSWER CATEGORIES)*

> 6.33 *Interviewer: ask this question also when the respondent did not answer the previous question (INSTRI).*
> *Is the monthly income of your household higher than $10,000 (REQ)?*
> *1. Yes*
> *2. No (ANSWER CATEGORIES)*

> 6.34 *We need the information about income of your household to be able to analyze the survey results for different types of households. An income group is enough.*
> *It would help us a lot if you would be able to state what income group your household belongs to (MOTIV).*
> *Please tell me to which one of the following income groups your household belongs after tax and other deductions (REQ).*
> *1. $20,000–$50,000*
> *2. $50,000–$100,000*
> *3. Higher than $100,000 (ANSWER CATEGORIES)*

One also can find more complex structures with three components by inserting more components before the request for an answer. Typical examples are as follows:

> 6.35 *The following request for an answer deals with your work (INTRO).*
> *Some people think that work is necessary to support themselves and their families (INFOC).*

Do you like your work or do you do it as a necessity to earn money
(REQ + answer categories)?
1. Likes his/her work.
2. Work is a necessity to earn money (ANSWER CATEGORIES).

6.36 *Now, I would like to talk about abortion (INTRO).*
"Abortion" means the deliberate termination of a pregnancy by a
physician (INFOD).
Are there in your opinion situations that justify an abortion (REQ)?
1. Yes
2. No (ANSWER CATEGORIES)

In the same way, an instruction for the respondent or a motivation for an answer can be used.

Other structures with four components are also possible. For example, starting with an introduction followed by a request for an answer, an extra instruction or information and answer categories at the end. An example could be:

6.37 *In the following requests for an answer, we would like to ask you*
about your leisure activities (INTRO).
Please tell me which of the following activities you prefer most in your
spare time (REQ).
Indicate only one activity (INSTRR).
1. Sports
2. Watching TV
3. Reading
4. Going shopping
5. Talking with people
6. Something else (ANSWER CATEGORIES)

Finally, even more complex structures can be found in the literature such as inserting several components before and after the request for an answer, for example:

6.38 *We want to ask you about your education (INTRO).*
It is very important for us to get a good picture of the education of
our citizens (MOTIV).
By "education," we understand the schools you finished for a degree,
and we want to know the highest education you finished with a degree
(INFOD).
What was the highest level of educational training that you finished
with a degree (REQ)?
1. Primary school
2. Lower vocational training
3. High school
4. Higher vocational training
5. University
6. Others (ANSWER CATEGORIES)

Here follows another example:

> 6.39 *Now, we would like to ask you about issues that are frequently discussed in the media (INTRO).*
> *When a physician assists a patient at his/her own request to die, we call this euthanasia (INFOD).*
> *Some people and political parties wish that euthanasia should be forbidden. Others are of the opinion that a physician always must comply with a patient's request to die. However, there are also people whose opinion lies in between (INFOC).*
> *What is your opinion about euthanasia (REQ)?*
> **Interviewer: present the card (INSTRI).**
> *People who favor euthanasia should choose the number 7, which means that "a physician has to comply with a patient's request." People who are against euthanasia should choose the number 1, which means that "euthanasia should be forbidden." People who are neither for or against it should choose the number 4 (INSTRR)*
> *Where would you place yourself on the scale (REQ)?*

1_____4_____7

Euthanasia	*neither for nor*	*A physician has to comply*
should be forbidden	*against it*	*with a patient's wish*
(ANSWER scale on card)		

This type of request is not used often, because the text becomes very complex and it is unclear whether respondents can answer these questions at all.

6.2.3 The Frequency of Occurrence

After having introduced various structures of closed survey items on the basis of selected data, we can now investigate the frequency of occurrence of the different structures of survey items, as in Table 6.1.

The frequency of occurrence of survey items relating to subjective requests for an answer is studied on the basis of Molenaar's sample. Table 6.1 summarizes the structures of closed survey items that we encountered in this data set. This table shows clearly that structures where answer categories are embedded in the request for an answer are more frequent than structures without embedding them. This is because most interviews are still interviewer-administered in the Netherlands. With the increase in the number of self-administered interviews like mail and Web surveys, this distribution might change quite rapidly.

The table also shows that researchers avoid highly complex survey items. Although complex items are possible as we have shown, they are seldom used in market and opinion research. Most frequently, the items consist of two components. An inspection of the English and German survey items we had collected also confirmed that the structures mentioned in Table 6.1 were similar. Survey items consisting of more than three components were infrequent. Also, the most common extension of a basic structure of a survey item is to start with some information about the content of the

TABLE 6.1 Overview of structures of closed survey items encountered in the sample of requests for answers for subjective variables with closed response categories

Structure of closed survey items relating to subjective requests	Number of components	Frequency	
		%	Absolute
REQ + answer categories	2	8	14
REQ + embedded answer categories	2	39	73
REQ + embedded answer categories + answer categories	2	3	5
INFOC + REQ + embedded answer categories	3	30	53
INFOC + REQ + answer categories	3	5	9
INTRO + REQ + answer categories	3	1	1
INTRO + REQ + embedded answer categories	3	2	3
INTRO + REQ + embedded answer categories + answer categories	3	10	19
INTRO + REQ + REQ + answer categories	4	1	2
INTRO + INFOD + REQ + embedded answer categories	4	1	2
Total		*100*	*181*

following survey item. Another possibility is the use of an introduction. These two structures may be substitutes of one another as they seldom occur simultaneously.

6.2.4   The Complexity of Survey Items

From the respondents' perspective, survey items consisting of various components are more difficult to understand than requests with only one component.

In the literature (Molenaar 1986; Graesser et al. 2000b; Tourangeau et al. 2000), different measures for complexity are used that coincide partially with the ones we make use of. In our research (see Chapter 12), we register which components are present in a survey item. In addition, we determine the complexity of the introduction and the request separately. For both parts, the complexity is studied with indices that will be discussed in more detail.

One of these indicators for complexity is the *number of interrogative sentences*. If there is more than one interrogative sentence in a request, the respondent has to decide which one should be answered, which in turn complicates the comprehension of the whole request.

Another characteristic that increases the difficulty of comprehension relates to the *number of subordinated clauses*. If a component contains one or more subordinate clauses that are embedded in the main clause, it can be assumed that the respondent needs several mental operations before fully understanding the sentence. As an example, we mention the following three requests:

6.40   *Do you think, although there is no certainty, that your financial situation will improve in the future?*

6.41   *Do you think that your financial situation will improve in the future?*

6.42   *In your opinion, will your financial situation improve in the future?*

Example 6.40 contains two subordinate clauses, where the second contains the proper request. Example 6.41 consists of only one subordinate clause containing a request.

The third example (6.42) has no subordinate clauses, and the request is stated in the main clause. This example is the easiest to comprehend.

The number of words of a component also contributes to its complexity. The more words it contains, the more difficult it is to understand. This also can be studied by means of *the average number of words for each sentence.*

Still another characteristic that adds to the complexity of a sentence is the mean *number of syllables* in a sentence. It is assumed that the more syllables a sentence contains, the more difficult it is to understand.

The last characteristic relating to complexity is *the number of abstract nouns on the total number of nouns. Abstract nouns* indicate objects that in principle cannot be touched, which means that they do not refer to living beings or physical objects, while concrete nouns refer to the latter categories. We assume that the comprehension becomes more difficult with the increase of abstract nouns in comparison to the number of concrete nouns.

This overview of complexity characteristics suggests reducing complexity by using only one interrogative sentence in the request, few subordinate clauses, and short sentences with a minimal amount of abstract nouns. Most survey researchers agree with these recommendations (see literature).

6.3 WHAT FORM OF SURVEY ITEMS SHOULD BE RECOMMENDED?

Our knowledge about the optimal form of survey items is still rather limited. Some new results on this topic will be mentioned in Part III of this book. However, about the use of some components, some research has already been done. First of all, Belson (1981) studied different forms of media items. For example, after respondents answered the question *"How often did you watch TV during the last week?,"* he asked them how they had interpreted the terms "watch TV," "you," and "last week." He wanted to see how many people interpreted the question according to the wishes of the researcher. It turned out that many people interpreted the terms differently as expected. For example, "watch TV" for some people meant that they were in the room while the TV was on. "You" could mean the respondent or his/her family. "Last week" was by some people interpreted as in the evening only, ignoring daytime viewing. Also weekend viewing was occasionally ignored.

In order to improve the question, Belson (1981) tried to include definitions of what was meant. This led to the following formulation:

How often did you watch TV during the last week?

By TV watching, we mean that you, yourself, are really watching the TV

while we would like to ask you to include day viewing and weekend viewing.

Somewhat surprising was that the number of misinterpretations of the request for an answer in the new form was not much lower than for the former. Belson's explanation was that the question was too long and people had already made up their minds before the definitions were given.

This does not mean that the length of the survey item always has negative effects on the quality of the responses. Schuman and Presser (1981) and Sudman and Bradburn (1983) suggest that the length of the survey item can have a positive effect if the topic is announced early and no further substantial information is given. For example, the question

6.43   *Should abortion be legalized?*
       0. No
       1. Yes

can be extended without adding new information as follows:

6.44   *The next question concerns the legalization of abortion. People have*
       different opinions about this issue. Therefore, we would like to know
       your personal opinion.
       Could you tell me what your opinion is, should abortion be legalized?
       0. No
       1. Yes

The important difference between the longer form of Belson and the last example is that no relevant information is added after the request for an answer has been made clear. In the last form, the respondents have more time to think and thereby improve their answers.

The other side of the coin is that one has to give extra information if the object of the question is not known to many respondents. For example, the meaning of the terms euthanasia, democracy, globalization, the WTO, and so on may be unknown to large portions of the population. In that case, an explanation of the term is necessary if they are to be used in a survey.

The findings of Belson suggest that these definitions definitely should not be given after the question, but before the request. We suggest starting a request with a definition of the concept. For example, if we want to know the opinion about a policy of the WTO with respect to free trade, we could use the following survey item:

6.45   *In order to regulate world trade, an organization of the UN, called*
       WTO, develops rules for the world trade to reduce the protection of
       countries of their own products and therefore to promote free trade in
       the whole world. What do you think of these activities of the UN?
       Express your opinion in a number between 0 and 10 where
       0 = completely against, 5 = neutral, and 10 = completely in favor.

In this case, a definition of the concept of interest is needed, and it should be given in relative simple words *before* the question is asked. This is to ensure that the respondents listen to the explanation before they decide their response.

6.4   SUMMARY

In this chapter, we introduced different components of survey items and described the combinations of components that occur in open-ended and closed survey items. Given the data in Table 6.1, we can conclude that closed survey items consist of two (50%) or three (49%) components. The most often encountered structure of a closed survey item with two components consisted of a request for an answer with an answer component. Since some requests require an introduction or more information or an instruction, these components were sometimes added (introductions 15%, information regarding the content 35%). However, there is always a tradeoff between precision and complexity. We have mentioned the following aspects of a survey item that increase its complexity:

- The number of components in the survey item
- The presence of more than one interrogative sentence
- The number of subordinated clauses
- The number of words per sentence
- The mean number of words in a sentence
- The mean number of syllables per word
- The ratio of abstract and concrete nouns

Although it is in general suggested that complex sentences should be avoided, there is also research suggesting that increasing the length of the questions improves the quality of the answers. Some new results with respect to the effects of these features of survey items on their quality will be discussed in Part III when we present the effects of these aforementioned choices on data quality.

EXERCISES

In these exercises several examples of survey items from empirical research are presented.

Decompose the different survey items into their components.

1. *Before we proceed to the main topic of the questionnaire, we would like to ask the following question:*
   How long have you lived in your present neighborhood?
   Number of years   _____
   Don't know   _____
   Not answered   _____
   ENTER YEAR ROUNDED TO NEAREST YEAR.
   PROBE FOR BEST ESTIMATE.
   IF LESS THAN 1 YEAR, CODE 0.

2. *How far would you say you discuss politics and current affairs?*
   SHOW CARD
   1. *A few times a week*   ☐
   2. *A few times a month*   ☐
   3. *A few times a year*   ☐
   4. *Never or almost never*   ☐
   5. *Don't know*   ☐

3. *Do you <u>actively</u> provide any support for ill people, elderly neighbors, acquaintances, or other people <u>without</u> doing it through an organization or club?*
    REGISTER ONLY UNPAID, VOLUNTARY ACTIVITY. INCLUDE ANY FINANCIAL SUPPORT GIVEN BY THE RESPONDENT TO ILL, ELDERLY, ETC.
    1. *Weekly*   ☐
    2. *Monthly*   ☐
    3. *Yearly*   ☐
    4. *Never, or almost never*   ☐

4. *We all know that no political system is perfect but some may be better than others. Therefore, we would like to ask you the following about the functioning of democracy in our country. How satisfied are you with the way democracy functions in our country?*
    a. *Very dissatisfied*
    b. *Quite dissatisfied*
    c. *Neither satisfied nor dissatisfied*
    d. *Quite satisfied*
    e. *Very satisfied*

5. In your questionnaire, did you also use components other than requests for answers and answer categories? If so, check whether they are in line with common practice or whether you did something unusual? According to your judgment, is that good or bad?

7

SURVEY ITEMS IN BATTERIES

In the last chapter, we discussed the forms that single survey items can take. However, in Chapter 4, we mentioned that researchers in the social sciences often bring items together in batteries. In that case, the different survey items do not stand alone any-more but often are connected by one introduction, instruction, and one request for an answer and a set of answer categories. Since we treat each text unit that requires one response as a survey item, we have to give special attention to the definition of survey items of batteries. The problem is that the different survey items in a battery contain very different text components even though they are often assumed to be equal and treated the same.

What distinguishes batteries is the mode of data collection in which they have been placed. Therefore, we start this chapter with batteries that are used in oral inter-views, followed by a discussion about batteries in mail surveys, and finally batteries employed in computer-assisted self-interviews (CASI) are discussed. In each case, we will discuss which components should be seen as belonging to each survey item. In the summary and discussion, we also will give some recommendations.

We will discuss the different battery types, not because we think that batteries are a good tool for survey research, but because they are so popular. As we have indi-cated in Chapter 4, we think that the standard batteries of agree/disagree responses have done more harm than good for the social sciences. Having given our words of caution and advice, let us start with battery forms employed in oral interviews.

Design, Evaluation, and Analysis of Questionnaires for Survey Research, Second Edition.
Willem E. Saris and Irmtraud N. Gallhofer.
© 2014 John Wiley & Sons, Inc. Published 2014 by John Wiley & Sons, Inc.

7.1 BATTERIES IN ORAL INTERVIEWS

Typical for batteries in oral interviews is that the interviewer reads the items for the respondent. Within this battery class, there is a difference between the face-to-face interview with show cards containing information and the telephone interview, where show cards cannot be used. Let us start with an example of a battery without show cards. A typical example of an oral battery without show cards has been presented as oral battery 1.

In this example, information about the content is read first, and then a request for an answer with implied "yes" or "no" answers is read. Next, the interviewer has to read the first item, wait for the answer, present the next item, and so on. Hence, the introduction and the request are read only before the first survey item and then not anymore. As a consequence, we assume that each survey item after the first one consists only of the statement (or stimulus) and the response categories.

Given the interview process we have suggested earlier, the information about the content and the request for an answer belong to the first item, while all other items consist only of a stimulus since the interviewer does not repeat the answer categories for each item. We think that this is formally correct even though it may not be in agreement with the intention of the original designer of the battery. Moreover, we cannot be certain that the introduction and the question retain a strong presence in the mind of the respondent when they are not repeated for each item.

Oral battery 1

There are different ways of attempting to bring about improvements or counteract deterioration within society.

During the last 12 months, have you done any of the following?
Firstly … READ OUT

	Yes	No
A. *Contacted a politician*		
B. *Contacted an association or organization*		
C. *Contacted a national, regional, or local civil servant*		
D. *Worked for a political party*		
E. *Worked for a (political) action group*		
F. *Worked for another organization or association*		
G. *Worn or displayed a campaign badge/sticker*		
H. *Signed a petition*		
I. *Taken part in a public demonstration*		
J. *Taken part in a strike*		
K. *Boycotted certain products*		

This kind of battery can be used only for very simple response categories as in this example. For more complex response categories, the quality of the responses will improve if the respondent is provided with visual aids. Visual aids help the respondent in two ways: (1) to provide the response alternatives on a card so that they can answer each item consistently and (2) to provide the respondent with the statements. The latter method makes sense if the statements are complex or for emphasis. We will give an example of both types from research practice.

In the following example presented in oral battery 2, the respondents are provided with information about the response alternatives on card D1. First, we present the form provided to the interviewer and then card D1.

In this case, the introduction and the question belong to the first item, and the next items all contain a stimulus and response categories because the respondents have these answer categories always in front of them.

Oral battery 2

CARD D1: *Policies are decided at various different levels. Using this card, at which level do you think policies should be decided mainly about…*

READ OUT AND CODE ONE ON EACH LINE.

	International level	European level	National level	Regional or local level	(DK)
D1…*protecting the environment*	1	2	3	4	
D2…*fighting against organized crime*	1	2	3	4	
D3…*agriculture*	1	2	3	4	
D4…*defense*	1	2	3	4	
D5…*social welfare*	1	2	3	4	
D6…*aid to developing countries*	1	2	3	4	
D7…*immigration and refugees*	1	2	3	4	
D8…*interest rates*	1	2	3	4	

Card D1

Card D1

International level

European level

National level

Regional or local level

If the answer categories are simple but the statements are complex or important, the content of the card can be changed. The next example, oral battery 3, demonstrates this point. First, we present the card for the respondents and after that the form provided to the interviewer.

CARD for oral battery 3

The woman became pregnant because she was raped.

*The woman became pregnant even though she used
a contraceptive pill.*

.

.

.

*The woman became pregnant although there are
already enough children in the family.*

Oral battery 3

*An issue often discussed nowadays is abortion. By abortion, we understand the purposeful
termination of a pregnancy.*

*On this card, some circumstances are indicated under which an abortion might be
carried out.*
*Could you tell me for each circumstance mentioned on the card whether you think that an
abortion is permissible?*

READ OUT

	Permissible	Not permissible
The woman became pregnant because she was raped.		
The woman became pregnant even though she used a contraceptive pill.		
.		
.		
The woman became pregnant although there are already enough children in the family.		

In this case, the response alternatives are rather simple, but the researcher wants the respondents to carefully consider the different possible situations, and therefore, the show card presents the different conditions that have to be evaluated. All the information before the first item belongs to the first item, while the second to the last item contains the statement because the response alternatives are not repeated.

It is even possible that both the stimuli and the answer categories are provided on a show card. An example is given (oral battery 4). The first form is the interviewer version of the battery. Card K2 contains the same information for the respondents.

Oral battery 4

Looking at card K2, how important are each of these things in your life?

Firstly ... **READ OUT**

	Not important at all									Very important	(Don't know)
A. *...family?*	00	01	02	03	04	05	06	07	08	09	10	88
B. *...friends?*	00	01	02	03	04	05	06	07	08	09	10	88
C. *...leisure time?*	00	01	02	03	04	05	06	07	08	09	10	88
D. *...politics?*	00	01	02	03	04	05	06	07	08	09	10	88
E. *...work?*	00	01	02	03	04	05	06	07	08	09	10	88
F. *...religion?*	00	01	02	03	04	05	06	07	08	09	10	88
G. *...voluntary organizations?*	00	01	02	03	04	05	06	07	08	09	10	88

Card K2

	Not important at all									Very important
A. *...family?*	00	01	02	03	04	05	06	07	08	09	10
B. *...friends?*	00	01	02	03	04	05	06	07	08	09	10
C. *...leisure time?*	00	01	02	03	04	05	06	07	08	09	10
D. *...politics?*	00	01	02	03	04	05	06	07	08	09	10
E. *...work?*	00	01	02	03	04	05	06	07	08	09	10
F. *...religion?*	00	01	02	03	04	05	06	07	08	09	10
G. *...voluntary organizations?*	00	01	02	03	04	05	06	07	08	09	10

In this case, the show card (K2) contains the stimuli as well as the response alternatives. For item A, the introduction to the question, the request for an answer, and the answer categories belong to the survey item. Items B–G consist of the stimulus and the response categories.

7.2 BATTERIES IN MAIL SURVEYS

Batteries are also used in mail surveys. We provide common structures of this type in the following examples starting with mail battery 1.

Mail battery 1

For each statement, could you tell me which answer seems to fit your situation?

	Agree strongly 1	Agree moderately 2	In the middle 3	Disagree moderately 4	Disagree strongly 5
My financial situation has improved over the past year.					

My career prospects have
improved over the past year.

.

.

.

My relational problems have
worsened over the past year.

The difference between the mail battery and the oral battery is that the respondents have to do all the work themselves. They have to read the question, the first survey item, and the answer categories. Then, they have to fill in an answer and read the next statement and look for a proper answer again. Hence, the question is read only before the first survey item and then not again. As a consequence, we assume that each survey item after the first one consists of the statement (or stimulus) and the response categories.

A slightly more complex battery is presented in mail battery 2. The battery of requests begins with an introduction or an information regarding the content, after which the request for an answer with answer categories is given, followed by the statements.

Mail battery 2

Here are some statements about our society.

Do you agree or disagree with the following statements?

	Agree	Disagree
People in our society still have great ideals.		
The government should improve gun control.		
.		
Most of our citizens are interested only in making money.		

In this case, an introduction and the question are given before the first survey item. As we have suggested in the preceding text, we assume that these two belong to the first item, while the next items consist of only a statement and answer categories.

In the next example (mail battery 3), the complexity of the battery is increased by adding other components.

Mail battery 3

There are different ways people can express their disapproval about the measures of authorities.

I will mention some to you. Indicate whether you approve of them.
How much do you approve or disapprove of this action?

	Approve completely 1	Approve 2	Disapprove 3	Disapprove completely 4
At a school, some teachers are in danger of losing their jobs. They therefore organize a strike without the approval of the trade union.				
Suppose that people are against a new law and therefore they occupy the Parliament in order to hamper the work of the representatives.				
.				
.				
.				
Suppose that the government had decided to increase the number of pupils in elementary school classes. Some teachers do not accept this and threaten to go on strike.				

This battery starts with information about the content in two sentences. Next, a request for an answer without answer categories is provided in the form of an imperative. However, this is not the real request for an answer because the answer categories do not match with the answer categories suggested later. Furthermore, the next sentence is also a request for an answer. The former request should be seen as an introduction, and the latter question is the real request for an answer. The first item and the answer categories that follow are presented in a table. As we mentioned previously, we assume that the first item contains all the information including the item, while the second and subsequent items consist only of the statement plus their answer categories.

A thoroughly developed and tested mail battery is the *choice questionnaire*. This procedure has been developed to collect a *well-considered opinion* of the respondent (Saris et al. 1984; Neijens 1987; Bütschi 1997). The reason for this development was that it was realized that people may understand a question, like the one about the policy of the WTO, after an explanation is given. However, this does not mean that they have a well-informed opinion about it (Converse 1964). Saris et al. (1984) suggested the use of a procedure called the *choice questionnaire* to collect a well-considered public opinion. Typical for the choice questionnaire is that respondents are provided with arguments on paper both in favor and against the choice they have to make. The procedure is rather elaborate and cannot be shown in detail here; for more information, we refer the reader to Neijens (1987). The problem mentioned in Chapter 6, about the free trade policy of the WTO, was developed in line with the choice questionnaire approach as presented in Table 7.1.

TABLE 7.1   An illustration of a simple choice questionnaire

The possible consequences of the reduction of the protection of national products and the promotion of free trade in the whole world are presented as follows.

Please evaluate the following consequences of this policy by first determining whether a consequence is an advantage or a disadvantage and consequently evaluate the size of the advantage or disadvantage with a number, where a neither large nor small advantage or disadvantage is 400.

How large is the advantage or disadvantage of the following?	*Advantage neither large nor small = 400*	*Disadvantage neither large nor small = 400*
The bankruptcy of some local companies in some underdeveloped countries		
The investments of international companies		
More efficiency in the companies so that they can compete internationally, etc.		

In order to regulate world trade, an organization of the UN, called the WTO, develops rules for world trade.
To reduce the protection of countries of their own products and to promote free trade in the whole world.
Are you in favor or against free trade in the world?
0 against   1 in favor

The table shows that the respondents are provided with information about the choice they have to make. This information is provided in the form of questions concerning evaluations of consequences for the possible options. In this example, only one option has been discussed; however, the option that no free trade policy is introduced could be treated in the same way. People are asked to give numeric evaluations of the size of the possible advantages and disadvantages because in this way it is possible to get a total of the advantages and disadvantages for each option and to make a choice on the basis of these total evaluations of the options.

Saris et al. (1984) and Neijens (1987) have demonstrated that with this approach, the final choice of the respondents was consistent with their judgments for approximately 70% of the cases. On the other hand, the consistency was around 30% if the judgments were asked after the choice was made without the information provided by this approach. We conclude that the information aids in creating a well-informed choice. The choice questionnaire is discussed here because it is a very elaborate procedure to provide the respondents with information before they are asked to make their choice. This approach differs from the previously discussed batteries of survey questions because all battery items were prepared for the final choice. However, most of the time, the items are supposed to measure different aspects of an object and are not aimed at preparation for a later judgment.

The final result of choice questionnaires can be very different from that of the naïve opinion of the respondent without the given information. Therefore, the choice questionnaire should be used only to measure well-informed opinion and not to measure naïve opinions that are based mainly on the first ideas that come to the mind of the respondent (Zaller 1992).

7.3   BATTERIES IN CASI

In the early development of computer-assisted data collection, the CASI mode often contained a series of identical requests for an answer and answer categories for a series of stimuli or statements. Typical for such series of survey items is that the formulation is exactly the same for each item and that only one introduction with other possible components is given before the first survey item is mentioned. The items are treated equally because the interview programs use substitution procedures. An example of such an instruction to an interview program could look as follows:

> *#Casibattery 10 1*
> *# item 1*
> *healthcare*
> *#item 2*
> *social services*
> *#item 3*
> .
>
> .
> *# item 10*
> *social security*
> *#*
> *#Question with 5 answer categories*
> *What is your opinion about our "S"?*
> >   *1. Very satisfactory*
> >   *2. Satisfactory*
> >   *3. Neither satisfactory nor unsatisfactory*
> >   *4. Unsatisfactory*
> >   *5. Very unsatisfactory*

The first line indicates that this battery consists of 10 stimuli and only one question; then, follow the 10 stimuli and after that, the request follows with the answer categories. In the request, "S" is mentioned, which is substituted by the different stimuli or statements in the presentation of the questions to the respondents. Using this interview program, the following computer screens will be presented:

Screen 1 of CASI battery 1

What is your opinion about our healthcare?
>   *1. Very satisfactory*
>   *2. Satisfactory*

   3. Neither satisfactory nor unsatisfactory
   4. Unsatisfactory
   5. Very unsatisfactory

Screen 2 of CASI battery 1

What is your opinion about our social services?
   1. Very satisfactory
   2. Satisfactory
   3. Neither satisfactory nor unsatisfactory
   4. Unsatisfactory
   5. Very unsatisfactory

Screen 10 of CASI battery 1

What is your opinion about our social security?
   1. Very satisfactory
   2. Satisfactory
   3. Neither satisfactory nor unsatisfactory
   4. Unsatisfactory
   5. Very unsatisfactory

In contrast to the previous batteries, all survey items contain exactly the same information and therefore have the same complexity.

This kind of battery has been used not only for stimuli but also for statements as the next example shows. The screens of CASI battery 2 look as follows:

Screen 1 of CASI battery 2

Which of the following statements is true or false?

Screen 2 of CASI battery 2

The European Monetary Union will be governed by a government composed of all participant nations.
Is this true or false?
   1. True
   2. False

Screen 3 of CASI battery 2

The language of the European government will be French.
Is this true or false?
   1. True
   2. False

Screen *n* of CASI battery 2

Decisions in the European government are made by majority.

Is this true or false?
  1. True
  2. False

There is one difference with battery 1, namely, that before the first item, an intro-duction (in question form) is presented (screen 1 of CASI battery 2). Although the introduction is aimed at all items, it will be read only once before the first items. Therefore, we have decided that in this case, only the first item has an introduction and the other items have no introduction.

Both examples are very simple. Far more complex examples can be found in the research literature. In CASI battery 3, we provide an example of a rather complex case.

Screen 1 of CASI battery 3

We would like to ask you how serious you find some illnesses.

You can indicate the seriousness of the illnesses with a number between 0 and 100, where 100 means very serious and 0 means not at all serious. Thus, the more serious the illness, the larger the number.

Now comes the illness:

Screen 2 of CASI battery 3

Cancer

How serious do you consider this illness?
0 = not at all serious, 100 = very serious

Which number indicates the seriousness of this illness?

.

.

Screen *n* of CASI battery 3

Aids

How serious do you consider this illness?
0 = not at all serious, 100 = very serious

Which number indicates the seriousness of this illness?

This example is more complex because an introduction is presented and then an instruction including a definition, followed by another instruction and, finally, a sec-ond introduction. The first item comes after all this information, and all items are treated equally. But, the first item is very different because of the large amount of information provided before it. It is doubtful how much of the information provided will be available in the mind of the respondent when the second and following items are answered. It is for this reason that we suggest connecting all information to the first item and not to the other items. This would lead to the second and the following survey items consisting of information that is placed on the screen for that item such as a stimulus, a request with answer categories, and a second rephrasing of the basic question. Repetition of the main request for an answer after each stimulus is more typical for computerized questionnaires than in other modes of data collection.

Screen of CASI battery 4

There are different ways of trying to improve things in this country or help prevent things from going wrong. During the last 12 months, have you done any of the following? Tick all that apply.

Contacted a politician, government, or local government official	☐
Worked in a political party or action group	☐
Worked in another organization or association	☐
Worn or displayed a campaign badge/sticker	☐
Signed a petition	☐
Taken part in a lawful public demonstration	☐
Boycotted certain products	☐
Deliberately bought certain products for political, ethical, or environmental reasons	☐
Donated money to a political organization or group	☐
Participated in illegal protest activities	☐

In the later days of computer-assisted data collection and in the present Web surveys, many survey items are presented on one computer screen. An example is given in CASI battery 4. It is the measurement of "political action" in Web survey format, which was also presented in oral battery 1. We think that the respondents will read the introduction, the instruction, and the request first. Then, they will proceed to read the first statement and click on the box. Next, they will read the subsequent statement and decide whether to select the box. Having finished one, they will proceed to the next statement and complete it in the same manner, until the whole list is completed. Here, we assume that the first item contains much more information than the second to the last item, which consists only of the statement itself. A new element employed in this battery form is that the respondents can click on the boxes to indicate their answers.

CASI battery 5 illustrates that an 11-point scale can be used in this manner with many items being displayed on the same computer screen. The information belonging to the different items can also be determined in the same way as was described earlier.

Screen of CASI battery 5

Please indicate how much you personally trust each of the institutions below. Selecting the leftmost button means you do not trust an institution at all, and the rightmost button means you have complete trust.

	No trust at all									Complete trust	(Don't Know)
The parliament	O	O	O	O	O	O	O	O	O	O	O	O
The legal system	O	O	O	O	O	O	O	O	O	O	O	O
The police	O	O	O	O	O	O	O	O	O	O	O	O
Politicians	O	O	O	O	O	O	O	O	O	O	O	O
Political parties	O	O	O	O	O	O	O	O	O	O	O	O
The European Parliament	O	O	O	O	O	O	O	O	O	O	O	O
The United Nations	O	O	O	O	O	O	O	O	O	O	O	O

7.4   SUMMARY AND DISCUSSION

In this chapter, we have presented the most common ways in which batteries are used in survey research. One issue we have emphasized here and that has not been discussed in the literature thus far is the difference between the batteries as they are operationalized in their respective modes of data collection. We have also shown that this has distinct implications for the description of the survey items in the battery. We suggested including those components in the different survey items that are explicitly provided to the respondent when the survey item is introduced. That means that the first item in the battery has a form very different than the other items in the battery, because the information given before the first item belongs to the first item and not to the others. This is particularly relevant because it affects the estimation of complexity (as discussed in the codebook) of the survey items within a battery, especially between the first and the other items, and across batteries between items in different modes of data collection. There are no studies showing that the use of stimuli has negative effects on the quality of the data collected. Batteries with statements have been more frequently criticized. These issues were covered in Chapter 4 and will not be repeated here. We will only note again that there is sufficient empirical evidence pointing in the direction that in the majority of cases trait-specific questions are of better quality than are battery items.

The battery method is an obvious choice asking for a reaction to many different statements. Often, information has to be added to the battery about concepts or the procedure of response; however, there are limits to these possibilities. These limits depend on the topic discussed and the responses that are asked, which points to a difference between batteries with stimuli and batteries with statements.

Although we have discussed batteries with stimuli and statements together, they have their differences. On the basis of a random sample of survey items mentioned in the last chapter (Molenaar 1986), we have found that both types of batteries have in more than 90% of the cases an introduction before the first item is presented. But it was also found that they differ in that batteries with statements need extra instructions (78% of the batteries) versus those with stimuli (less than 2% of the cases). This may be a consequence of the length of the statements used, since otherwise there is no difference between these two types of batteries.

Let us give a final example (complex battery) to emphasize that there are limits to the use of batteries and that these limits vary for the different modes of data collection.

Complex battery with two requests

CARD K1
For each of the voluntary organizations I will now mention, please use this card to tell me whether any of these things apply to you now or in the last 12 months and, if so, which.

READ OUT EACH ORGANIZATION IN TURN. PROMPT*: Which others.*
ASK K2 FOR EACH ORGANIZATION CODED 1 TO 4 AT K1*.*

> *K2*   *Do you have <u>personal friends</u> within this organization?*
> ***READ OUT EACH ORGANIZATION CODED 1 TO 4****.*

	K1		*CODE ALL THAT APPLY FOR EACH ORGANIZATION*			*K2*		
	None	*Member*	*Participated?*	*Donated money*	*Voluntary work*		*Personal friends?*	
							Yes	*No*
A. ...firstly, a sports club or club for outdoor activities, do any of the descriptions on the card apply to you?	0	1	2	3	4	810–813	1	2
B. ...an organization or cultural or hobby activities?	0	1	2	3	4	814–817	1	2
C. ...a trade union?	0	1	2	3	4	818–821	1	2
D. ...a business, professional, or farmers' organization?	0	1	2	3	4	822–825	1	2

NOW ASK THE AFOREMENTIONED K2 FOR EACH ORGANIZATION CODED 1

In this example, two questions are presented in one table. Instructions are given to the interviewer above the table about how the questioning should be conducted. In this case, a show card is provided with the response alternatives to the first question:

Show card for the complex battery

- *A member of such an organization*
- *Participated in an activity arranged by such an organization*
- *Donated money to such an organization*
- *Have done voluntary (unpaid) work for such an organization*

It should be clear that this combination of two batteries, which in fact requires responses for five requests per organization, is rather complex. This format is recommended only in the case where the interviewers have been trained well, and it is not for a mail questionnaire or a CASI.

With this illustration, we finish the discussion about batteries of survey items; much more research is needed in order to determine what the effects of the different forms on the quality of the responses are. In Part III, results of such research will be presented.

EXERCISES

1.   Two interviewer forms of batteries are presented as follows.

Example 1   *How important is each of the following in your life? Here, I have a card with a scale of 0–10 where 10 means "very important" and 0 means "not important at all." Where would you place yourself on this scale? SHOW CARD.*

	Not important at all										Very important	Don't know
A. Family and friends	0	1	2	3	4	5	6	7	8	9	10	88
B. Leisure time	0	1	2	3	4	5	6	7	8	9	10	88
C. Politics	0	1	2	3	4	5	6	7	8	9	10	88
D. Work	0	1	2	3	4	5	6	7	8	9	10	88
E. Clubs	0	1	2	3	4	5	6	7	8	9	10	88
F. Community organization	0	1	2	3	4	5	6	7	8	9	10	88

Example 2   *As you know, there are different opinions as to what it takes to be a good citizen. I would like to ask you to examine the characteristics listed on the card. Thinking about what you personally think, how important is it:*

SHOW CARD

	Not at all important										Very important	Don't know
	0		1	2	3	4	5	6	7	8	9	10	88
A. To show solidarity with people who are worse off than yourself													
B. To vote in public elections												
C. Never to try to evade taxes												
D. To form your own opinion, independently of others												
E. Always to obey laws and regulations												
F. To be active in organizations												

Answer the following questions:

a. What would you put on the card for these two examples?

b. Given the choice in a., indicate for both batteries what text belongs to which survey item.

c. What kind of components is presented before the first item in each scale?

2. Is there a reason to use batteries in your own questionnaire?

    a. If so, why?

    b. What is your proposal?

8

MODE OF DATA COLLECTION AND OTHER CHOICES

The first part of this chapter is dedicated to the discussion of some choices over which the researcher has little control but which can have considerable influence over the survey results. The first choice to be discussed is the mode of data collection. In principle, the researcher is free to choose any mode; however, in reality, the options are restricted by the budget that is available for the study. Anyway, the mode of data collection affects the quality of the survey and also its costs. We will spend considerable attention to the mode of data collection since it is a very important decision and because its possibilities increase rapidly.

The second choice we will discuss concerns the position of the question in the questionnaire. Not all questions can be placed in the best place of the questionnaire. Therefore, a policy should be developed that deals with how to place the questions in the questionnaire.

A third issue to discuss is the layout of the questionnaires. Unfortunately, there is still very little known about the effects of layout; however, we will give references where relevant information about this issue can be found.

Finally, there is the choice of the language for the questionnaire. This is, of course, not a real choice. Nevertheless, a limited choice exists with respect to the language related to minority groups in a country. For example, should the Moroccan people in France be interviewed in French or in Moroccan Arabic? It is also important for comparative research to know whether it makes a difference on the substantive conclusions if the questions are asked in a given language.

Design, Evaluation, and Analysis of Questionnaires for Survey Research, Second Edition.
Willem E. Saris and Irmtraud N. Gallhofer.
© 2014 John Wiley & Sons, Inc. Published 2014 by John Wiley & Sons, Inc.

8.1 THE CHOICE OF THE MODE OF DATA COLLECTION

Data collection is developing rapidly. In the 1960s and 1970s, there were only three procedures for data collection: paper-and-pencil interviewing (PAPI) by an interviewer in the home of the respondent; traditional telephone interviewing, where the interview was done by telephone; and, finally, mail questionnaires, which were done without the presence of an interviewer and where respondents had to fill in the forms themselves.

In the mid-1980s, the computer made its entry into survey research. First, the telephone interview was computerized. The computer-assisted telephone interview (CATI) is now a very common form of data collection for survey research. Also, in the beginning of the 1990s, the first experiments with computer-assisted personal interviewing (CAPI), as a substitute for PAPI, were done even though at that time the portable computer was not yet available (Danielson and Maarstad 1982; Saris et al. 1982). In 1985, the first experiments with computer-assisted self-interviews (CASI) as a substitute for mail questionnaires (Kiesler and Sproull 1986; Saris and de Pijper 1986) were conducted.

The first two forms of computer-assisted interviewing systems did not change much for the respondents; they caused mainly a change for the interviewer. The last form really made an impact on the respondents because they had to answer questions on a computer and not on paper. There were two forms to this approach. The form that resembled mail questionnaires was the disk-by-mail (DBM) approach. In this case, a diskette with the interview and interview program was sent to the respondents, who would answer the questions on their own computers and send the disks back to the research agency. The second approach was the *telepanel* (Saris and De Pijper 1986; Saris 1991, 1998b). In this approach, a random sample of the population was provided with computers and modems (if necessary). With the equipment, interviews could be sent to the households via cable and the answers could be returned without intervention of an interviewer. These two approaches required the respondents to have a computer. Using the DBM system, one could only study populations with computers like businesses or physicians, doctors, etc. The telepanel system required a large investment in computers in order to allow for studies of representative samples of the population. The telepanel approach also required the use of the same households for many studies because of the large startup investment. As a consequence, this research design became automatically a panel survey design.

The next developmental phase was the further automatization of the telephone interview. Two new forms have been developed for large-scale surveys: touch-tone data entry (TDE) and voice recognition entry (VRE). In both cases, the questions are presented to the respondent by a device that can present a question in natural language via a recorder or a computer. The respondent is asked to choose an answer by pressing keys on the handset of the telephone (TDE) or to mention the answer loudly; these answers are interpreted by a computer and coded (VRE). These two approaches were developed for very large surveys in the United States (Harrel and Clayton 1991; Phipps and Tupek 1991).

A new phase in data collection has been the development of the World Wide Web with its possibilities to reach very many people with relatively low costs. This approach can become the alternative for mail questionnaires, DBM, and the telepanel if the Web facilities are sufficiently widespread to allow research with representative samples of the population. As long as this is not the case, the telepanel procedure should be used by providing a representative sample of respondents with the necessary equipment to participate in research. This is done by Knowledge Networks (Couper 2000) in Palo Alto (United States) and by Centerdata in Tilburg (the Netherlands). The development of the Web survey is not a fundamental change compared with DBM or the telepanel, but is an improvement in efficiency and in cost reduction.

The most recent development is that experiments are done with audio self-administered questionnaires (ASAQ) and audio computer-assisted self-interviewing (ACASI). The purpose of these approaches is to make it possible for illiterate people to fill in self-administered questionnaires because the questions are read to them via a recorder (ASAQ) or a computer (ACASI). On the other hand, it also provides the respondents with an environment where they can answer questions without being influenced by people in their surrounding (an interviewer or people in the household). This is especially important for sensitive questions where socially desirable answers can be expected. In order to provide such an environment, the questions are presented through a headphone, and the respondents can mark their answers on a response form (ASAQ) or by pressing keys on a computer (ACASI). This approach leads to results for sensitive issues that are very different from results obtained by the standard data collection methods (Jobe et al. 1997; Tourangeau and Smith 1986; Turner et al. 1998). A more detailed overview of the historical development of data collection can be found in Couper et al. (1998).[1]

8.1.1 Relevant Characteristics of the Different Modes

More important than the historical sequence of events are the differences in characteristics of the different modes of data collection for:

1. The presence of an interviewer
2. The mode of presentation (oral or visual)
3. The role of the computer

Tourangeau et al. (2000) make a fourth distinction between oral responses: written and keyed responses. Evidently more cognitive skills are required for writing than for keying answers and oral responses, although this has not led to different data collection methods. Most methods use oral responses if an interviewer is present but keyed

---

[1] Our overview deviates on only one point of the report of Couper et al. (1998), which is the early development of CAPI. Although they were informed that experiments with CAPI were done as early as 1980, they did not mention this in their overview. They thought that it was not possible at that time because the PC was not yet available. But these experiments were not done with PCs but with the first Apple computer, which appeared on the market in 1979. This computer was placed in a box with a screen. With this box, interviewers went to the houses of respondents for interviews. These experiments have been described in Saris et al. (1982).

TABLE 8.1 Different methods for data collection distinguished on the basis of the role of the interviewer, the mode of presentation, and the role of the computer

Presence of interviewer	Role of computer	Presented		
		ORAL	ORAL/VISUAL	VISUAL
Present	CAI	CAPI	CAPI+	CASI-IP
	NO	PAPI	PAPI+	—
Distant	CAI	CATI	—	TP, Web-IP
	NO	Tel. In	—	—
Absent	CAI	ACASI	ACASI+	Web, DBM
	NO		ASAQ	ASAQ+
		TDE/VRE		Mail

CAI, computer-assisted interviewing; CAPI, computer-assisted personal interviewing; CAPI+, CAPI plus show cards; PAPI, paper-and-pencil interviewing; PAPI+, PAPI plus show cards; CASI-IP, computer-assisted self-interviewing with an interviewer present; CATI, computer-assisted telephone interview; TP, telepanel; Web-IP, Web survey plus interviewer present at a distance; Tel. In, telephone interviewing; ACASI, audio computer-assisted self-interviewing; ACASI+, ACASI plus possibility to read the questions; Web, Web survey; DBM, disk by mail; ASAQ, audio self-administered questionnaire; ASAQ+, ASAQ plus possibility of reading the questions; TDE, touch-tone data entry; VRE, voice recognition entry; Mail, standard mail interview.

or written answers if no interviewer is available. Therefore, this distinction is completely confounded with the role of the interviewer, although this is not absolutely necessary.

The other three distinctions mentioned previously have led to different procedures, displayed in Table 8.1. The majority of methods use oral presentation, but the newer methods increasingly use a visual presentation of the questions, with the exception of different experiments with audio systems. The procedures employing oral and visual presentations started with show cards in oral procedures or by reading facilities next to audio facilities. In the following, we will briefly discuss the possible consequences of the different choices.

8.1.2 The Presence of the Interviewer

A distinction that has always existed in survey research concerns the role of the interviewer. In personal interviewing, the interviewer is present during the interview and normally has to ask the questions and record the answers. In telephone interviewing, the task of the interviewer is the same, but the interviewer is not physically present at the interview and can ask the questions from a long distance. Finally, in all self-administered interviews, there is no interviewer present at all.

The advantage of the presence of the interviewer during the interview, either in person or at a distance, is that the respondents do not need to have reading and writing abilities. Normally, the interviewer will read the requests for an answer to the respondent. Another advantage is that the interviewer can help with difficult questions. Some people suggest that the interviewer in a personal interview can also see the nonverbal reaction of the respondent, which is not possible in telephone interviews. Another distinction between the two approaches with an interviewer is that a limited number of people do not have a telephone and therefore cannot be reached by telephone. Another point is that people are more inclined to participate if an interviewer appears at the door than when the interviewer asks for cooperation through the telephone. Finally, the item nonresponse may be lower in a personal interview if the interviewer can build up a good relationship with the respondent during the interview.

However, the personal interview with an interviewer is costly because of the travel expenses and the interview takes relatively more time. The presence of an interviewer might lead to more socially desirable answers to sensitive questions in a personal interview than over those administered over the telephone because of the "closeness" of the interviewer. However, although several studies showed this effect, others did not, as was summarized by Tourangeau et al. (2000).

A major difference between using interviewers or the approaches without an interviewer is that in the latter it is not possible for an interviewer to reformulate the questions as often happens in personal and telephone interviews (Van der Zouwen and Dijkstra 1996). This might be useful in helping the respondent but it also makes the answers incomparable if the question asked is not the same.

Another major difference is that the interviewer is not present, which in turn affects the answer results for sensitive issue questions. In general, there is a lesser social desirability effect in self-administration procedures. A series of studies by Aquilino and LoSciuto (1990), Aquilino (1994), Gfröerer and Hughes (1992), Turner et al. (1998), and Schober and Conrad (1997) have demonstrated these effects. Tourangeau and Smith (1996) and Tourangeau et al. (1997) show that these effects are even larger if ACASI is used instead of simple CASI.

The last findings are certainly of major importance and suggest that for sensitive issues it is much better to opt for self-administered methods instead of interviewer-administered questionnaires. Otherwise, the differences are not significant. Even the coverage error in telephone interviewing does not have a dramatic effect because the number of people without a telephone is rather small in most countries and therefore the effect of this deviant group on the total result is in general rather insignificant (Lass et al. 1997).

That does not exclude the possibility that in certain cases one mode can be more effectively used than another. For example, Kalfs (1993) demonstrated that the time spent on transport is more suited for telephone than for self-administered interviewing because the interviewer can be instructed to check the sequence of events better than respondents can. On the other hand, she also found that self-administered interviewing worked better for recording TV watching than telephone interviewing because higher-educated people were not willing to report their total TV viewing time in the CATI mode. They reported approximately half an hour more TV viewing time in the CASI mode.

So, we have to conclude that the quality of the measures depends very heavily on the topic studied. If the topic is rather complex, an interviewer can be helpful. If the topic is simple but sensitive, self-completion is advisable. However, if the topic is simple and not sensitive, the presence or absence of the interviewer will not make much difference.

8.1.3 The Mode of Presentation

The second characteristic to be discussed is the presentation mode of the requests for answers. In personal and telephone interviewing, it is natural to make an oral presentation. For self-administered methods, the most natural procedure of presenting is visual.

More recently, mixed procedures have also been developed. For example, a very important tool in personal interviewing nowadays is the show card. While the interviewer reads the question, the respondent receives a card with the response alternatives. In case of a battery of statements, the respondents can also be provided with a card representing the different statements about which they have to give a judgment.

Another new possibility in self-administered surveys with a computer is that the text on the screen is presented and read to the respondent. This is typically the case for TDE, VRE, ASAQ, and ACASI.

Tourangeau et al. (2000) point out that the cognitive requirement is higher for a visual presentation than for an oral because people have to be able to read. The problem of illiteracy received more attention in the United States than in Europe. Therefore the ASAQ and ACASI methods are more popular there than in Europe. Much of the research in Europe is still administered by personal or telephone interviews, and so the illiteracy problem does not play such an important role.

However, the visual or mixed presentation is very helpful because it appears that people in general have a limited capability to remember the information with which they are provided. For example, for the oral interview, the respondent is unable to go back and check information. However, if visual information is provided in the form of response categories or statements, then the respondent can quickly go back and recapture the information to give a better response. That is why show cards have gained in popularity in personal interviews, which is not possible for telephone interviews.

Normally, survey researchers try to include the response categories in the requests for an answer. For example:

8.1   *Are you very much in favor, much in favor, in favor, against, much against, or very much against the extension of the EU to include the Central European countries?*

This request for an answer can be much simpler and more naturally formulated in a completely visual presentation:

8.2   *How much are you in favor or against the extension of the EU to include Central European countries?*
  1. *Very much in favor*
  2. *Much in favor*
  3. *In favor*
  4. *Against*
  5. *Much against*
  6. *Very much against*

This difference does not have to exist, of course, because the second form of the question could also be used in oral presentations if the interviewer is instructed to read all the response categories or if a show card with the response categories is used. However, looking at the survey literature, questions like the first one are rather common even though their formulation is not very attractive.

A major disadvantage of the completely visual paper representation is the lack of process control about the way the respondent answers the questions. A respondent can skip questions that should not be skipped or answer questions in a standard way without giving due diligence for the specific characteristic of each question. Dillman (2000), who has focused his attention on addressing this problem, suggests designing the layout of the pages in order to guide the respondent in the desired sequence to answer the questions. We will come back to this issue when we talk about layout design. On the other hand, lack of individual attention for each separate question also occurs during oral presentations in the case of the interviewer who would like to advance the interview as quickly as possible. In both cases, this can lead to the problem of response set (Krosnick 1991). However, it should be mentioned that visual representations used in self-completion surveys often take more time than do personal or telephone interviews and are often done at a time chosen by the respondent. From this, it can be inferred that there is less time pressure in visual presentations than in oral presentations, and this could improve the overall response quality.

8.1.4   The Role of the Computer

Although the use of computers in survey research has revolutionized data collection, the difference between procedures with and without a computer is not always as great as expected. For example, from the respondents' perspective, personal interviewing and telephone interviewing has not fundamentally changed because of the introduction of the computer (Bemelmans-Spork and Sikkel 1986). The procedural difference has occurred mainly for the interviewers. In fact, their task is simplified because in a good CAPI and CATI the routing in the questionnaire is controlled by the computer and the interviewer no longer has a need to plan the next question anymore and can focus on the interview process.

Moreover, the computer could perform more of the interviewer's tasks. For example, the computer can:

- Check whether the answers are appropriate
- Provide the respondents with more information, if necessary
- Code open answers into predetermined categories
- Store the given answers in a systematic way

The computer has the potential to reduce the interviewer's tasks considerably. However, the interviewer is still better at:

- Obtaining cooperation
- Motivating the respondent to answer
- Building up sufficient confidence for honest answers

But the computer can do certain tasks quicker than the interviewer, such as:

- Calculations for routing
- Substitution of previous answers
- Randomization of questions and answer categories
- Validation of responses
- Complex branching
- Complex coding
- Exact formulation of questions and information
- Provision of help

For a detailed discussion of how these tasks can be performed by computer programs, we refer the reader to Saris (1991). However, these extra facilities have their price. In order to obtain all of them, a considerable amount of time to develop computer-assisted interviews has to be invested. In addition, more time is needed to check whether all these tasks have been formulated properly. Programs are available to check automatically that a question can be reached and does not lead to a dead end, but all sequences need to be checked on substantive correctness. This cannot be done by computer programs. Wrong routings in a paper questionnaire can be corrected rather easily by the interviewer; however, this is not the case in computer-assisted interviews. Therefore, checking the questionnaire is an essential task of development.

Nevertheless, checking the structure of the questionnaire is not enough. The respondent or interviewer can also make mistakes, and as a consequence, the routing might be wrong. For example, a respondent who does not have a job can be coded as having one. In that case, the respondent may get all kinds of questions about labor activities that do not apply. To prevent such confusions, Saris (1991) has suggested *summary and correction* (SC) screens, which summarize prior provided information to be checked by the respondent before the program makes the next branching decision. This SC screen turns out to be a very efficient tool in computer-assisted data collection. An interviewer can be taught complex computer tasks that are too much for respondents. Therefore, for self-administered interviews, without an interviewer present, the SC screen is even more important than for CAPI and CATI. How this tool and others can be developed can be found in Saris (1991).

The role of the computer in self-administered questionnaire applications can be most helpful. First, the interview program:

- Can perform all the tasks we have previously mentioned, which can be used to substitute the interviewer except for obtaining cooperation
- Can ensure that the respondent will not skip a question by mistake
- Provides automatic visual questions and can also show pictures or short movies
- Can read the text of the questions as has been developed in ACASI
- Can use line production scales that have turned out to be very effective to obtain judgments on a continuous scale
- Can be used to present complex instructions that are not possible in oral presentations

In the case of a panel of respondents for the data collection, it is possible to use previous answers to reduce the effort of the respondent and to verify the answers on unlikely responses. An example is the "dynamic SC screen," where respondents can check the correctness of previous answers without filling in the forms again. If a change has occurred, they can restrict their work to the necessary corrections on the screen. Another example is the dynamic range and consistency check, where previous answers with respect to income, prices, or quantities are used in later interviews, that take into account the variability in these quantities by using a dynamic confidence interval around the point estimator (Hartman and Saris 1991).

This brief overview of the potential of computer-assisted interviewing indicates that a considerable increase in data quality can be obtained by efficient use of the computer. Tortora (1985) conducted an experiment that demonstrated that checks available in CATI could prevent nearly 80% of all the corrections that were normally needed after the data collection. As we have seen, even more advantages can be obtained in self-administered questionnaires improving mail questionnaires, including the commonly used diaries (Kalfs 1993). So far, these data quality improvements have been obtained mainly with respect to factual information. For subjective variables, the results are more difficult to improve because it is more difficult to specify rules for inconsistent answers. An interesting possibility is the development of procedures to detect unlikely answers elaborated by Martini (2001).

We have to mention that this quality improvement by CAI also has its costs, not only in terms of hardware but also in the time one has to spend for developing computer-assisted interviews, followed by checking the correctness of questionnaire routing. The mere fact that a computer is used is not enough to improve data quality. Considerable time has to be spent on the development of computer-assisted questionnaires, which makes sense only for very expensive or often repeated studies. Therefore, improved quality is not guaranteed by using computer-assisted data collection. Only if special attention is given to the development of the questionnaire, these advantages are obtained.

This point is important to emphasize because a mistaken assumption with the development of the Web survey is that it does do survey research at very low costs. However, there are three problems with this approach: (1) it is impossible to get a representative sample of a national population through the Web at this moment—there are still too few people connected to the Web, and this group is rather deviant from the population as a whole; (2) the cooperation with Web surveys is not such that one can expect to get a representative sample from the population; and (3) few researchers incorporate the possibilities of computer-assisted interviewing that were mentioned previously. As a consequence, formulation of the questionnaires is most usually on the level of the standard mail survey. In this case, a step backward is taken in terms of survey quality, in comparison with what is possible, and even though computer-assisted data collection is used. Better procedures are used by Centerdata and Knowledge Network (Couper 2000), which follow the telepanel procedure, providing a random sample of the population with the necessary equipment to use real computer-assisted data collection with consistency checks and other possible tools for data collection.

8.1.5 Procedures without Asking Questions

So far, we have discussed procedures that present requests for an answer to the respondents. In marketing research, procedures have been developed where no questions are asked at all anymore. Two popular approaches are the "people meter" and the "barcode scanner." The former is used for recording the amount of time that people spend watching different TV programs. It is a box placed on the TV that can record any program that is viewed on it. In order to know who is watching the program, the viewers have to indicate on a remote control whether they are watching. In this way, their TV watching behavior is registered without asking any question.

The barcode scanner is used for recording the purchases of consumers. The most efficient procedure uses the barcode reader at the cash register. In order to be able to connect the purchases to a person, the consumer is first asked to present an identity card with a barcode. The code is registered, followed by all goods bought to account for all purchases without asking any question in person.

These systems also have their shortcomings, but they are much more efficient than the traditional procedures using diaries or asking questions. For a discussion around issues concerning these topics, we can refer to Belson (1981) and Kalfs (1993), for TV viewing to Silberstein and Scott (1991), and to Kaper (1999) for consumer behavior. We will not elaborate on further details here, because we will concentrate on approaches using requests for an answer. We will conclude by stating that automatic registration is more efficient but also much more expensive.

8.1.6 Mixed-Mode Data Collection

The most commonly used procedures have different response rates and different costs associated with them. A general trend for all modes of data collection is that the response rate decreases. In order to improve the response rate, mixed-mode data collection has been suggested, which employs several data collection methods for one study. The mixed-mode design is developed on the basis of the knowledge that different methods lead to different levels of nonresponse for different groups. These unequal response probabilities make mixed-mode data collection attractive, even more so when the (financial) costs of the different data collection methods are taken into account. Mail questionnaires are by far the most economical of the three traditional data collection modes, followed by telephone interviewing, while face-to-face interviewing is the most expensive one (Dillman 1991; Pruchno and Hayden 2000). Because the response rates of these three data collection methods are inversely related to financial expenses, it seems to pay off to start with the cheapest data collection method (mail), follow up the nonrespondents by telephone, and approach those who still are not reached or not willing to participate, by a personal interview. In this way, the highest response level at the lowest cost can be achieved, and the differences of the selection process between the three data collection modes are turned to the advantage of the survey researcher.

The main disadvantage of mixed-mode data collection is the possibility of mode effects. The results with respect to mode effects are not so clear yet. It is most likely that it depends on the topic being studied. This means that pilot studies are needed before the mixed-mode data collection can be used in large-scale survey research. An example of such a study is done by Saris and Kaase (1997) with respect to a comparison of telephone and face-to-face research for the Eurobarometer. They found significant differences between the methods but indicated as well procedures for how to overcome these problems. A study has been done by Voogt and Saris (2003) dealing with election studies. They did not find mode effects comparing mail and telephone interviewing, but the personal interviewing gave results different from the other two mentioned earlier. A lot of useful suggestions to minimize the effects of the mode can be found in Dillman's work (2000).

8.2   THE POSITION IN THE QUESTIONNAIRE

A next issue that requires attention is the construction of the questionnaire. So far, we have discussed only single requests for an answer or batteries but not the ordering of the different requests or batteries in a questionnaire. In this context, there are four principles that require attention. The first is that a prior request for an answer can have an effect on a later request for an answer (Schuman and Presser 1981). The second principle is that one should not mix all requests randomly with each other as is often done in omnibus surveys. Dillman (2000) is a strong supporter of ordering the question by topic. However, this increases the risk of ordering effects. He also suggests a third principle to start questionnaires with the topic that has been mentioned to the respondents to get their cooperation. These questions should be relatively simple, apply to all respondents, and also be interesting in order to increase the cooperation to respond. Contrary to this rule is a fourth principle that suggests that the answers to the first requests are probably not as good as later responses because the respondents have to learn how to answer and to gain confidence with the interviewer and the interview process. According to this principle, it is advisable not to ask about the major topic of the survey immediately at the beginning of the interview.

The first two rules concern the choice to order the survey items by topic or not. In psychological tests, the items are normally randomly ordered going from one topic to another without any connecting statement between the different questions. This is done to avoid sequence effects. Dillman (2000) argues that the respondents get the impression that the interviewer does not listen to their answers because the next question has nothing to do with their prior answer. Another problem is that it also puts a heavy cognitive burden on the respondents because they have to search for answers in a different part of their memories and it is questionable whether most respondents are willing to do so. This might lead to *satisfying* behavior as suggested by Krosnick (1991), which means that the respondent does not look for the optimal answer anymore, but only for an acceptable answer with minimal effort, as, for example, a "don't know" answer or the same response.

The other side of the coin is that grouping the questions by topic can lead to order effects. An example of an order effect is called the "evenhandedness effect" (Hyman and Sheatsley 1950). The number of "yes" answers to "Should a communist reporter be allowed to report on a visit to America as he/she saw it?" increases when the respondents were first asked "Should an American reporter be allowed to report on a visit to the Soviet Union as he/she saw it?" Dillman (2000) mentions several other similar examples of effects of one question on the next for distinct reasons. One is called the "anchoring effect" and suggests that a distant object is evaluated differently if asked first than if it is evaluated after a more familiar local object. Another is called the "carryover effect," which means that, for example, happiness judgments are more positive after a question about marriage than before the question. It has also been shown that overall evaluations become lower after the evaluations of some specific aspects. For more examples, we refer the reader to Schuman and Presser (1981); their examples suggest watching out for the possibility of order effects if requests for answers have an obvious relationship. Nevertheless, we think that it is better to order the questions by topic instead of using a random order of the request for answers because it improves the cooperation of the people and reduces their cognitive burden, which is in general quite high.

The second pair of contradictory rules suggests (1) starting with the main topic of the questionnaire using simple questions that apply to all and are interesting for the respondents and (2) not starting with the main theme of the study until the people are more familiar with the procedure and the interview process and have confidence in the interviewer. According to rule 1, starting with questions about the background of the respondent or household would be a mistake, while according to rule 2, it could be acceptable.

Although we prefer the rule to start as soon as possible with the topic announced in order to get the cooperation of the respondent and to start with simple questions that apply to all people and are interesting, we don't suggest that the main topic should be in the first question. However, in the second part of this book, we will demonstrate that the respondents continuously learn how to answer and as a consequence their answers improve. Therefore, it is preferable to delay the most important questions in order to elicit better responses. In general, respondents will not be surprised to answer some more general questions about their background before the main topic questions. Since these questions are simple and general, they serve to familiarize the respondent with the procedure and the interviewer.

After a general introduction, topics that are clearly related to the main topic should be introduced. The order of the different topics should be partially determined by the complexity of the questions, because the later that complex questions are asked, the better the respondents will be able to answer the questions. However, these complex questions should not be asked at a moment that the respondents are starting to get bored with the interview. Therefore, the best ordering of the topics should be based on the logical sequence of the topics and the quality of the responses, which will be discussed in Part III of this book.

8.3   THE LAYOUT OF THE QUESTIONNAIRE

Although the layout of the questionnaire is important for any questionnaire, the quality of the layout is more important in self-administered questionnaires than in interviewer-administered questionnaires. Interviewers can be taught the meaning of the different signs and procedures in the layout. This is normally not possible in self-administered questionnaires; therefore, the layout of such a questionnaire should be completely self-evident.

This is even more important for paper questionnaires than for computer-assisted self-administered questionnaires. For paper questionnaires, the respondent has to find the routing of the questionnaire that skips the fewest requests for an answer by mistake and that answers the requests in the proper order. For computer-administered questionnaires, the computer can take over this task. However, in the new Web surveys, often, this routing task is not taken over by the program, therefore producing rather incomplete response files.

Unfortunately, very little is known about the optimum rules for designing questionnaires. The best source for this information is Dillman (2000), who formulated a number of general rules mainly for mail questionnaires and in his recent work also for questionnaires presented on computer screens. For more detailed information, we refer the reader to his work. A general rule of thumb is to choose one system for the layout and to be consistent throughout the whole questionnaire so that the respondent and interviewer will know what to expect.

8.4   DIFFERENCES DUE TO USE OF DIFFERENT LANGUAGES

The researcher does not have much choice over the language to be used in the questionnaire. Normally, a country has one national language, and that is the language to be used to formulate the questions; for large minority groups, a questionnaire in the language of the minority group can also be considered. Such translation tasks are rather complex because a functionally equivalent version of the questionnaire needs to be generated (Harkness et al. 2003) wherein the different translations of the requests for answers have an equivalent meaning.

In the European Social Survey (ESS), which is conducted in 23 countries in Europe, a translation into more than 20 languages was necessary. This translation process was organized in the following way. First, a source questionnaire was developed in several steps in English (ESS 2002). This questionnaire, with annotations with respect to the concepts used, was sent to the different countries. In each country, two translators were asked to make a translation independently of each other. The resulting questionnaires were provided to a reviewer who looked at the differences between the two translations. Finally, the national coordinator of the study together with the reviewer made the final decisions with respect to the definite formulation of the questionnaire in each respective national language (Harkness et al. 2003). After this complex and laborious process, a set of questionnaires with as similar as possible requests for answers and response categories were developed.

Although all due diligence is taken in cross-national research, there is no guarantee that the results from the different countries or cultural groups are comparable. Even if all the texts produced are completely functionally equivalent, it cannot be excluded that the same request for an answer generates more random errors and/or systematic errors in one country than in another. Illustrations have been provided by Saris (1997) for the Eurobarometer and in Saris and Gallhofer (2002) for the ESS. How such tests can be conducted will be discussed in further detail in Parts III and IV. Given these findings, their argument is to compare results of different countries after correcting for measurement error; otherwise, it is unknown whether the differences between the countries are due to measurement error or represent real differences.

8.5 SUMMARY AND DISCUSSION

In this last chapter of Part II, an overview of the developments in data collection methods was provided. In that context, we have discussed three choices that in combination determine the data collection method to select:

1. The presence or absence of an interviewer
2. The mode of presentation (oral/visual)
3. The role of the computer

With respect to the presence or absence of an interviewer, we have to conclude that the quality of the measures depends heavily on the topic studied. Depending on whether the topic is rather complex, an interviewer can be helpful. However, if the topic is simple but sensitive, self-completion is advisable. When the topic is simple and not sensitive, the presence or absence of the interviewer will not make much difference.

The mode of presentation can make quite a difference for sensitive topics. We have also mentioned that the mode of data collection will change the way the questions are formulated; for this reason, it is difficult to study pure mode effects.

The role of the computer in survey research is very interesting as it can take over many tasks of the interviewer. The computer can also do certain tasks that an interviewer cannot do, leading to considerable improvement of a questionnaire. On the other hand, designing computer-assisted questionnaires is a very time-intensive process and therefore also rather expensive. Because of the costs involved, we recommend this method for large-scale studies and longitudinal studies. However, computer-assisted data collection does not always improve the quality of the surveys. To date, the improvements that have been noted are mainly for objective variables. For subjective variables, it is more difficult to specify efficient consistency checks.

The second issue discussed was the ordering of the questions within the questionnaire; we advise the reader to be aware of the possibility of order effects in case of requests for answers that have an obvious relationship. Nevertheless, we think that it is better to order the questions by topic instead of using a random order of the requests for an answer. This will improve the cooperation of the respondents and reduce their cognitive burden.

The third issue discussed concerned the layout of questionnaires. In general, it is recommended that the researcher chooses only one system for the layout and adheres to it throughout the whole questionnaire. This is important not only for self-administered questionnaires but also for interviewer-administered ones, because then the respondent and interviewer both know what to expect.

The last topic was the effect of the language used. Normally, the official language cannot be chosen, but often, there is a choice in using a different language for large minority groups. In such cases, the questionnaires are not just translations but have to be made as functionally equivalent as possible. However, this does not guarantee that the results for different groups or countries can be compared. It is possible that in the different groups the reactions on the optimally equivalent requests create different types of errors that lead to different results. Therefore, it was suggested that one can never compare results directly across groups without correction for measurement error, which will be further discussed in Part IV.

EXERCISES

Imagine that we want to design a questionnaire to determine the satisfaction of the people with the government and the reasons for this satisfaction. This could be done in different ways, and the next questions deal with these different options.

1. The most important question, of course, is about satisfaction with the government with a 7-point completely labeled response scale.
    Formulate such a request for the following data collection modes:
    a. For a mail survey
    b. For a personal interview in two ways: one with a show card and one without
    c. For telephone interviewing, also in two ways: one in two steps and one direct
    d. Which one would you prefer?

2. Next, we would like to know which policies influence the satisfaction with the government the most. We want to ask about the economy, health service, education system, security, and so on.
    a. Formulate such a block of requests in different ways (and in detail).
    b. Should these requests be asked before or after the general satisfaction request mentioned in question 1?

3. Routing in paper questionnaires requires special attention. Take as an example the two-step procedure asked as request 1c.
    a. Indicate in detail how you would make the layout of the page to avoid problems.
    b. How would this routing be done in CAI?

4. Other questions concern the sources of income (salary, pension, unemployment money, savings, alimentation, etc.), especially the amount of money they get from each source.

    a. Formulate a question block for a face-to-face study to ask these questions.

    b. How can a computer-assisted data collection simplify the task taking into account the position of the people on the labor market?

    c. In computer-assisted panel research, this can be made even simpler using prior information. How?

5. As a last part in this questionnaire, we want to formulate questions about satisfaction with income.

    a. Formulate a question asking for an absolute judgment with a labeled 11-point scale for a face-to-face study.

    b. Formulate also a question asking a relative judgment with a labeled 11-point scale for a face-to-face study.

    c. Can we ask the same questions by telephone? If not, how should the procedure be adjusted?

    d. Do you think that the order of these questions will make a difference?

    e. Can we use the answers to these two questions to check for consistency? If so, how can that be done?

6. For the study you are designing, choose the most appropriate mode of data collection. Determine the order of the questions and specify the layout of the questionnaire.

PART III

ESTIMATION AND PREDICTION OF THE QUALITY OF QUESTIONS

Until now, we have been discussing the different choices that have to be made in order to design survey items, a questionnaire, and a data collection instrument. The design of survey questionnaires can be a scientific activity if we know the quality of different possible forms of questions. A lot of research has been done to get this information.

In this part, we discuss the following steps in this process in sequence:

Chapter 9 discusses the criteria for the quality of survey questions: reliability, validity, and method effects.

In Chapter 10, we present the more classical design of MTMM experiments to estimate the reliability, validity, and method effects of survey questions.

Chapter 11 discusses the MTMM design that has been used in the European Social Survey (ESS) for estimation of the quality criteria.

In Chapter 12, we will describe the experiments that have been incorporated in the later analysis. Because the experiments of the period 1979–1997 have been described in Saris and Gallhofer (2007a), we will concentrate in this chapter on the experiments that have been done in the context of the ESS. We will pay attention to planned differences between questions and unplanned differences. We will also present some results with respect to the quality of the questions over all experiments carried out.

The quality of all possible questions cannot be studied by MTMM experiments. Therefore, the questions involved in the MTMM experiments are coded with respect to their characteristics, and after that, a study has been done to predict the quality of these questions involved in the experiments. This study has also been reported in Chapter 12.

Design, Evaluation, and Analysis of Questionnaires for Survey Research, Second Edition.
Willem E. Saris and Irmtraud N. Gallhofer.
© 2014 John Wiley & Sons, Inc. Published 2014 by John Wiley & Sons, Inc.

9

CRITERIA FOR THE QUALITY OF SURVEY MEASURES

In Parts I and II, we have seen that the development of a survey item demands making choices concerning the structure of the item and the data collection procedure. Some of these choices follow directly from the aim of the study, such as the choice of the topic of the survey item(s) (church attendance, neighborhood, etc.) and the concept measured by the request for an answer (evaluations, norms, etc.). But there are also many choices that are not fixed and these choices will influence the quality of the survey items. They have to do with the formulation of the request, the response scales, and any additional components such as introduction, motivation, position in the questionnaire, and the mode of data collection. Therefore, it is highly desirable to have some information about the quality of a survey item before it is used in the field.

Several procedures have been developed to evaluate survey items before they are used in the final survey. The oldest and most commonly used approach is, of course, the use of pretests and debriefing of the interviewers regarding any problems that may arise in the questionnaire. Another approach, suggested by Belson (1981), is to ask people during a pretest, after they have answered a request for an answer, how they interpreted the different concepts in the survey item while they were answering the requests. A third approach is the use of "think aloud" protocols during interviews. A fourth approach is to assess the cognitive difficulty of a request for an answer. This can be done by an expert panel (Presser and Blair 1994) or on the basis of a coding scheme (Forsyth et al. 1992; Van der Zouwen 2000) or by using a computer program (Graesser et al. 2000a, b). The latter authors developed a computer program to

Design, Evaluation, and Analysis of Questionnaires for Survey Research, Second Edition.
Willem E. Saris and Irmtraud N. Gallhofer.
© 2014 John Wiley & Sons, Inc. Published 2014 by John Wiley & Sons, Inc.

evaluate survey items in relation to their linguistic and cognitive difficulty. A fifth approach, which is now rather popular, is to present respondents with different formulations of a survey item in a laboratory setting in order to see what the effect of these wording changes is (Esposito et al. 1991; Esposito and Rothgeb 1997; Snijkers 2002). For an overview of the different possible cognitive approaches to the evaluation of requests, we recommend Sudman et al. (1996). A rather different approach is interaction or behavioral coding. This approach checks to see whether the interaction between the interviewer and the respondent follows a standard pattern or whether deviant interactions occur (Dijkstra and Van der Zouwen 1982). Such deviant interactions can indicate problems in the questionnaire related to specific concepts or the sequence of the survey items.

All these approaches are directed at detecting response problems. The hypothesis is that problems in the formulation of the survey item will reduce the quality of the responses of the respondents. However, the standard criteria for data quality, such as validity, reliability, method effect, and item nonresponse, are not directly evaluated. Campbell and Fiske (1959) suggested that validity, reliability, and method effects can be evaluated only if more than one method is used to measure the same trait that is, in our research, a concept-by-intuition. Their design is called the multitrait–multimethod (MTMM) design and is widely used in psychology and psychometrics (Wothke 1996). This approach has also attracted attention in marketing research (Bagozzi and Yi 1991). In survey research, it has been elaborated and applied by Andrews (1984), whose method has been used for different topics and request forms in several languages: English (Andrews 1984), German (Költringer 1995), and Dutch (Scherpenzeel and Saris 1997). Andrews (1984) suggested using a meta-analysis of the available MTMM studies to determine the effect of different choices made in the design of survey requests on the reliability, validity, and method effects. Following his suggestion, Saris and Gallhofer (2007a) conducted a meta-analysis of the available 87 MTMM studies to summarize the effects that different request characteristics have on reliability and validity. Since 2002 in every round of the European Social Survey, 4–6 MTMM experiments have been done in 25–30 European countries in more than 20 different languages. Chapter 12 describes the results of the meta-analysis of these experiments, and Chapter 13 and later chapters indicate how these results can be used to predict both the quality of survey items before they are used in practice and how formulations of survey items can be improved where the quality of the original formulation is insufficient. In this chapter, we will illustrate some effects of these choices, followed by indicating what criteria should be used to evaluate the quality of survey requests. In Chapters 10 and 11, we discuss procedures to evaluate questions with respect to the selected quality criteria.

9.1 DIFFERENT METHODS, DIFFERENT RESULTS

Normally, all variables are measured using a single method. Thus, one cannot see how much of the variance of the variables is random measurement error and how much is systematic method variance. Campbell and Fiske (1959) suggested that

TABLE 9.1   An MTMM study in the ESS pilot study (2002)

For the three traits, the following three requests were employed:

1. *On the whole, how satisfied are you with the present state of the economy in Britain?*
2. *Now think about the national government. How satisfied are you with the way it is doing its job?*
3. *And on the whole, how satisfied are you with the way democracy works in Britain?*

In this experiment, the following response scales were used to generate the three different methods:

Method 1: *(1) very satisfied, (2) fairly satisfied, (3) fairly dissatisfied, (4) very dissatisfied*
Method 2:

very dissatisfied   *very satisfied*

       0   *1*   *2*   *3*   *4*   *5*   *6*   *7*   *8*   *9*   *10*

Method 3: *(1) not at all satisfied, (2) satisfied, (3) rather satisfied, (4) very satisfied*

multiple methods for multiple traits should be employed in order to detect error components. The standard MTMM approach nowadays uses at least three traits that are measured by at least three different methods, leading to nine different observed variables. In this way, a correlation matrix of 9 × 9 is obtained. In order to illustrate this type of procedure, Table 9.1 presents a brief summary of an MTMM experiment conducted in a British pilot study for the first round of the European Social Survey (2002). Three different traits and three methods were used in the study.

In this study, the topic of the survey items (national politics/economy) remained the same across all methods. Also, the concept measured (a feeling of satisfaction) is held constant. Only the way in which the respondents are asked to express their feelings varies. The first and third methods use a 4-point scale, while the second method uses an 11-point scale. This also means that the second method provides a midpoint on the scale, while the other two do not. Furthermore, the first and second methods use a bipolar scale, while the third method uses a unipolar scale. In addition, the direction of the response categories changes in the first method compared to the second and the third methods. The first and third methods have completely labeled categories, while the second method has labels only at the endpoints of the scale.

There are other aspects in which the requests are similar, although they could have been different. For example, in Table 9.1, direct requests have been selected for the study. It is, however, very common in survey research to specify a general request such as "How satisfied are you with the following aspects in society?" followed by the provision of stimuli such as the present economic situation, the national government, and the way the democracy functions. Furthermore, all three requests are unbalanced, asking "how satisfied" people are without mentioning the possibility of dissatisfaction. They have no explicit "don't know" option, and all three have no introduction and subordinate clauses, making the

TABLE 9.2   The response distribution for the nine requests specified in Table 9.1

	Satisfaction with the								
	Economy method			Government method			Democracy method		
Responses	1	2	3	1	2	3	1	2	3
Dissatisfied	167	134	99	268	208	169	187	152	169
Neutral	—	—	102	—	—	100	—	—	100
Satisfied	273	193	320	191	128	258	223	176	258
Very satisfied	26	7	12	11	2	4	43	7	4
Missing	19	49	54	15	47	54	32	50	54

survey items relatively short. There is no need to discuss here other relevant characteristics of requests because they have already been covered in Parts I and II of this book.

Identical characteristics of the three requests cannot generate differences, except for random errors, but those aspects that do differ can generate differences in the responses. Many studies have looked at the differences in response distributions, for example, Schuman and Presser (1981). Table 9.2 presents a summary of the responses. We have made the responses as comparable as possible by summing up categories that can be clustered.[1]

Table 9.2 shows that quite different results are obtained depending on what method for the formulation of the answer categories is used. If we did not know that these answers come from the same 485 people for the same requests, we could conclude that these responses come from different populations or that the requests measure different opinions.

One obvious effect is the effect of the neutral category of the second method, which changes the overall distribution of the answers. Another phenomenon that seems to be systematic is that the third method generates much more "satisfied" responses than do the other two. This has to do with the unipolar character of the scale and the extreme label for the negative category, which is "not at all satisfied." This label seems to move some respondents to the category "satisfied" where they otherwise are "neutral" or "dissatisfied." It also appears that the number of people saying that they are "very satisfied" decreases if more response categories are available to express satisfaction: method 1 has only two, method 2 has five, and method 3 has three possibilities.

Finally, a very clear effect can be observed for the number of missing values. This might have to do with the positioning of the request for an answer in the questionnaire and other characteristics of the questionnaire.

Because the same people were asked for all nine requests, it is possible to look at the cross tables of the requests for the same topic, which demonstrates what

[1] In method 3, very dissatisfied=0, dissatisfied=1–4, neither satisfied nor dissatisfied=5, satisfied=6–9, very satisfied=10.

TABLE 9.3   The cross table of satisfaction with the economy measured with method 1 and method 3

Measured with	Method 1				
Method 3	Very satisfied	Fairly satisfied	Fairly dissatisfied	Very dissatisfied	Total
Not at all satisfied	0	19	42	36	97
Satisfied	10	173	44	7	234
Rather satisfied	6	54	13	4	77
Very satisfied	8	4	0	0	12
Total	*24*	*250*	*99*	*47*	*420*

Source: These data were taken from the British pilot study of the ESS.

TABLE 9.4   The cross table of satisfaction with the economy measured with methods 1 and 2

Measured with	Method 1				
Method 2	Very satisfied	Fairly satisfied	Fairly dissatisfied	Very dissatisfied	Total
Very dissatisfied					
0	0	1	5	12	18
1	0	3	2	8	13
2	0	4	8	8	20
3	0	12	22	7	41
4	0	19	18	1	38
5	3	61	24	9	97
6	0	41	12	1	54
7	2	60	6	1	69
8	7	43	4	1	55
9	5	9	0	0	14
10	7	0	0	0	7
Very satisfied					
Total	*24*	*253*	*101*	*48*	*426*

the link between the different responses is. In Table 9.3 and Table 9.4, we present the results for two of the three possible combinations for satisfaction with the economy.

Table 9.3 shows the following striking results:

- Sixty-eight respondents claim that they are dissatisfied with method 1 and satisfied[1] with method 3.
- Nineteen respondents claim that they are satisfied with method 1 and dissatisfied with method 3.

- Sixteen respondents claim that they are very satisfied with method 1 and are satisfied or rather satisfied with method 3.
- Four respondents claim that they are only fairly satisfied with method 1 and are very satisfied with method 3.

A similar analysis (see note 1) can be made for Table 9.4 on the following page in which:

- Twenty-five respondents claim that they are dissatisfied with method 1 and satisfied with method 2.
- Thirty-nine respondents claim that they are satisfied with method 1 and dissatisfied with method 2.
- Nine respondents claim that they are very satisfied with method 1 and only satisfied with method 2.
- Nine respondents claim to be only satisfied with method 1 and very satisfied with method 2.

In summary, Table 9.3 and Table 9.4 had 107 inconsistent answers out of approximately 420 respondents who answered both requests. Some of these inconsistent answers may be mistakes, but it is also possible that there is a systematic pattern in these inconsistencies due to the three methods. To clarify whether that is the case, we also looked at the same two tables for the topic "satisfaction with the government." The results are presented in Table 9.5 and Table 9.6.

Table 9.5 shows the following errors:

- Eighty-five respondents claim to be dissatisfied with method 1 and satisfied with method 3.
- Twelve respondents claim to be satisfied with method 1 and dissatisfied with method 3.
- Eight respondents claim to be very satisfied with method 1 and satisfied with method 3.
- Two respondents are only satisfied with method 1 and very satisfied with method 3.

In Table 9.6, it can be seen that:

- Twenty-six respondents claim to be dissatisfied with method 1 and satisfied with method 2.
- Thirty-one respondents claim to be satisfied with method 1 and dissatisfied with method 2.
- Seven respondents claim to be very satisfied with method 1 and only satisfied with method 2.
- One claims to be satisfied with method 1 and very satisfied with method 2.
- Thirty-six respondents claim to be very dissatisfied with method 1 and only dissatisfied with method 2.
- Five respondents claim to be only dissatisfied with method 1 and very dissatisfied with method 2.

TABLE 9.5   The cross table of satisfaction with the government measured with method 1 and method 3

Measured with	Method 1				
Method 3	Very satisfied	Fairly satisfied	Fairly dissatisfied	Very dissatisfied	Total
Not at all satisfied	0	12	89	67	168
Satisfied	4	124	67	2	197
Rather satisfied	4	35	15	1	55
Very satisfied	2	2	0	0	4
Total	*10*	*173*	*171*	*70*	*424*

TABLE 9.6   The cross table of satisfaction with the government measured with methods 1 and 2

Measured with	Method 1				
Method 2	Very satisfied	Fairly satisfied	Fairly dissatisfied	Very dissatisfied	Total
Very dissatisfied					
0	0	1	5	27	33
1	0	5	8	13	26
2	0	3	27	12	42
3	0	7	37	9	53
4	1	14	35	2	52
5	1	49	38	7	95
6	1	28	13	0	42
7	0	29	9	0	38
8	2	32	3	1	38
9	4	5	0	0	9
10	1	1	0	0	2
Very satisfied					
Total	*10*	*174*	*175*	*71*	*430*

In Table 9.5 and Table 9.6, we discovered 107 inconsistent answers out of the approximately 420 respondents who answered both requests. As mentioned previously, some inconsistency may be random errors, but there are also some evident patterns in these errors. For example, there are many more people claiming to be dissatisfied with method 1 and satisfied with method 3 rather than vice versa. This is also true for Table 9.3 and Table 9.5. This phenomenon cannot be found in Table 9.4 and Table 9.6, where the effect is rather reversed. But in the latter case, we see that there are many more extreme responses for method 1 than for method 2. These results seem to suggest that there are random errors due to mistakes but also systematic effects connected with the differences between the methods. This issue will be discussed in a later section.

TABLE 9.7   Correlations between the nine variables of the MTMM experiment with respect to satisfaction with political outcomes

	Method 1			Method 2			Method 3		
	Q1	Q2	Q3	Q1	Q2	Q3	Q1	Q2	Q3
Method 1									
Q1	1.00								
Q2	.481	1.00							
Q3	.373	.552	1.00						
Method 2									
Q1	−.626	−.422	−.410	1.00					
Q2	−.429	−.663	−.532	.642	1.00				
Q3	−.453	−.495	−.669	.612	.693	1.00			
Method 3									
Q1	−.502	−.347	−.332	.584	.436	.438	1.00		
Q2	−.370	−.608	−.399	.429	.653	.466	.556	1.00	
Q3	−.336	−.406	−.566	.406	.471	.638	.514	.558	1.00

Given the results, the general conclusion should be that the correspondence between the different measures for the same people is rather low. A measure commonly used to determine the correspondence of responses is the correlation coefficient. It does not come as a surprise that the correlations in these tables are not very high, even though the requests are supposed to measure the same traits. The correlations in these tables are as follows: Table 9.3, −.502; Table 9.4, −.626; Table 9.5, −.608; and Table 9.6, −.663. We have also calculated the same correlations for the concept "satisfaction with the democracy." The respective correlations are −.566 and −.669. We see that these relationships are rather weak since the proportion of similarity is equal to the correlation squared, which peaks around .40.

Let us now look at what happens to the correlations between the measures of the different traits. Table 9.7 presents the correlations between the nine measures.

These results clearly indicate the need for further investigation of the quality of the different measures, since the correlations between the three requests Q1–Q3 are very different for the different methods. For the first method, the correlations vary between .373 and .552; for the second method between .612 and .693; and for the third method between .514 and .558.

All these results raise questions such as:

- How can such differences be explained?
- What are the correct values?
- What is the best method?

To answer these requests, quality criteria for survey measures are required. This is the topic of one of the next sections.

9.2 HOW THESE DIFFERENCES CAN BE EXPLAINED

In order to explain the differences discussed earlier, something has to be said about relations between variables in general, namely, how such relations can be formulated and how these relationships and correlations are linked. After this general introduction, we will apply our knowledge of the measurement situation we were previously discussing.

9.2.1 Specifications of Relationships between Variables in General

In the literature, a distinction is made between direct effects, indirect effects, spurious relationships, and joint effects. The different relations are illustrated in Figure 9.1. The arrow in Figure 9.1a going from x to y indicates a *direct effect* of a variable x on a variable y. In Figure 9.1b, there is no arrow from x to y directly, so there is no direct effect, but there is an *indirect effect* because x influences z and z in turn influences y, and so x has an influence on y but only indirectly. This is not the case in Figure 9.1c. There, the relationship between y and x is called *spurious* because the two variables have neither a direct nor an indirect effect on each other, but there is a third variable z that influences both. Therefore, we can expect a relationship between x and y, but this relationship is not due to any effect of x on y. Finally, in Figure 9.1d, there is also no effect of x on y, but it is unclear where the relationship between x and y comes from, as the direction of the effect between z and w is unknown (indicated by a double-headed arrow). In this case, z could produce a spurious relation and w could do so as well, making the relationship unclear. This type of relationship is called a *joint effect* due to z and w.

In order to make the discussion less abstract, let us look at the example in Figure 9.2. Figure 9.2 represents a causal model explaining "political interest." It is assumed that this variable is directly influenced by education (b_3) and socioeconomic status ("SES") (b_4) and that "SES" is directly influenced by "income" (b_1) and "education" (b_2). We could continue with the explanation of the relationship between "income" and "education," but here, we are not interested in these details, so we are leaving it unspecified (ρ_{ie}).

The different b coefficients indicate the size of the effects (for an explanation, we refer the reader to Appendix 9.1), while ρ_{ie} represents the size of the correlation between the variables "income" and "education." Note that we only have to specify

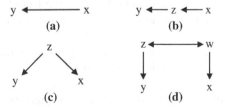

FIGURE 9.1 Different relationships between the variables x and y: (a) a direct effect, (b) an indirect effect, (c) a spurious relation, and (d) a joint effect.

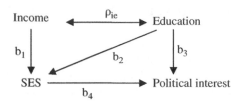

FIGURE 9.2   An example of causal relationships.

the direct effects because from these follow the indirect effects, spurious relations, and joint effects.

In order to be more precise about the size of the indirect effects, spurious relationships, and joint effects, we specify that:

Theorem 1   The indirect effect, spurious relations, and joint effects are equal to the product of the coefficients in the direction of the arrow from one variable to the other, without passing the same variable twice and going against the direction or the arrow.

From such causal models, predictions can be made concerning the size of the correlations between each pair of the variables, assuming that no other variables play a role in this process. These predictions are as follows:

Theorem 2   The correlation between two variables is equal to the sum of the direct effect, indirect effects, spurious relationships, and joint effects between these variables.

In the literature on structural equation modeling, these two theorems have been proven. For a simple introduction, we refer the reader to Saris and Stronkhorst (1984); for a more complete and an elaborate discussion, Bollen (1989) is recommended.

Let us now apply these theorems to derive the size of the different correlations between the variables mentioned in the causal model of Figure 9.2. The results are presented below, where the different correlations are denoted by $\rho(i,j)$.

ρ(income, education) $=$ joint effect $= \rho_{ie}$
ρ(income, SES) $=$ direct effect $+$ joint effect $= b_1 + \rho_{ie}b_2$
ρ(income, political interest) $=$ indirect effect $+$ joint effects $= b_1b_4 + \rho_{ie}b_3 + \rho_{ie}b_2b_4$
ρ(education, SES) $=$ direct effect $+$ joint effect $= b_2 + b_1\rho_{ie}$
ρ(education, political interest) $=$ direct effect $+$ indirect effect $+$ joint effect
                                $= b_3 + b_2b_4 + \rho_{ie}b_1b_4$
ρ(SES, political interest) $=$ direct effect $+$ spurious $+$ joint effect
                                $= b_4 + b_2b_3 + b_1\rho_{ie}b_3$

On the basis of the estimates of the different effects, predictions can be made about the size of the correlations of the variables. Note that in all cases, the correlation between any two variables is not equal to the direct effect between the two variables.

FIGURE 9.3 The classical model for random errors.

Sometimes, there is even no direct effect at all, while there will be a correlation due to other effects or relationships, for example, the correlation between "income" and "political interest."

In the next chapter, we will show that these relationships between the correlations and the effects (also called *parameters*) can be used to estimate the values of these parameters if the sizes of the correlations are known from research. But, in this chapter, we concentrate on the formulation of models. So, in the next section, we will use this approach to specify measurement models.

9.2.2 Specification of Measurement Models

In the first part of this chapter, we have shown that we can expect two types of errors: random and systematic. This means that the response variables we use in survey research will not be the same as the variables we want to measure. Looking only at the random errors, psychometricians (Lord and Novick 1968) have suggested the model of Figure 9.3.

This model suggests that the response variable (y) is determined directly by two other variables: the so-called true score (t) and the random errors represented by the random variable (e). The true score variable is the observed response variable corrected for random measurement error. For example, imagine that we measure "political interest" by a direct request "How interested are you in politics?" using an 11-point scale and we assume only random mistakes in the answers. For example, some people mistakenly pick a score 6 while their true score is 5, while others choose a 3 by mistake while their true score is 4, and so on. Therefore, y represents the observed answer to the request, e represents the random errors (here in both cases 1), and t is equal to the observed answer corrected for random errors.

However, as we have seen before, this model is too simple because systematic errors also occur. In that case, the variable we want to measure is not the same as the variable measured. This can be modeled by making a distinction between the variable we want to measure (f), the true score (t), and the response variable (y). This idea is presented in Figure 9.4.

To continue our example, if "political interest" is not measured by a direct request but by "the amount of time spent watching political programs on TV," then there is a difference between the variable we would like to measure and the variable really

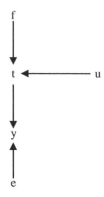

FIGURE 9.4   The measurement model with random (e) and systematic (u) errors.

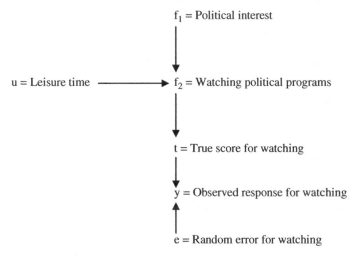

FIGURE 9.5   A measurement for political interest with a systematic error (u) and a random error (e).

measured. We certainly expect that "political interest" affects "the time people spend watching for political programs on TV"; however, the relationship will not be perfect, especially because "the amount of leisure time" a person has will have an effect on "the time this person will watch TV." This could be modeled as is done in Figure 9.5.

If the relationship between f_2 and t is perfect so that $f_2 = t$, then this model will be reduced to the model presented in Figure 9.4. However, in Figure 9.5, there is a systematic error because the variable to be measured is measured indirectly using an indicator that also contains another component (leisure time).

There can also be another reason why f is not necessarily equal to the true score. We have seen that just varying the answer categories for measuring opinion of

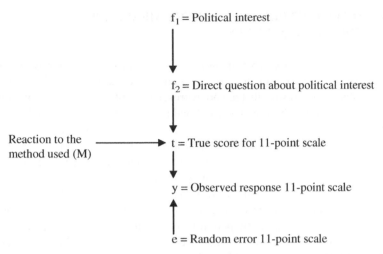

f_1 = Political interest

f_2 = Direct question about political interest

Reaction to the method used (M) ⟶ t = True score for 11-point scale

y = Observed response 11-point scale

e = Random error 11-point scale

FIGURE 9.6   A measurement for political interest with a systematic method factor (M) and a random error.

respondents can also produce distinct and systematically different results. Therefore, it appears that the method also has an effect on the response. For example, the direct request "How interested are you in politics?" will not be influenced by any other variables.[2] So we can assume that $f_1 = f_2$ in Figure 9.6. Even when we ignore this possible difference, we observe systematically different responses to the direct request depending on formulation choices of the direct request. We demonstrated that the reaction of respondents toward an 11-point scale can vary between respondents (with respect to the range of responses) and their individual reactions can be different for different response scales. The reaction of respondents to a specific method is called the *method factor*, which is modeled in Figure 9.6.

If we assume again that the link between f_2 and f_1 is perfect ($f_2 = f_1$), then this model reduces to the model of Figure 9.4 with a specific component affecting the true score "the reaction to the method used."

Although a combination of the two reasons for differences between f_1 and t is rather common, in the next sections, we will concentrate on the second kind of possible systematic error, because the first example represents a shift of concept that is discussed later in Part IV. Until then, we will always assume that the different indicators do not contain systematic errors due to a conceptual shift and only contain errors due to the reaction of the method used. This modeling of measurement processes can help us specify quality criteria for measures and explain the differences in the data.

---

[2] However, it can be argued that the answers to the direct request are indeed not influenced by any variables other than "political interest," although the possibility that "a tendency to social desirable answers" might also play a role.

9.3   QUALITY CRITERIA FOR SURVEY MEASURES AND THEIR CONSEQUENCES

The first quality criterion for survey items is to have as little *item nonresponse* as possible. This is an obvious criterion because missing values have a disrupting effect on the analysis, which can lead to results that are not representative of the population of interest.

A second criterion is *bias*, which is defined as a systematic difference between the real values of the variable of interest and the observed scores corrected for random measurement errors.[3] For objective variables, real values can be obtained, and thus, the method that provides responses, corrected for random errors (true scores), closest to the real values is preferable. A typical example comes from voting research. After the elections, the participation in the elections is known. This result can be compared with the result that is obtained by survey research performed using different methods. It is a well-known fact that participation is overestimated when using standard procedures. Therefore, a new method that does not overestimate the participation or produces a smaller bias is preferable to the standard procedures.

In the case of subjective variables, where the real values are not available, it is only possible to study different methods that generate different distributions of responses as we have done previously. If differences between two methods are observed, at least one method is biased; however, both can also be biased.

These two criteria have been given a lot of attention in split-ballot experiments (see Schuman and Presser (1981) for a summary). Molenaar (1986) has studied the same criteria focusing on nonexperimental research (1986). In summary, these criteria give a description of the observed differences of nonresponse and differences of response distributions.

There are also quality criteria that provide an explanation for the weak correlation between indicators that should measure the same variable and the differences in correlations between variables for different methods as we have seen in Table 9.7. To explain these observations, the concepts *reliability*, *validity*, and *method effect* need to be studied.

In order to do so, we extend the model of Figure 9.4 to two variables of interest, for example, "satisfaction with the government" and the "satisfaction with the economy." The measurement model for two variables is presented in Figure 9.7.

In this model, it is assumed that:

- f_i is the trait factor i of interest measured by a direct question.
- y_{ij} is the observed variable (variable or trait i measured by method j).
- t_{ij} is the "true score" of the response variable y_{ij}.
- M_j is the method factor that represents a specific reaction of respondents to a method and therefore generates a systematic error.
- e_{ij} is the random measurement error term for y_{ij}.

---

[3] This simple definition serves for the purpose of this text. However, a precise definition is found in Groves (1989).

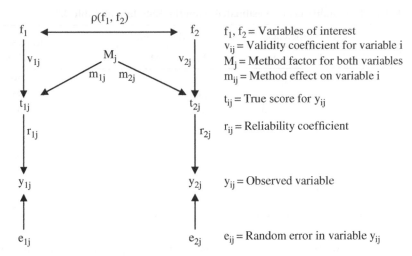

f_1, f_2 = Variables of interest
v_{ij} = Validity coefficient for variable i
M_j = Method factor for both variables
m_{ij} = Method effect on variable i

t_{ij} = True score for y_{ij}

r_{ij} = Reliability coefficient

y_{ij} = Observed variable

e_{ij} = Random error in variable y_{ij}

FIGURE 9.7   The measurement model for two traits measured with the same method.

The r_{ij} coefficients represent the standardized effects of the true scores on the observed scores. This effect is smaller if the random errors are larger. This coefficient is called the *reliability coefficient*.

The v_{ij} coefficients represent the standardized effects of the variables of interest on the true scores for the variables that are really measured. Therefore, this coefficient is called the *validity coefficient*.

The m_{ij} coefficients represent the standardized effects of the method factor on the true scores, called the *method effect*. An increase in the method effect results in a decrease in validity and vice versa. It can be shown that for this model, $m_{ij}^2 = 1 - v_{ij}^2$, and therefore, the method effect is equal to the invalidity due to the method used.

Reliability is defined as the strength of the relationship between the observed response (y_{ij}) and the true score (t_{ij}), that is, r_{ij}^2.

Validity is defined as the strength of the relationship between the variable of interest (f_i) and the true score (t_{ij}), that is, v_{ij}^2.

The *systematic method effect* is the strength of the relationship between the method factor (M_j) and the true score (t_{ij}) resulting in m_{ij}^2.

The *total quality of a measure* is defined as the strength of the relationship between the observed variable and the variable on interest, that is, $(r_{ij}v_{ij})^2$.

The *effect of the method on the correlations* is equal to $r_{1j}m_{1j}m_{2j}r_{2j}$.

The reason for employing these definitions and their criteria becomes evident after examining the effect of the characteristics of the measurement model on the correlations between observed variables.

Using the two theorems we have presented previously, it can be shown that the correlation between the observed variables $\rho(y_{1j}, y_{2j})$ is equal to the joint effect of

TABLE 9.8 The quality criteria estimated from the ESS data of Table 9.7

	Validity coefficients			Method effects			Reliability coefficients
	F_1	F_2	F_3	M_1	M_2	M_3	
t_{11}	.93			.36			.79
t_{21}		.94		.35			.85
t_{32}			.95	.33			.81
t_{12}	.91				.41		.91
t_{22}		.92			.39		.94
t_{32}			.93		.38		.93
t_{13}	.85					.52	.82
t_{23}		.87				.50	.87
t_{33}		.88				.48	.84

the variables that we want to measure (f_1 and f_2) plus the spurious correlation due to the method factor as demonstrated in formula 9.1:

$$\rho(y_{1j},y_{2j}) = r_{1j}v_{1j}\rho(f_1,f_2)v_{2j}r_{2j} + r_{1j}m_{1j}m_{2j}r_{2j} \qquad (9.1)$$

Note that r_{ij} and v_{ij}, which are always smaller than 1, will decrease the correlation (see first term), while the method effects, if they are not 0, can generate an increase in the correlation (see second term). This result suggests that it is possible that the low correlations for methods 1 and 3 in Table 9.7 are due to a lower reliability of methods 1 and 3 compared to method 2. However, it is also possible that the correlations of method 2 are higher because of greater systematic method effects of this method.

Using the specification of the model mentioned in Figure 9.7, the results presented in Table 9.8 are obtained for the variables of Table 9.7. How this result is obtained will be the topic of the next chapter. For the moment, we will concentrate on the meaning of the results. They suggest that method 2 has higher reliability coefficients than the other methods and that its method effects are intermediate.

If we know that the correlation between the first two traits was estimated at .69, it can be verified by substituting the values of the reliability, validity, and method coefficients in Equation 9.1 that such different observed correlations as .490 for method 1 and .631 method 2 can be obtained.

Equation 9.1 for method 1 gives

$$\rho_{y11,y12} = .79 \times .93 \times .69 \times .94 \times .85 + .79 \times .36 \times .35 \times .85$$
$$= .405 + .085 = .490$$

Equation 9.1 for method 2 gives

$$\rho_{y21,y22} = .91 \times .91 \times .69 \times .92 \times .94 + .91 \times .41 \times .39 \times .94$$
$$= .494 + .137 = .631$$

This result shows that the observed correlation between the same two variables was .141 higher for method 2 than for method 1 because its reliability was higher, while the method effect was higher for method 2. So, with these quality estimates, we can quite well explain the difference in correlations for the different methods. However, both correlations were not very good estimates of the correlation between the two variables because of the random and systematic errors in the data. Our best estimate of the correlation between these two variables corrected for measurement error is .69. So, both correlations were incorrect. How we obtained the estimate of the relationship corrected for measurement errors will be discussed in the next chapter.

Our results show that the differences in the correlations obtained can almost entirely be explained by differences in data quality between the different measurement procedures. It also illustrates how important, for social science research, reliability and validity are as defined. Therefore, it is also important to know how these quality criteria can be estimated. However, let us now turn to some other commonly used criteria for data quality.

9.4 ALTERNATIVE CRITERIA FOR DATA QUALITY

Out of the many possible criteria for data quality, we will discuss only the most common ones and indicate some problems associated with them.[4]

9.4.1 Test–Retest Reliability

A very popular idea is that reliability can be determined by repeating the same observation twice as in the model of Figure 9.8.

Here, f_i is the variable to be measured, and y_{i1} and y_{i2} are the responses to the request used to measure this variable. This approach requires that the same method be used on two occasions. If the model holds true, then the correlation between the two variables can be due only to the product of the two reliability coefficients of the two measures:

$$\rho_{yi1,yi2} = r_{i1} \cdot r_{i2}$$

But since the same measure is used twice, we can assume that $r_{i1} = r_{i2}$ and then it follows that the reliability $= r_{i1}^2 = r_{i2}^2 = \rho_{yi1,yi2}$. In this case, the reliability of the measure is equal to the test–retest correlation.

However, the aforementioned representation is too simple, and it is better to start with the model shown in Figure 9.9.

The difference with the previous model is that a distinction is made between the latent variable for the first and second measure, accounting for a change that might have occurred while conducting the two observations. Additionally, the possibility is

---

[4] This section gives a wider perspective on the way reliability and validity can be defined; however, it is not essential for understanding the approach discussed in this book.

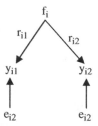

FIGURE 9.8   The standard test–retest model.

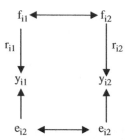

FIGURE 9.9   A more realistic test–retest model.

left open that respondents remember their first answers, indicated by a correlation between the error terms. In order to arrive from this model to the earlier model, the following assumptions were made:

1. No change in opinion between the first and the second measurements
2. No memory effects
3. No method effects
4. Equal reliability for the different measures of the same trait

This approach is unrealistic because it assumes that the measurement procedure can be repeated in exactly the same way (assumption 4).

Furthermore, if the time between the repetitions is too short, we can expect a memory effect (assumption 2), and if the time is too long, the opinion may be changed (assumption 1). Finally, possible method effects cannot be detected, while they may play an important role (assumption 3).

Therefore, this approach is not an accurate representation of reality. Although many people think that it is a robust procedure, it is based on a number of unattainable assumptions and a less restricted approach is needed.

9.4.2   The Quasi-simplex Approach

The previously specified approach can be made more manageable by using three observations instead of two. This approach has been suggested by Heise (1969), improved by Wiley and Wiley (1970), and used by Alwin and Krosnick (1991) to

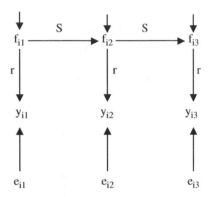

FIGURE 9.10   The quasi-simplex model for three repeated observations.

evaluate measurement instruments for survey research. Its advantage is that it is no longer necessary to assume that no change has occurred, and it is suggested that the memory effect can be avoided by making the time gap between the observations so long that a memory effect can no longer be expected. Figure 9.10 displays the suggested model.

In Figure 9.10, s is the stability coefficient and r is the reliability coefficient. This approach has two major problems. First, it assumes that it is not possible that considerations that are associated with the variable of interest are forgotten at time 2 but return at time 3. This would suppose that there is an effect of f_{i1} on f_{i3} that is not possible for technical reasons. However, because of these effects, wrong estimates of the quality of the measures will be obtained as discussed by Coenders et al. (1999).

The second problem is that any temporary component in the variables that is not present at the next occasion will be treated as error, while it might be a substantive part of the latent variable at a given point in time. For example, if we ask about "life satisfaction" and the respondent is in a bad mood, that person's score will be lower than if the same respondent is in a good mood on a different occasion. The mood component is a real part of the satisfaction variable, but because the mood changes rapidly, this component will end up in the error term. Therefore, the error term increases and the reliability decreases: not because of lack of data quality but also because of the instability of a component within the variable of interest. For further discussion of this point, we refer the reader to Van der Veld (2006). However, this would not occur if the measures were conducted quickly in the same survey, but then memory effect might emerge again. For these reasons, this approach is not preferable for defining the reliability coefficient.

9.4.3   Correlations with Other Variables

In order to evaluate the validity of different measures for the same variable, it has been suggested to use the correlation with other variables that are known to correlate with the variable of interest. The measure with the highest correlation is then the best estimate. Following this line of reasoning, this approach is modeled in Figure 9.11.

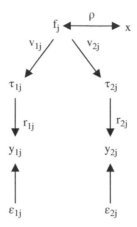

FIGURE 9.11 A standard model to evaluate validity.

In this figure, ρ is the correlation between the variable of interest and the external criterion variable (x). The other coefficients have their previously discussed meanings.

From this model, it follows that

$$\rho_{y1j,xi} = r_{1j}v_{1j}\rho$$

and

$$\rho_{y2j,xi} = r_{2j}v_{2j}\rho$$

This demonstrates that correlations can be different because of differences in validity, differences in reliability, or both. It also suggests that these correlations are not the proper criteria to evaluate the validity of measures. The validity of a measure should be evaluated by comparing the validity coefficients that we have presented in Section 9.3, in order to avoid confusion between reliability and validity, as is the case when using the correlation with a criterion variable.

9.5 SUMMARY AND DISCUSSION

In this chapter, we have demonstrated that the influence of different choices in the development of a survey item can have a significant effect on the item nonresponse, the distribution of the variables, and the correlation between different traits measured by different methods. These results led us to the conclusion that the first two quality criteria are:

1. Item nonresponse
2. Bias in the response

Furthermore, we have shown that the differences in the correlations between the variables for the different methods can be explained by the size of the random errors and systematic errors or the reliability and the validity and method effects, which were defined as follows: reliability as the strength of the relationship between the observed variable and the true score and validity as the strength of the relationship between the true score and the latent trait of interest.

Other measures that are often used and their several critiques were also elaborated. Given that we have clearly defined the quality criteria in the next two chapters, we will discuss how these quality criteria can be and have been estimated in practice.

EXERCISES

1. Asking people for information about their salary is a problem because many people refuse to answer the request. Therefore, alternative procedures are employed and compared with factual information collected from the employers.

   In the following, we provide two requests used in practice and the results that were obtained are compared with the factual information.
   Q1: What is your net monthly income?
   Q2: Could you tell me to what category your net monthly income belongs to?

In euros	Factual info (%)	Q1 (%)	Q2 (%)
<1000	5	9	5
1000–1500	10	12	11
1500–2000	30	35	32
2000–2500	30	32	33
2500–3000	10	8	11
3000–3500	5	2	4
3500–4000	4	1	2
4000–4500	3	1	1
>4500	3	0	1
100%=no. of responses	1000	700	850

   a. What, do you think, is the best measure for income?

   b. What was the criterion for answering "a"?

2. To see the effect of reliability on the observed correlation, we evaluate the following cases of the model of Figure 9.7:

   If the correlation between the variables corrected for measurement error $\rho(f_1,f_2)=.9$, the validity$=1$, and the method effect$=0$, what is the correlation between the observed variables in the following cases?

Reliability	Reliability	Observed correlation
y_{11}	y_{12}	$r_{y11,y12}$
1.0	1.0	
.9	.9	
.8	.8	
.7	.7	
.6	.6	

To answer these questions, use Equation 9.1.

3. The effect of the validity and method effect on the correlation between the observed variables is studied, while the correlation between the variables corrected for measurement error = .4 and the reliability = .8 for both measures.
   What is the correlation between the observed variables in the following cases?

Validity	Validity	Method effect	Method effect	Correlation
y_{11}	y_{12}	y_{11}	y_{12}	$r_{y11,y12}$
1.0	1.0	0.0	0.0	
.9	.9	.43	.43	
.8	.8	.6	.6	
.7	.7	.71	.71	

4. Is there any reason to assume that one or more of the requests of your own questionnaire are without random and/or systematic measurement error? If so, why?

5. Can you think of a measure that is without measurement error?

6. For the test–retest approach, we have specified the following model[5]:

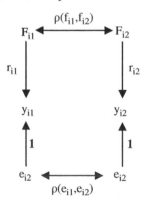

   a. Express the correlation between the observed variables in the parameters of the model.
   b. Show that the specified restrictions actually lead to the simple test–retest model.

---

[5] This question requires reading of Section 9.4.

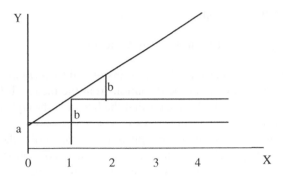

FIGURE 9A.1   The linear regression model.

APPENDIX 9.1   THE SPECIFICATION OF STRUCTURAL EQUATION MODELS

Structural equation models (SEMs) have been discussed extensively in the literature. For a simple introduction, we refer the reader to Saris and Stronkhorst (1984), and for a more elaborate text, we suggest the text of Bollen (1989). Therefore, our introduction to this topic will be very brief, and we mention only those aspects that are relevant for the models discussed in this and the next chapter.

The simplest case is a model with only two variables: a cause X and an effect variable Y. Assuming a linear relationship, we can formulate

$$Y = a + bX + u \tag{9A.1}$$

where u represent all variables affecting Y that are not explicitly mentioned. If we assume that the mean of u is 0, then the mean value of $Y = a$, if $X = 0$. This coefficient is called the *intercept* of the equation. The mean value of Y will increase by b for any increase of X of 1 unit on its scale. This effect b of X, which is called the *unstandardized regression coefficient*, is always the same. Therefore, the relationship between the two variables is linear, as presented in Figure 9A.1.

From equation (9A.1), it follows for the mean of Y indicated by $\mu(Y)$ that

$$\mu(Y) = \left(\frac{1}{n}\right)\sum(Y) = a + b\left(\frac{1}{n}\right)\sum(X) + \left(\frac{1}{n}\right)\sum(u) \tag{9A.2}$$

where the summation is done over the scores of all respondents. Since the mean of u is 0, we derive that

$$\mu(Y) = a + b\mu(X) \tag{9A.3}$$

From (9A.1) and (9A.3), it follows that

$$(Y - \mu(Y)) = b(X - \mu(X)) + u \tag{9A.4}$$

In this case, the scores of the variables are transformed to *deviation scores* because they are a numeric expression of the deviation from the means. It follows that the intercept of the equation is 0. This means that the line goes through the zero point or that $Y = 0$ if $X = 0$. We also see that the regression coefficient remains the same.

It is very common and useful in our application to standardize the variables. This can be done by dividing each of the variables, expressed in deviation scores, by their standard deviation $(\sigma(i))$, and from 9A.5, we get formula 9A.6:

$$\frac{Y - \mu(Y)}{\sigma(Y)} = b\left(\frac{\sigma(x)}{\sigma(y)}\right)\left(\frac{X - \mu(X)}{\sigma(X)}\right) + \left(\frac{u}{\sigma(Y)}\right) \tag{9A.5}$$

It can easily be verified that if 9A.5 is true, then 9A.6 is also true. The last equation can be rewritten as

$$y = \beta x + u \tag{9A.6}$$

Now y and x are standardized variables, while the effect of x on y is β, and the relationship with the previous coefficient is

$$\beta = \frac{b\sigma(X)}{\sigma(Y)} \tag{9A.7}$$

This effect should be interpreted as the effect on y of an increase of x with 1 standard deviation and is therefore called the *standardized regression coefficient*. In order to indicate the *strength of the effect* of x on y, it is squared:

$$\beta^2 = \text{strength of the relationship} \tag{9A.8}$$

Note that $\beta = 0$ if and only if $b = 0$ because the standard deviations are always larger than 0. The standardized coefficient as well as the unstandardized coefficient indicates whether a variable has an effect on another variable.

This model can be extended by introducing more causal variables. In case we assume that the effects of the different variables are additive, model 9A.9 becomes a model of multiple regression with standardized variables:

$$y = \beta_1 x_1 + \beta_2 x_2 + \cdots + \beta_n x_n + u \tag{9A.9}$$

In this case, the strength of the effects of the different variables on the effect variables can be compared by comparing the β_i coefficients.

Models can be further extended by introducing more effect variables. In that case, for each effect variable, a separate equation like 9A.9 should be formulated, which includes in the equation only those variables that supposedly have a direct causal effect on the effect variable. Such a set of equations forms a causal model, and these models are compatible with the graphical models used in this text. Theorems 1 and 2 discussed previously can be proved using the causal models in algebraic form that were introduced in this appendix. For more detail, we refer the reader to Saris and Stronkhorst (1984) and Bollen (1989).

10

ESTIMATION OF RELIABILITY, VALIDITY, AND METHOD EFFECTS

In the last chapter, we discussed several criteria to consider while thinking about the quality of survey measures. Two criteria, item nonresponse and bias, do not require much further discussion. If different methods have been used for the same trait, item nonresponse can be observed directly from collected data. The same holds true for bias if factual information is available. However, this is not the case if an estimate of the relative bias of measures can be derived by comparing distributions of responses with each other. Molenaar (1986) has made useful suggestions for measures of relative bias.

More complicated is the estimation of the quality criteria of reliability, validity, and method effect. Therefore, this chapter and the next will concentrate on their estimation. In order to discuss the estimation of reliability and validity, we have to introduce the basic idea behind estimation of coefficients of measurement models. So far, it is not so evident that the effect of an unmeasured variable on a measured variable, let alone the effect of an unmeasured variable on another unmeasured variable, can be estimated.

Here, we start by addressing the problem of identifying the parameters of models with unmeasured variables. Next, we will discuss the estimation of the parameters. Only after we have introduced these basic principles we will concentrate on the estimation of the reliability and validity, demonstrating the kind of designs that have been used to estimate them, as was defined in the last chapter.

Design, Evaluation, and Analysis of Questionnaires for Survey Research, Second Edition.
Willem E. Saris and Irmtraud N. Gallhofer.
© 2014 John Wiley & Sons, Inc. Published 2014 by John Wiley & Sons, Inc.

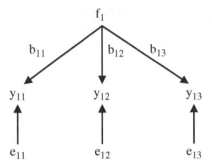

FIGURE 10.1   A simple measurement model assuming only random measurement errors.

10.1   IDENTIFICATION OF THE PARAMETERS OF A MEASUREMENT MODEL

In the last chapter, we introduced the formulation of causal models and two theorems to derive the relationship between the correlations of pairs of variables and the parameters of a model. Figure 10.1 represents a causal model for "satisfaction with the economy" where f_1 is a general measure of "satisfaction with the economy," while y_{11} is a measure of "satisfaction" using a bipolar 4-point scale. Measure y_{12} is a bipolar measure of satisfaction on an 11-point scale, and y_{13} is a measure of satisfaction on a unipolar 4-point scale. The formulation of these measures has been presented in Table 9.1.

In this model, f_1 is the variable that we would like to measure but that cannot be directly observed, except by asking questions. The variables y_{11} through y_{13} represent answers to these different questions. For the sake of simplicity, we assume that there are only random measurement errors (or $t_1 = f_1$). Using the two theorems mentioned in the last chapter, the following relationships of the correlations between the variables and the parameters of the model can be derived:

$$\rho(y_{11}, f_1) = b_{11} \tag{10.1}$$

$$\rho(y_{12}, f_1) = b_{12} \tag{10.2}$$

$$\rho(y_{13}, f_1) = b_{13} \tag{10.3}$$

$$\rho(y_{12}, y_{11}) = b_{12} b_{11} \tag{10.4}$$

$$\rho(y_{13}, y_{11}) = b_{13} b_{11} \tag{10.5}$$

$$\rho(y_{13}, y_{12}) = b_{13} b_{12} \tag{10.6}$$

If the correlations between these variables were known, the effects of the general judgment (f_1) on the specific observed answers would be easily determined on the basis of the first three relationships. However, the problem with measurement models is that the general variable (f_1) is not a measured variable, and by definition, its correlations with the observed variables are also unknown. So, the effects have to be estimated from the relationships between the observed variables from y_{11} to y_{13} (Equations 10.4–10.6). Formally, it can be shown that this is possible. From (10.4) and (10.5), we can derive

$$\frac{\rho(y_{12}, y_{11})}{\rho(y_{13}, y_{11})} = \frac{b_{12} b_{11}}{b_{13} b_{11}} = \frac{b_{12}}{b_{13}}$$

and therefore

$$b_{12} = b_{13} \left(\frac{\rho(y_{12}, y_{11})}{\rho(y_{13}, y_{11})} \right)$$

If we substitute this result in (10.6), we get the equation

$$\rho(y_{13}, y_{12}) = b_{13} \left(\frac{\rho(y_{12}, y_{11})}{\rho(y_{13}, y_{11})} \right) b_{13}$$

or

$$b_{13}^2 = \frac{\rho(y_{13}, y_{11}) \rho(y_{13}, y_{12})}{\rho(y_{12}, y_{11})} \tag{10.7}$$

From this, it can be seen that if these three correlations are known, the effect b_{13} is also known. Also, if b_{13} is known, the other effects can be obtained from Equations (10.5) and (10.6).

The proof shows that the estimate of an effect of an unmeasured variable on measured variables can be obtained from the correlations between the observed variables. Let us check this assumption for the given example. In Table 9.7, the correlations for the three observed variables can be found. Assuming that the values mentioned are the correct values for the population correlations, we can derive the following:

$$\rho(y_{12}, y_{11}) = -.626 = b_{12} b_{11}$$

$$\rho(y_{13}, y_{11}) = -.502 = b_{13} b_{11}$$

$$\rho(y_{13}, y_{12}) = .584 = b_{13} b_{12}$$

Applying the previously indicated procedure, we first get that $b_{13}^2 = -.502 \times .584 / -.626 = .475$ or[1] $-.475$, making $b_{13} = .689$; it follows from Equation (10.6) that $b_{12} = .584/.689 = .847$ and from (10.5) that $b_{11} = -.502/.689 = -.729$.

---

[1] That a positive or negative value is possible is not just a numeric result but also a logical one, given that the scale of the latent variable is not fixed and can go from low to high or from high to low. Therefore, the sign of the effect of this variable can be positive or negative. After the sign for one effect has been chosen, all others are also determined.

The results show that "satisfaction with the economy" (f_1) has the strongest relationship with the observed scores of method 2 (y_{12}). Therefore, this measure seems preferable, because it contains the smallest amount of random measurement errors. We also see that the effect for method 1 is negative. This is because the scale of the observed variable goes from positive to negative, while for the other two variables, the scale goes in the opposite direction.

This example demonstrates that it is possible to estimate the size of the effect of an unmeasured variable on a measured variable. The question is, however, whether this is always the case and, if not, when it is not the case. It can easily be verified that the necessary condition is that there should be at least as many correlations as unknown parameters. If there are only two observed variables, there is only one correlation; however, two coefficients need to be estimated, which is impossible.

Even if the necessary condition is fulfilled, the values of the parameters cannot always be obtained. The sufficient condition for "*identification*" is difficult to formulate. For a more complete analysis of sufficient conditions, we refer the reader to the literature on structural equation modeling (Bollen 1989). A practical approach is that one uses programs for the estimation of structural equation models to determine whether the parameters of a model can actually be estimated. The programs will indicate whether the model is not identified and even indicate which parameter cannot be uniquely estimated.[2]

A requirement for the quality of the estimates is that the model be correctly specified, because the relationships that are derived for the correlations and their parameters are based on the assumption that it is so. While the estimation of the values of the parameters is based on these relationships, the correctness of the specification of a model is determined by a test. Such a test for structural equations models requires that there be more correlations than parameters to be estimated. The difference between the number of correlations and the number of parameters is called the *degree of freedom* (df). So the necessary condition for any testing is that $df > 0$.

In the previous example (Figure 10.1), a test is not possible because there are only three correlations for three parameters that have to be estimated, having $df = 0$. There is a perfect solution presented, but there is no information for a test because all information has been used to determine the values of the parameters. Adding one more measure for "satisfaction with the economy," we have the model presented in Figure 10.2.

In Figure 10.2, there are three more correlations between the observed variables and only one extra effect parameter that needs to be estimated. In this case, $df = 2$ and a test is possible. Let us illustrate it again by extending the previously given example: Suppose for y_{14} that we have found the following correlations with the other variables:

---

[2] The programs mention the first parameter that cannot be estimated, but there are often more possible candidates. Therefore, one should use such a suggestion only as one of the possible suggestions.

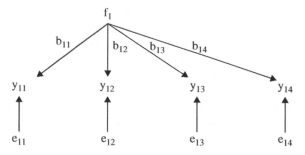

FIGURE 10.2 A simple measurement model with one unobserved and four observed variables assuming only random measurement errors.

$$\rho(y_{14}, y_{11}) = -.474$$

$$\rho(y_{14}, y_{12}) = .551$$

$$\rho(y_{14}, y_{13}) = .309$$

Now, we have six correlations of observed variables and only four effects to estimate as can be seen from the following sets of equations. The first three are the same as we have used before, but we can add three more for the three extra correlations:

$$\rho(y_{12}, y_{11}) = -.626 = b_{12}b_{11} \tag{10.8}$$

$$\rho(y_{13}, y_{11}) = -.502 = b_{13}b_{11} \tag{10.9}$$

$$\rho(y_{13}, y_{12}) = .584 = b_{13}b_{12} \tag{10.10}$$

$$\rho(y_{14}, y_{11}) = -.474 = b_{14}b_{11} \tag{10.11}$$

$$\rho(y_{14}, y_{12}) = .551 = b_{14}b_{12} \tag{10.12}$$

$$\rho(y_{14}, y_{13}) = .309 = b_{14}b_{13} \tag{10.13}$$

From the first three Equations (10.8)–(10.10), b_{11}, b_{12}, and b_{13} can be estimated as we did before, while from the last three Equations (10.11)–(10.13), only one is needed to estimate b_{14}. So there are indeed 2 df. Keeping in mind that $b_{11} = -.729$, it follows from Equation (10.11) that $b_{14} = -.474/-.729 = .650$. Now, all parameters are determined, but we have two equations left that have not been used for determining the values of the parameters.

Equations (10.12) and (10.13) can be used for a test because now all coefficients are known and we can control whether the correlations in these two equations can be reproduced by the values obtained for the effect parameters. For these two equations, we expect that

$$\rho(y_{14}, y_{12}) = b_{14}b_{12} = .847 \times .650 = .551$$

$$\rho(y_{14}, y_{13}) = b_{14}b_{13} = .475 \times .650 = .309$$

In this constructed example, the test would indicate that the model is correct because the effect parameters produce exactly the values of the two correlations that were not used for estimation. If, however, the correlation between the variables y_{14} and y_{13} had been .560, then one would think that there may be something wrong with the model, because we would expect, on the basis of the estimated values of the coefficients, the correlation to be .309, while in reality it is much larger (.560).

The differences between the correlations of the observed variables and their predicted values are called the *residuals*. It will be clear that the model has to be rejected if the residuals are too large. In the next section, we will discuss this. However, keep in mind that such a test is possible only if df > 0. If df = 0, the residuals are normally[3] 0 by definition, because there is only one perfect solution.

10.2 ESTIMATION OF PARAMETERS OF MODELS WITH UNMEASURED VARIABLES

So far, we have discussed only under what conditions the parameters of measurement models are identifiable and the fact that the models can be tested. We have not yet discussed how the parameters can be estimated. The procedure we have used so far to show that the parameters can be identified is not satisfactory for estimation for two reasons. The first reason is that the selection of the Equations (10.8)–(10.11) to determine the values of the parameters was arbitrary. We could have used four other equations, and in general, different estimates of the coefficients will be derived for each of the chosen sets of four equations, making this procedure unacceptable.

A second reason is that the correlations in the population are not known. Only estimates of these correlations can be obtained using a sample from the population, which will be denoted with r_{ij} in order to indicate very clearly that they represent *sample correlations*. The relationship between the correlations and the effect parameters holds perfectly only for the population correlations and not for the sample correlations. Here, we simply cannot use the previously mentioned Equations (10.8)–(10.13) to estimate coefficients.

There are several general principles to derive estimators for structural equations models. We will discuss the *unweighted least squares* (ULS) procedure and the *weighted least squares* (WLS) procedure. Both procedures are based on the residuals between the sample correlations and the expected values of these correlations that

---

[3] There is the possibility that the necessary condition for identification is satisfied, but that the model is not identified anyway, and in that case, the residuals can be larger than 0.

are a function of the parameters $f_{ij}(\mathbf{p})$. In this formulation, $\mathbf{p}$ represents the vector of parameters of the model, and f_{ij} is the specific function that gives the link between the population correlations and the parameters for the variables i and j.

The ULS procedure suggests looking for the parameter values that minimize the unweighted sum of squared residuals[4]:

$$F_{ULS} = \sum \left(r_{ij} - f_{ij}(\mathbf{p}) \right)^2 \tag{10.14}$$

where r_{ij} is the correlation in the sample between variables y_i and y_j. The summation is computed on all unique elements of the correlation matrix.

The WLS procedure suggests looking for the parameter values that minimize the weighted sum of squared residuals:

$$F_{WLS} = \sum w_{ij} \left(r_{ij} - f_{ij}(\mathbf{p}) \right)^2 \tag{10.15}$$

where w_{ij} is the weight for the term with the residual for the correlation r_{ij}. The summation is also computed on all unique elements of the correlation matrix.

The difference between the two methods is only that the ULS procedure gives all correlations an equal weight, while the WLS procedure varies the weights for the different correlations. These weights can be chosen in different ways. The most commonly used procedure is the maximum likelihood (ML) estimator, which can be seen as a WLS estimator with specific values for the weights (w_{ij}). The ML estimator provides standard errors for the parameters and a test statistic for the fit of the model. However, it was developed under the assumption that the observed variables have a multivariate normal distribution. More recently, it was demonstrated that the ML estimator is robust under very general conditions (Satorra 1990, 1992). There are also other estimators using different weighting techniques; for further information about this topic, we recommend the books on SEM by Bollen (1989) and Kaplan (2000).

Also, how the values of the parameters are determined is an issue that is beyond the scope of this text. However, the ULS procedure can easily be illustrated by the example with the four observed "satisfaction" variables discussed earlier. In this case, the minimum of the ULS criterion has to be found for the following function:

$$
\begin{aligned}
F_{ULS} = & \left(-.626 - b_{12}b_{11} \right)^2 + \left(-.502 - b_{13}b_{11} \right)^2 + \left(.584 - b_{12}b_{13} \right)^2 \\
& + \left(-.474 - b_{14}b_{11} \right)^2 + \left(.551 - b_{14}b_{12} \right)^2 + \left(.309 = b_{14}b_{13} \right)^2 \\
& + \left(1 - \left(b_{11}^2 + \text{variance}(e_{11}) \right) \right)^2 + \left(1 - \left(b_{12}^2 + \text{variance}(e_{12}) \right) \right)^2 \\
& + \left(1 - \left(b_{13}^2 + \text{variance}(e_{13}) \right) \right)^2 + \left(1 - \left(b_{14}^2 + \text{variance}(e_{14}) \right) \right)^2
\end{aligned}
\tag{10.16}
$$

---

[4] For simplicity sake, we specified the functions to be minimized for the correlation coefficients; however, it is recommended to use the covariance matrix as data. For details on this issue, we recommend the work of Bollen (1989).

TABLE 10.1   The estimated values of the parameters obtained by a hand calculation compared to the ULS procedure of the LISREL program

Parameter	Values of the parameters obtained	
	By hand Equations (10.8)–(10.11)	By LISREL/ULS method
b_{11}	−.729	−.75
b_{12}	.847	.89
b_{13}	.689	.63
b_{14}	.650	.59
Fit of the model: RMSR	.080	.029

Each term on the right-hand side of (10.16) is a residual. The first six terms come directly from the Equations (10.8)–(10.13). The first value is always the observed sample correlation, and the second term of the function of the parameters is according to the model equal to the population correlation. The last four terms are introduced to obtain an estimate of the variance of the measurement errors. They are based on the diagonal elements of the correlation matrix representing the variances of the standardized variables, which are by definition equal to 1. The variances should be equal to the explained variance (b_{ij}^2) plus the unexplained variance or the variance of e_{ij}.

To obtain the optimal values of the parameters, an iterative procedure is used that looks for those values that minimize the function as a whole. In the case of a perfect fit, we get exactly the same solution as we have found earlier. This is, however, not the case anymore if the sample correlation between y_{13} and y_{14} or $r(y_{13}, y_{14}) = .560$ as we suggested before. Using the program LISREL to estimate the parameters and applying the ULS estimation method, we find the result presented in Table 10.1.

The ULS method tries to find an optimal solution for all cells of the correlation matrix, while employing the hand calculation, only the first four equations have been used. When using ULS, the *root mean squared residuals* (RMSR) are considerably smaller and can be used as a measure for the fit of the model to the data. However, it is difficult to determine when the residuals are too large. Some people suggest to reject the model if the RMSR is larger than .1. If that criterion would have been used, this model would not be rejected. An alternative is to look for possible misspecifications in the model. The programs that can estimate these models provide estimates of the expected *parameter change* or EPC and provide also a test statistic for these estimates, the *modification index*. For this issue, we refer the reader to the literature (Saris et al. 1987).

We could discuss about estimation and testing procedures of SEM much more, but we refer the reader to the literature mentioned in this section.

10.3   ESTIMATING RELIABILITY, VALIDITY, AND METHOD EFFECTS

Now that the general principles of identification, estimation, and testing of a measurement model have been introduced, we will discuss the estimation of the three criteria: reliability, validity, and method effects. Figure 10.3 presents the same model for two traits from the last chapter.

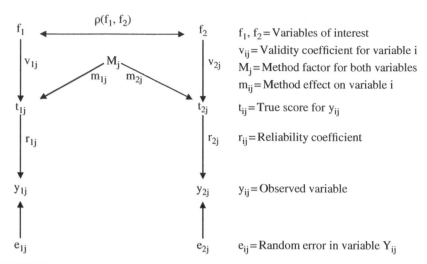

$f_1, f_2 =$ Variables of interest

$v_{ij} =$ Validity coefficient for variable i

$M_j =$ Method factor for both variables

$m_{ij} =$ Method effect on variable i

$t_{ij} =$ True score for y_{ij}

$r_{ij} =$ Reliability coefficient

$y_{ij} =$ Observed variable

$e_{ij} =$ Random error in variable Y_{ij}

FIGURE 10.3 The measurement model for two traits measured with the same method.

This model differs from the models presented in Figure 10.1 and Figure 10.2 in that method-specific systematic errors are also introduced. This makes the model more realistic while not changing the general approach.

Using the two theorems presented in Chapter 9, it was demonstrated that the correlation between the observed variables, $\rho(y_{1j}, y_{2j})$, is equal to the joint effect of the variables that we want to measure (f_1 and f_2) plus the spurious correlation due to the method effects, as follows:

$$\rho(y_{1j}, y_{2j}) = r_{1j} v_{1j} \rho(f_1, f_2) v_{2j} r_{2j} + r_{1j} m_{1j} m_{2j} r_{2j} \qquad (10.17)$$

We have shown in the preceding text that the reliability, validity, and method effects are the parameters of this model. The issue within this model is that there are two reliability coefficients, two validity coefficients, two method effects, and one correlation between the two latent traits, leaving us with seven unknown parameters, while only one correlation can be obtained from the data. It is impossible to estimate these seven parameters from just one correlation. Therefore, in the following section, we will discuss more complex designs to estimate the parameters.

Campbell and Fiske (1959) suggested using multiple traits and multiple methods (MTMM). The classical MTMM approach recommends the use of a minimum of three traits that are measured with three different methods leading to nine different observed variables. The example of Table 10.2 was discussed in Table 9.1.

Collecting data using this MTMM design, the data for nine variables are obtained, and from that data, a correlation matrix of 9×9 is obtained. The model formulated to estimate the reliability, validity, and method effects is an extension of the model presented in Figure 10.3. Figure 10.4 illustrates the relationships between the true scores (t) and their general factors of interest. Figure 10.4 shows that each trait (f_i)

TABLE 10.2   The classic MTMM design used in the ESS pilot study

The three traits were presented by the following three requests:

- *On the whole, how satisfied are you with the present state of the economy in Britain?*
- *Now, think about the national government. How satisfied are you with the way it is doing its job?*
- *And on the whole, how satisfied are you with the way democracy works in Britain?*

The three methods are specified by the following response scales:

(1) Very satisfied, (2) fairly satisfied, (3) fairly dissatisfied, (4) very dissatisfied

Very dissatisfied   *very satisfied*

| 0 | 1 | 2 | 3 | 4 | 5 | 6 | 7 | 8 | 9 | 10 |

(1) Not at all satisfied, (2) satisfied, (3) rather satisfied, (4) very satisfied

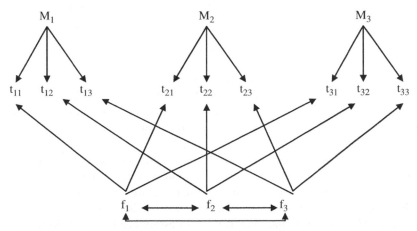

FIGURE 10.4   MTMM model illustrating the TS and their factors of interest.

is measured in three ways. It is assumed that the traits are correlated but that the method factors (M_1, M_2, M_3) are not correlated. To reduce the complexity of the figure, it is not indicated that for each t, there is an observed response variable that is affected by the t and a random error as was previously introduced in the model in Figure 10.3. However, these relationships, although not made explicit, are implied.

It is normally assumed that the correlations between the factors and the error terms are 0, but there is debate about the actual specification of the correlations between the different factors. Some researchers allow for all possible correlations between the factors while mentioning estimation problems[5] (Marsh and Bailey 1991; Kenny and Kashy 1992; Eid 2000). Andrews (1984) and Saris (1990) suggest that the trait factors can be

[5] This approach lends itself to nonconvergence in the iterative estimation procedure or improper solutions such as negative variances.

TABLE 10.3   **The correlations between the nine variables of the MTMM experiment with respect to satisfaction with political outcomes**

	Method 1			Method 2			Method 3		
	Q1	Q2	Q3	Q1	Q2	Q3	Q1	Q2	Q3
Method 1									
Q1	1.00								
Q2	.481	1.00							
Q3	.373	.552	1.00						
Method 2									
Q1	−.626	−.422	−.410	1.00					
Q2	−.429	−.663	−.532	.642	1.00				
Q3	−.453	−.495	−.669	.612	.693	1.00			
Method 3									
Q1	−.502	−.374	−.332	.584	.436	.438	1.00		
Q2	−.370	−.608	−.399	.429	.653	.466	.556	1.00	
Q3	−.336	−.406	−.566	.406	.471	.638	.514	.558	1.00
Means	2.42	2.71	2.45	5.26	4.37	5.13	2.01	1.75	2.01
sd	.77	.76	.84	2.29	2.37	2.44	.72	.71	.77

allowed to correlate, but should be uncorrelated with the method factors, while the method factors themselves are uncorrelated. Using this latter specification, combined with the assumption of equal method effects for each method, almost no estimation problems occur in the analysis. This was demonstrated by Corten et al. (2002) in a study in which 79 MTMM experiments were reanalyzed.

The MTMM design of three traits and three methods generates 45 correlations and variances. In turn, these 45 pieces of information provide sufficient information to estimate nine reliability and nine validity coefficients, three method effect coefficients, and three correlations between the traits. In total, there are 24 parameters to be estimated. This leaves $45 - 24 = 21$ df, meaning that the necessary condition for identification is fulfilled. It also can be shown that the sufficient condition for identification is satisfied, and given that df=21, a test of the model is possible.

Table 10.3 presents again the correlations that we derived between the nine measures obtained from a sample of 481 people in the British population. Using the specifications of the model indicated previously and the ML estimator to estimate the quality indicators, the results presented in Table 10.4 are obtained.[6] (The input for the LISREL program that estimates the parameters of the model is presented in Appendix 10.1.)

No important misspecifications in this model were detected. Therefore, the model does not have to be rejected, and the estimated values of the parameters are probably a good approximation of the true values of the parameters. The parameter

---

[6] In this case, the ML estimator is used. The estimation is done using the covariance matrix as the input matrix and not the correlation matrix (see Appendix 10.1). Thereafter, the estimates are standardized to obtain the requested coefficients. A result of this is that the standardized method effects are not exactly equal to each other.

TABLE 10.4   Standardized estimates of the MTMM model specified for the ESS data of Table 10.3

	Validity coefficients			Method effects			Reliability coefficients
	f_1	f_2	f_3	m_1	m_2	m_3	
t_{11}	.93			.36			.79
t_{21}		.94		.35			.85
t_{31}			.95	.33			.81
t_{12}	.91				.41		.91
t_{22}		.92			.39		.94
t_{32}			.93		.38		.93
t_{13}	.85					.52	.82
t_{23}		.87				.50	.87
t_{33}			.88			.48	.84

values point to method 2 having the highest reliability for these traits. With respect to validity, the first two methods have the highest scores and are approximately equal. When considering all estimates, method 2 is preferable to the other methods.

Note that the validity and the method effects do not have to be evaluated separately because they complement each other, as was mentioned previously: $v_{ij}^2 = 1 - m_{1j}^2$. With this example, we have shown how the MTMM approach can be used to evaluate the quality of several survey items with respect to validity and reliability.

10.4   SUMMARY AND DISCUSSION

The reliability, validity coefficients, and the method effects are defined as parameters of a measurement model and indicate the effects of unobserved variables on observed variables or even on unobserved variables. This chapter showed that these coefficients can be estimated from the data that can be obtained through research. After an introduction to the identification problem, general procedures for the estimation of the parameters and testing of the models were discussed.

Furthermore, it was demonstrated that the classic MTMM design suggested by Campbell and Fiske (1959) can be used to estimate the data quality criteria of reliability, validity, and method effects. This proved that the design can evaluate specific forms of requests for an answer with respect to the specified quality criteria.

There are many alternative models suggested for MTMM data. A review of some of the older models can be found in Wothke (1996). Among them is the *confirmatory factor analysis* model for MTMM data (Werts and Linn 1970; Althauser et al. 1971; Alwin 1974). An alternative parameterization of this model was proposed as the TS model by Saris and Andrews (1991), while the *correlated uniqueness* model has been suggested by Kenny (1976), Marsh (1989), and Marsh and Bailey (1991). Saris and Aalberts (2003) compared models presenting different explanations for the correlated uniqueness. Models with *multiplicative method effects* have been suggested by Campbell and O'Connell (1967), Browne (1984), and Cudeck (1988). Coenders and Saris (1998, 2000) showed that the multiplicative model can be formulated as a special case of the correlated uniqueness model of Marsh (1989). We suggest the use of the *TS MTMM model* specified by Saris and Andrews (1991) because Corten et al. (2002) and Saris and Aalberts (2003) have shown that this model has the best fit for large series of data sets for MTMM experiments. The classic MTMM model is locally equivalent with the TS model, meaning that the difference is only in its parameterization. For more details on why we prefer this model, see Appendix 10.2.

The MTMM approach also has its disadvantages. If each researcher performed MTMM experiments for all the variables of his/her model, it would be very inefficient and expensive, because he/she would have to ask six more requests to evaluate three original measures. In other words, the respondents would have to answer the requests about the same topic on three different occasions and in three different ways. This raises the questions of whether this type of research can be avoided, whether this research is really necessary, and whether or not the task of the respondents can be reduced.

So far, all MTMM experiments have employed the classical MTMM design or a panel design with two waves where each wave had only two observations for the same trait while at the same time the order of the requests was random for the different respondents (Scherpenzeel and Saris 1997). The advantage within the latter method is that the response burden of each wave is reduced and the strength of opinion can be estimated (Scherpenzeel and Saris 2006). The disadvantages are that the total response burden is increased by one extra measure and that a frequently observed panel is needed to apply this design. Although this MTMM design has been used in a large number of studies because of the presence of a frequently observed panel (Scherpenzeel 1995), we think that this is not a solution that can be recommended in general. Therefore, given the limited possibilities of this particular design, other types of designs have been elaborated, such as the split-ballot MTMM design (Saris et al. 2004), which will be discussed in the next chapter. We recommend this chapter only if you are interested in going into the details of this design; otherwise, please skip Chapter 11 and move directly to Chapter 12, where a solution of how to avoid MTMM research in applied research is presented.

EXERCISES

1. A study evaluating the quality of requests measuring "political efficacy" was conducted using following requests for an answer:

   How far do you agree or disagree with the following statements?

   a. *Sometimes politics and government seem so complicated that I cannot really understand what is going on.*

b. *I think I can take an active role in a group that is focused on political issues.*

c. *I understand and judge important political questions very well.*

The response categories were:

1. *Strongly disagree*
2. *Disagree*
3. *Neither disagree nor agree*
4. *Agree*
5. *Strongly agree*

The 5-point category scale was used twice: at the very beginning of the questionnaire and once at the end. Therefore, the only difference between the two sets of requests was the positioning in the questionnaire. We call these requests "agree/disagree" (A/D) requests. One other method was used to measure "political efficacy." Instead of the A/D format, a "trait-specific method" or TSM request format was employed. The requests were:

1. *How often do politics and government seem so complicated that you cannot really understand what is going on?*
   1. *Never*
   2. *Seldom*
   3. *Occasionally*
   4. *Regularly*
   5. *Frequently*

2. *Do you think that you could take an active role in a group that is focused on political issues?*
   1. *Definitely not*
   2. *Probably not*
   3. *Not sure either way*
   4. *Probably*
   5. *Definitely*

3. *How good are you at understanding and judging political questions?*
   1. *Very bad*
   2. *Bad*
   3. *Neither good nor bad*
   4. *Good*
   5. *Very good*

An MTMM study evaluating these requests led to the following results: first, we represent a response distribution for the different requests presenting the means, standard deviations (sd), and the missing values of the distributions of the responses.

	First A/D		Second A/D		TSM	
	Mean	sd	Mean	sd	Mean	sd
Item 1	2.91	1.21	2.87	1.12	2.90	1.10
Item 2	2.28	1.24	2.38	1.21	2.17	1.21
Item 3	2.94	1.12	3.06	1.08	3.23	.99
Missing	34		82			55

In the following, we provide results of the estimation of the reliability, validity, and method effects:

	Request 1	Request 2	Request 3
Reliability coeff.			
A/D core	.69	.76	.76
A/D drop-off	.82	.91	.79
TSM drop-off	.88	.92	.87
Validity coeff.			
A/D core	.84	.88	.87
A/D drop-off	1	1	1
TSM drop-off	1	1	1
Method effect[7]			
A/D core	.55	.48	.49
A/D drop-off	0	0	0
TSM drop-off	0	0	0

Please answer the following questions on the basis of the findings of the MTMM study:

a. What are, according to you, the best measures for the different traits?

b. Why are there differences between the measures?

c. Can these hypotheses be generalized to other requests?

2. In Figure 10.4, an MTMM model specifies the relationships between the true scores and their factors of interest:

a. Express the correlations between the true scores in the parameters of the model. Do this only for those correlations that generate a different expression.

b. Assuming that each true score has an observed variable that is not affected by any other variable except random measurement error, what do the correlations between the observed variables look like?

c. Do you have any suggestion about whether the parameters can be estimated from the correlations between the observed variables? (Solving the equations is too complicated.)

---

[7] In this table, we have ignored the signs. Only absolute values are presented in order to prevent confusion.

APPENDIX 10.1 INPUT OF LISREL FOR DATA ANALYSIS OF A CLASSIC MTMM STUDY

Analysis of the British satisfaction data for ESS

Data ng=1 ni=9 no=428 ma=cm
km
*
1.00
.481 1.00
.373 .552 1.00
−.626 −.422 −.410 1.00
−.429 −.663 −.532 .642 1.00
−.453 −.495 −.669 .612 .693 1.00
−.502 −.374 −.332 .584 .436 .438 1.00
−.370 −.608 −.399 .429 .653 .466 .556 1.00
−.336 −.406 −.566 .406 .471 .638 .514 .558 1.00
mean

*

2.42 2.71 2.45 5.26 4.37 5.13 2.01 1.75 2.01
sd

*

.77 .76 .84 2.29 2.37 2.44 .72 .71 .77

model ny=9 ne=9 nk=6 ly=fu,fi te=di,fr ps=di,fi be=fu,fi ga=fu,fi ph=sy,fi
value −1 ly 1 1 ly 2 2 ly 3 3
value 1 ly 4 4 ly 5 5 ly 6 6 ly 7 7 ly 8 8 ly 9 9

free ga 1 1 ga 4 1 ga 7 1 ga 2 2 ga 5 2 ga 8 2 ga 3 3 ga 6 3 ga 9 3

value 1 ga 1 4 ga 2 4 ga 3 4
value 1 ga 4 5 ga 5 5 ga 6 5 ga 7 6 ga 8 6 ga 9 6
free ph 2 1 ph 3 1 ph 3 2   ph 6 6 ph 5 5 ph 4 4
value 1 ph 1 1 ph 2 2 ph 3 3
start .5 all
out rs pc iter=200 adm=off sc

APPENDIX 10.2 RELATIONSHIP BETWEEN THE TS AND THE CLASSIC MTMM MODEL

The structure of the classical MTMM model follows directly from the basic characteristics of the TS model that can be specified in Equations (10.2A.1) and (10.2A.2):

$$y_{ij} = r_{ij} t + e_{ij} \qquad (10.2A.1)$$

$$t_{ij} = v_{ij} f_i + m_{ij} m_j \qquad (10.2A.2)$$

From this model, one can derive the most commonly used MTMM model by substitution of Equation (10.A.2) into Equation (10.2A1). It results in the models (10.2A.3) or (10.2A.4):

$$y_{ij} = r_{ij} v_{ij} f_i + r_{ij} m_{ij} m_j + e_{ij} \qquad (10.2A.3)$$

or

$$y_{ij} = q_{ij} f_i + s_{ij} m_j + e_{ij} \qquad (10.2A.4)$$

where

$$q_{ij} = r_{ij} v_{ij} \quad \text{and} \quad s_{ij} = r_{ij} m_{ij}.$$

One advantage of this formulation is that q_{ij} represents the strength of the relationship between the variable of interest and the observed variable and is an important indicator of the total quality of an instrument. Besides, s_{ij} represents the systematic effect of method j on response y_{ij}. Another advantage is that it simplifies Equation (9.1) to (10.2A5):

$$r(y_{1j}, y_{2j}) = q_{jr}(f_1, f_2) q_{2j} + s_{1j} s_{2j} \qquad (10.2A.5)$$

Although this model is quite instrumental, some limitations are connected with it. One of these is that the parameters themselves are products of more fundamental parameters. This creates problems because the estimates for the data quality of any model are derived only after the MTMM experiment is completed and the data analyzed. Therefore, in order to apply this approach for each item in the survey, two more requests have to be asked to estimate the item quality. The cost of doing this makes this approach unrealistic for standard survey research.

An alternative is to study the effects in terms of how different questionnaire design choices affect the quality of the criteria and to use the results for predicting the data quality before and after the data are collected. By making a meta-analysis to determine the effects of the question design choices on the quality criteria, we would be eliminating the additional survey items needed in substantive surveys. It is an approach that has been suggested by Andrews (1984) and has been applied in several other studies (Költringer 1995; Scherpenzeel and Saris 1997; Corten et al. 2002; Saris and Gallhofer 2007b).

In such a meta-analysis, it is desirable that the parameters to be estimated represent only one criterion and not mixtures of different criteria, in order to keep the explanation clear. It is for this particular reason that Saris and Andrews (1991) have suggested an alternative parameterization of the classical model: the TS model, presented in Equations (10.2A.1) and (10.2A.2), where the reliability and validity coefficients are separated and hence can be estimated independently from each other. Both coefficients can also vary between 0 and 1, which does not occur if one employs the reliability and the validity coefficient as Andrews (1984) did, starting with the classical model (10.2A.5). In agreement with Saris and Andrews (1991), we suggested that for the meta-analysis the TS MTMM model has major advantages, and therefore, we have presented the TS model in this chapter.

11

SPLIT-BALLOT MULTITRAIT–MULTIMETHOD DESIGNS[1]

Although the classical MTMM design is effective, it has one major problem, namely, that one has to ask the respondents three times nearly the same questions. As a consequence, people can become bored and answer with less seriousness as questions are repeated. It is also possible that they remember what they have said before, which means that the observed responses are not independent of each other.

In order to cope with this problem, there are two possible strategies: (1) to increase the time between the observations so that the previous answers cannot be remembered anymore or (2) to reduce the number of repeated observations. The first approach has been tried in the past. It was discussed that one can after 25 minutes repeat the same questions if similar questions are asked in between. This will be a solution for the second measures but not for the third ones. So, we have mentioned that Scherpenzeel (1995) have used in many experiments a panel design where at each point in time, only two observations of the same questions are made. However, this approach requires a panel that is commonly not available. The second strategy is to ask each respondent fewer questions while compensating for the "missing data by design" by collecting data from different subsamples of the population. In doing so, the designs look very similar to the frequently used split-ballot experiments and hence are called the "split-ballot MTMM design" or SB-MTMM design. This design

[1] This chapter is based on a paper by Saris W. E., A. Satorra, and G. Coenders 2004. A new approach to evaluating the quality of measurement instruments: The split-ballot MTMM design. *Sociological Methodology*, complemented with the results recently published by Revilla and Saris (2013).

Design, Evaluation, and Analysis of Questionnaires for Survey Research, Second Edition. Willem E. Saris and Irmtraud N. Gallhofer.

will be discussed in this chapter because it is the design that has been used for all MTMM experiments in the European Social Survey (ESS).

11.1 THE SPLIT-BALLOT MTMM DESIGN

In the commonly used split-ballot experiments, random samples from the same population receive different versions of the same requests. In other words, each respondent group gets one method. The split-ballot design makes it possible to compare the response distributions of the different requests across their forms and to assess their possible relative biases (Schuman and Presser 1981; Billiet et al. 1986).

In the SB-MTMM design, random samples of the same population are also used but with the difference that these groups get two different forms of the same request. In total, it is one less repetition than in the classical MTMM design and one more than in the commonly used split-ballot designs. We will show that the design, suggested by Saris (1998c), combines the benefits of the split-ballot approach and the MTMM approach in that it enables researchers to evaluate measurement bias, reliability, and validity simultaneously, and that it does so, while reducing response burden. Applications of this approach can also be found in Saris (1998c) and Kogovšek et al. (2001). A more complex alternative design has been suggested by Bunting et al. (2002). The suggestion to use split-ballot designs for structural equation models (SEM) can be traced back to Arminger and Sobel (1991).

11.1.1 The Two-Group Design

The two-group SB-MTMM design is structured as follows. The sample is split randomly into two groups. One group has to answer three survey items formulated by method 1, while the other group is given the same survey items presented in a second form, in the MTMM literature called "method 2." In the last part of the questionnaire, all respondents are presented with the three items, which are now formulated in method 3 format. The design can be summarized as tabulated in Figure 11.1.

In summary, under the two-group design, the researcher draws two comparable random samples from the same population and asks three requests about at least three traits in each sample: one time with the same and the other time with another form (method) of the same requests (traits) after sufficient time has elapsed. Van Meurs and Saris (1990) have demonstrated that after 25 minutes the memory effects are negligible if similar questions have been asked in between the two sets of questions. This time gap is enough to obtain independent measures in most circumstances.

	Time 1	Time 2
Sample 1	Form 1	Form 3
Sample 2	Form 2	Form 3

FIGURE 11.1   The two-group SB-MTMM design.

TABLE 11.1   Samples providing data for correlation estimation

	Method 1	Method 2	Method 3
Method 1	Sample 1		
Method 2	None	Sample 2	
Method 3	Sample 1	Sample 2	Sample 1 + 2

The design in Figure 11.1 matches the standard split-ballot design at time 1 and provides information about differences in response distributions between the methods. Combined with the information obtained at time 2, this design provides extra information. The question still remains whether the reliability, validity, and method effects can be estimated from this data, since each respondent answers only two requests about the same trait and not three, as is required from the classical MTMM design. The answer is not immediately evident since the necessary information for the 9×9 correlation matrix comes from different groups and is by design incomplete (see Table 11.1). Table 11.1 shows the groups that provide data for estimating variances and correlations between requests using either the same or different forms (methods).

In contrast to the classical design, no correlations are obtained for form 1 and form 2 requests, as they are missing by design. Otherwise, all correlations in the 9×9 matrix can be obtained on the basis of one or two samples, but the data come from different samples.

Each respondent is given the same requests only twice, reducing the response burden considerably. However, in large surveys, the sample can be split into more subsamples and hence evaluate more than one set of requests. However, the correlations between forms 1 and 2 cannot be estimated, resulting in a loss of degrees of freedom (df) when estimating the model on the now incomplete correlation matrix. This might make the estimation less effective than the standard design where all correlations are available, as in the three-group design.

11.1.2 The Three-Group Design

The three-group design proceeds as the previous design except that three groups or samples are used instead of two, leaving us with the following scheme (Fig. 11.2):

Using this design, all request forms are treated equally: they are measured once at the first and later at a second point in time. Therefore, there are also no missing correlations in the correlation matrix, as shown in Table 11.2.

Evidently, the major advantage of this approach is that all correlations can be obtained. A second advantage is that the order effects are canceled out because each measure comes once at the first position and another time at the second position within the questionnaire.

A major disadvantage, however, is that the main questionnaire has to be prepared in three different formats for the three different groups. In addition, the same measures are not obtained from all respondents. This may raise a serious issue in the

	Time 1	Time 2
Sample 1	Form 1	Form 2
Sample 2	Form 2	Form 3
Sample 3	Form 3	Form 1

FIGURE 11.2   The three-group SB-MTMM design.

TABLE 11.2   Samples providing data for correlation estimation

	Method 1	Method 2	Method 3
Method 1	Samples 1 and 3		
Method 2	Sample 1	Samples 1 and 2	
Method 3	Sample 3	Sample 2	Samples 2 and 3

analysis because the sample size is reduced with respect to its relationships with the other variables.[2] This design was for the first time used by Kogovšek et al. (2001).

11.1.3   Other SB-MTMM Designs

Other methods that are guided by the principles discussed previously can also be designed. The effects of different factors can be studied simultaneously, and interaction effects can be estimated. However, an alternative to this type of study is to employ a meta-analysis of many separate MTMM experiments under different conditions, which will be elaborated in the next chapter.

There is one other design that deserves special attention, the SB-MTMM design, which makes use of an exact replication of methods. In doing so, the occasion effects can be studied without placing an extra response burden on respondents. A possible design is illustrated in Figure 11.3.

Figure 11.3 models a complete four-group design for two methods and their replications. The advantage of this design is that the same information as with the other two designs is obtained, and in addition to the previous design, the occasion-specific variance can be estimated. This is only possible if exact repetition of the same measures is included in the design. In order to estimate these effects, the model specified in Chapter 10 has to be extended with an occasion-specific factor (Saris et al. 2004). This design can be reduced to a three-group design by leaving out sample 2 or 3 or alternatively sample 1 or 4, assuming that the order effects are negligible or that the occasion effects are the same for the different methods.

---

[2] A possible alternative would be to add to the study a relatively small subsample. For the whole sample, one would use method 1, the method expected to give the best results, in the main questionnaire; method 2 for one subgroup; and method 3 for another subgroup in an additional part of the questionnaire that relates to methodology. With the subsample, one would use method 2 for the main questionnaire and method 3 in the methodological part. In this way, method 1 is available for all people, and all three combinations of the forms are also available. Also, one could get an estimate of the complete covariance matrix for the MTMM analysis without harming the substantive analysis. But this design would cost extra money for the additional subsample. The appropriate size of the subsamples is a matter for further research.

	Time 1	Time 2
Sample 1	Form 1	Form 1
Sample 2	Form 1	Form 2
Sample 3	Form 2	Form 1
Sample 4	Form 2	Form 2

FIGURE 11.3   A four-group SB-MTMM design with exact replications.

Another similar design can be developed including three different methods; however, it is beyond the scope of this chapter to discuss further possibilities. For further information, we refer to Saris et al. (2004) and the first two large-scale applications of this design in the ESS (2002).

We hope that we clarified that the major advantage of these designs is the reduction of the response burden from three to two observations. Furthermore, in order to show that these designs can be applied in practice, we need to discuss, based on the collected data, the estimation of the parameters.

11.2   ESTIMATING AND TESTING MODELS FOR SPLIT-BALLOT MTMM EXPERIMENTS

The split-ballot MTMM experiment differs from the standard approach in that different equivalent samples of the same population are studied instead of just one. Given the random samples are drawn from the same populations, it is natural to assume that the model is exactly the same for all respondents and equal to the model we have specified in Figure 10.4, which includes the restrictions on the parameters suggested by Saris and Andrews (1991). The only difference is that not all requests have been asked in every group.

Since the assignment of individuals to groups has been made at random, and there is a large sample in each group, the most natural approach for estimating is the multiple-group SEM method (Jöreskog 1971). It is available in most of the SEM software packages. We refer to this approach as multiple-group structural equation model or MGSEM.[3] As indicated in the previous section, a common model is fitted across the samples, with equality constraints for all the parameters across groups. With the current software and applying the theory for multiple-group analysis, estimation can be made by using the maximum likelihood (ML) method or any other standard estimation procedure in SEM. In the case of non-normal data, robust standard errors and test statistics are available in the standard software packages.

---

[3] Because each group will be confronted with partially different measures of the same traits, certain software for multiple-group analysis will require some small tricks to be applied. This is the case for LISREL, where the standard approach expects the same set of observable variables in each group. Simple tricks to handle such a situation of the set of observable variables differing across groups were already described in the early work of Jöreskog (1971) and in the manual of the early versions of the LISREL program; such tricks are also described in Allison (1987). Multiple-group analysis with the software EQS, for example, does not require the same number of variables in the different groups.

For a review of multiple-group analysis in SEM as applied to all the designs enumerated in the present chapter, see Satorra (2000).

The incomplete data setup we are facing could also be considered as a missing data problem (Muthen et al. 1987). However, the approach for missing data assumes normality, while this design does not provide the theoretical basis for robust standard errors and corrected test statistics that are currently available in MGSEM software. Thus, since the multiple-group option offers the possibility of standard errors and test statistics that are protected from non-normality, we suggest that the multiple-group approach is preferable.

Given this situation, we suggest the MGSEM approach for estimating and testing the model on SB-MTMM data. In doing so, the correlation matrices are analyzed, while the data quality criteria (reliability, validity coefficients, and method effects) are obtained by standardizing the solution.

Although the statistical literature suggests that data quality indicators can be estimated using the SB-MTMM designs, we need to be careful while using the two-group designs with incomplete data, because they may lead to empirical underidentification problems. Before addressing this issue, we will illustrate an application of the two designs based on data from the same study discussed in the previous chapters.

11.3 EMPIRICAL EXAMPLES

In Chapters 9 and 10, an empirical example of the classical MTMM experiment was discussed. In order to illustrate the difference between this design and the SB-MTMM designs, we have randomly split the total sample of that study (n=428) into two (n=210) and three groups (n=140). Thereafter, we took only those variables that would have been collected had the two- or three-group MTMM design been used, for each group. In this way, we obtained incomplete correlation matrices for each group. Next, we estimated the model, using the multiple-group approach. Now, we will investigate the results, starting with the three-group design, where a complete correlation matrix is available for all groups. Later, we discuss the results for the two-group design, where the correlation information is incomplete.

11.3.1 Results for the Three-Group Design

The random sampling of the different groups and selection of the variables according to the three-group design has led to the results summarized in Table 11.3. First, this table indicates that in each sample incomplete data are obtained for the MTMM matrix. The correlations for the unobserved variables are represented by 0s and the variances by 1s. This presentation is necessary for the multiple-group analysis with incomplete data in LISREL but does not have to be used in general.

Keep in mind that these correlation matrices are incomplete because at each time interval, one set of variables is missing. We see also that we have summarized the response distributions in means and standard deviations, which can be compared across groups as is done in the standard split-ballot experiments. However, in this

TABLE 11.3 Data for three-group SB-MTMM analysis on the basis of three random samples from the British pilot study of the ESS

Correlations, means, and standard deviations of the first subsample

Correlations
1.00
.469 1.00
.250 .415 1.00
.0 .0 .0 1.00
.0 .0 .0 .0 1.00
.0 .0 .0 .0 .0 1.00
−.524 −.322 −.212 .0 .0 .0 1.00
−.313 −.523 −.273 .0 .0 .0 .509 1.00
−.244 −.313 −.517 .0 .0 .0 .442 .461 1.00
Means
2.39 2.69 2.41 .0 .0 .0 2.09 1.77 2.02
Standard deviations
.70 .71 .78 1.0 1.0 1.0 .71 .68 .73

Correlations, means, and standard deviations of the second subsample

Correlations
1.00
.0 1.00
.0 .0 1.00
.0 .0 .0 1.00
.0 .0 .0 .598 1.00
.0 .0 .0 .601 .694 1.00
.0 .0 .0 .588 .398 .517 1.00
.0 .0 .0 .395 .690 .504 .547 1.00
.0 .0 .0 .397 .462 .571 .545 .564 1.00
Means
.0 .0 .0 5.22 4.30 4.98 1.91 1.69 2.00
Standard deviations
1.0 1.0 1.0 2.27 2.51 2.47 .69 .65 .71

Correlations, means, and standard deviations of the third subsample

Correlations
1.00
.469 1.00
.393 .605 1.00
−.669 −.454 −.489 1.00
−.512 −.669 −.564 .707 1.00
−.495 −.508 −.742 .693 .729 1.00
.0 .0 .0 .0 .0 .0 1.00
.0 .0 .0 .0 .0 .0 .0 1.00
.0 .0 .0 .0 .0 .0 .0 .0 1.00
Means
2.41 2.65 2.50 5.18 4.32 4.99 .0 .0 .0
Standard deviations
.78 .77 .90 2.39 2.39 2.53 1.0 1.0 1.0

TABLE 11.4 Estimates of parameters for the full sample using three methods and for the three-group design with incomplete data in each group

	Full sample			Three-group SB-MTMM design		
	M1	M2	M3	M1	M2	M3
Reliability coefficient for						
Q1	.79	.91	.82	.78	.91	.84
Q2	.85	.94	.87	.82	.97	.86
Q3	.81	.93	.84	.83	.95	.77
Validity coefficient for						
Q1	.93	.91	.85	.94	.91	.86
Q2	.94	.92	.87	.94	.93	.85
Q3	.95	.93	.88	.96	.93	.84
Method variance	.05	.73	.09	.04[a]	.73	.09

[a] This coefficient is not significantly different from 0, while all others are significantly different from 0.

case, we also want estimates for the reliability, validity, and method effects. In estimating these coefficients from the data for the three randomly selected groups simultaneously, we have assumed that the model is the same for all groups except for the specification of variables selected for the three groups. The technical details of this analysis are given in Appendix 11.1, where the LISREL input is presented.

In Table 11.4, we provide the results of the estimation as provided by LISREL using the ML estimator.[4] The table also contains the full sample estimates for comparison. Given that on the basis of sampling fluctuations, one can expect differences between the different groups, the similarity between the results for the two designs indicates that the three-group SB-MTMM design can provide estimates for the parameters of the MTMM model that are very close to the estimates of the classical design. At the same time, the correlation matrices are rather incomplete since the respondents are asked to answer fewer requests about the same topic.

Moreover, the fact that the program did not indicate identification problems suggests that the model is identified even though the correlation matrices in the different subgroups are incomplete. Let us now investigate the same example in an identical manner assuming that a two-group design has been used.

11.3.2   Two-Group SB-MTMM Design

Using the two-group design, the same model is assumed to apply for the whole group, and the analysis is carried out in exactly the same manner. The data for this design are presented in Table 11.5. The procedure for filling in the empty cells in the

[4] In this case, LISREL reports a chi^2 of 54.7 with df = 111. However, the number of df is incorrect because in each matrix 24 correlations and variances were missing, so the df should be reduced by $3 \times 24 = 72$, and the correct df are 39.

TABLE 11.5   Data for two-group SB-MTMM analysis on the basis of two random samples from the British pilot study of the ESS

Correlations, means, and standard deviations of the first subsample

Correlations
1.00
.457 1.00
.347 .478 1.00
.0 .0 .0   1.00
.0 .0 .0   .0 1.00
.0 .0 .0   .0 .0 1.00
−.564 −.365 −.344 .0 .0 .0 1.00
−.366 −.597 −.359 .0 .0 .0 .546 1.00
−.350 −.386 −.530 .0 .0 .0 .512 .498 1.00
Means
2.42 2.75 2.43 .0 .0 .0 2.01 1.70 1.99
Standard deviations
74 .76 .83 1.0 1.0 1.0 .71 .67 .73

Correlations, means, and standard deviations of the second subsample

Correlations
1.00
.0 1.00
.0 .0 1.00
.0 .0 .0 1.00
.0 .0 .0 .686 1.00
.0 .0 .0 .669 .742 1.00
.0 .0 .0 .585 .449 .441 1.00
.0 .0 .0 .464 .684 .546 .568 1.00
.0 .0 .0 .397 .516 .674 .516 .607 1.00
Means
.0 .0 .0 5.26 4.49 5.10 2.01 1.80 2.02
Standard deviations
1.0 1.0 1.0 2.38 2.40 2.51 .74 .73 .81

table was the same in Table 11.5 as in Table 11.3. An important difference between the two designs is that in the two-group design, no correlations between the first and the second methods are available, and so, the coefficients have to be estimated on the basis of incomplete data.

The first analysis of these matrices did converge, but the variance of the first method factor was negative. This issue may also arise in the classical MTMM approach when a method factor has a variance very close to 0.

In Table 11.4, we have seen that the method variance for the first factor was not significantly different from 0 and rather small even though the estimate was based on two groups of 140 or 280 cases. In the two-group design, the variance has to be estimated on the basis of 210 cases, and the program does not provide a proper solution. A common remedy is to fix one parameter on a value close to 0. If we fix

TABLE 11.6 Estimates of the parameters for the full sample using three methods and for the two-group design with incomplete data

	Full sample			Two-group SB-MTMM design		
	M1	M2	M3	M1	M2	M3
Reliability for						
Q1	.79	.91	.82	.80	.93	.83
Q2	.85	.94	.87	.87	.96	.86
Q3	.81	.93	.84	.83	.98	.82
Validity for						
Q1	.93	.91	.85	.99	.90	.85
Q2	.94	.92	.87	.99	.91	.86
Q3	.95	.93	.88	.99	.92	.87
Method variances	.05	.73	.09	.01[a]	.86	.10

[a] This coefficient was fixed on the value .01 in order to avoid an improper solution.

the variance on .01, we get the result presented in Table 11.6.[5] With this restriction, the estimates provided by the program are close to the estimates obtained in the classical MTMM design. The largest differences in the validity coefficients for the first method are a direct consequence of the restriction introduced.

On the whole, the estimates are such that regarding the reliability coefficients, the conclusion drawn from the estimates obtained by the two-group design would not differ from those using the estimates of the one-group design where the second method has the highest reliability. Given the restriction introduced on the method variance, one should be very cautious to draw a definite conclusion about the validity coefficients and hence about the method effects.

Clearly, the fact that we had to introduce this restriction raises the question of whether the two-group design is identified and robust enough to be useful in practice. On the one hand, it would seem that the most natural approach is to reduce the response burden. On the other hand, when this approach is not robust enough to provide the same estimates as the classical or the three-group SB-MTMM design, then one of the other designs should be preferred.

With regard to the identification, we assert that the model is indeed identified under normal circumstances and the specified estimation procedure will provide consistent estimates of the population parameters.

Before proceeding to the next section, we should mention that the aforementioned example did not give a correct impression of the true quality of the different designs. The reason is that the quantity of data on which the parameters are based differed for the parameters in the different designs. The parameters of the classical design were based on approximately 420 cases. The parameter estimates in the three-group design are based on 280 respondents, while some parameter estimates in the two-group

[5] In this case, LISREL reports a chi^2 value of 12.7 with df=67, but also now, the df has to be corrected in the way discussed previously (footnote 4), and the correct df are 19.

design are based on either 210 or 420 cases. This consideration provides us with one explanation for the difference in performance between the designs. Hence, the topic of efficiency of the different designs is covered in the next section.

11.4   THE EMPIRICAL IDENTIFIABILITY AND EFFICIENCY OF THE DIFFERENT SB-MTMM DESIGNS

In order to study the robustness of these different designs, two different problems have to be evaluated. The first is that we would like to determine under what conditions the procedures break down even though the correct model has been specified. The second issue is what we can say about the efficiency of the different designs to estimate the parameters of the MTMM model. We will begin with addressing the first issue.

11.4.1   The Empirical Identifiability of the SB-MTMM Model

Three aspects of these models require special attention after the model has been correctly specified:

1. Minimal variance of one of the method factors
2. Lack of correlation between the latent traits
3. Equal correlations between the latent traits

The first problem of the minimal method variance is a problem of overfitting. In this case, a parameter is estimated that is not needed for the fit of the model to the data. If the model had been estimated with this coefficient fixed on 0, the fit would be equally good. This problem is not just an issue for SB-MTMM designs; it also occurs in the classical MTMM design. Normally, this problem is solved by fixing on zero the parameter that is not needed for the model. However, the challenge is to detect where the actual problem in the model lies. Our experience with the analyses of MTMM data is that negative variances for the method variances are obtained in unrestricted estimation procedures. Such solutions are of course inacceptable and are called in the literature "Heywood cases" or "improper solutions." In the case where an estimation procedure including constraints on the parameter values is used, the value 0 will automatically be obtained for the problematic method variance in order to avoid improper solutions.

The second condition, lack of correlations between the traits, can raise a problem because we know that the loadings of a factor model are identified if each trait has three indicators or two but then the traits have to be correlated with each other. If each trait has only two indicators and the correlation between the traits is 0, the situation is the same as for a model with one trait and two indicators, which is not identified. Applying this rule to the MTMM models, we can see that in the classical MTMM model, each trait has three indicators and is therefore identified under normal circumstances even if the correlations between the traits are 0. In the different groups

TABLE 11.7   Results obtained when running 180 SB-MTMM models for ESS rounds 1 and 4

	Experiments	NC	HC	PS	Total cases
Round 1	Media use	15	4	0	19
	Pol. efficacy	1	11	7	19
	Pol. orientation	4	8	7	19
	Satisfaction	3	9	7	19
	Social trust	3	13	3	19
	Political trust	2	10	7	19
Round 4	Media use	16	6	0	22
	Satisfaction	9	10	3	22
	Political trust	1	13	8	22
Total across experiments (total in %)		*54 (30.0%)*	*84 (46.7%)*	*42 (23.3%)*	*180 (100%)*

Note: NC=not convergent, HC=Heywood case, PS=proper solution.

of the SB-MTMM designs, each trait has only two indicators. Therefore, if the correlation between two traits is 0, the model in the different subgroups will not be identified. In such cases, the estimation procedure will not converge.

The third condition that can cause problems was detected by chance while studying the identification of the two-group MTMM design. It was discovered that the basic model of the two-group SB-MTMM design is not identified if the correlations between the traits are identical. Also, in these cases, the estimation procedure will not converge.

In the ESS, the two-group design has been chosen because otherwise at least some people would get in the main questionnaire questions using a different method than the other people in the samples. This would reduce the number of cases in the analysis.

Not many problems were expected in estimation of the models of the ESS experiments given the special conditions under which problems were expected; however, the reality was different. Table 11.7 reports the number of nonconvergence (NC), Heywood cases (HC), and proper solutions (PS) for the different topics that were used for the SB-MTMM experiments. The table shows that the analysis leads in 30.0% of the data sets to nonconvergence and 46.7% to Heywood cases. In only 23% of the analyses, a proper solution is obtained. Differences between experiments may be observed: the media use experiment seems particularly problematic in both rounds, with no proper solutions at all. In this case, the result is understandable because the correlations between the reported time spent watching television, listening to the radio, and reading newspapers are almost 0.

This may explain the problems encountered for the variables media use. For the other topics, Revilla and Saris (2013) found out that often the correlations between the latent variables are so close that this will be the reason for these problems. For more details of this issue, we refer to the mentioned paper.

Eid (2000) and others have suggested that the problem of identification in MTMM models could be solved by reducing the number of method factors with 1. This means that one method factor is the standard, having no method effects, and the other methods are compared with this method. This is also in line with the classical way of dealing with Heywood cases, which consists in fixing to 0 nonsignificant negative estimates that should not be negative in theory. However, in many experiments, one can see that different method variances fixed to 0 led to models that could not be rejected but generated very different estimates. Especially the quality estimates obtained depended very much on the selected method as the reference method. It will be clear that this solution could not be used for our purposes because we needed not a relative estimate of quality but an absolute one.

Revilla and Saris (2011) have shown that the problems of the two-group design are solved in case of very large samples. In the ESS, each experiment has been done in at least 25 countries with samples of around 1500 cases. If each country is analyzed separately, there are only 1500 cases and one would have problems. But if we assume for a moment that the model is the same in all countries, this means that for each experiment we have at least 37,500 cases taking all countries together. Such a sample would be enough to avoid nonconvergence for most of the experiments. Using multiple-group analysis of any SEM program, one can estimate such a model. We have used in this case the program LISREL 8.5.

It should be clear that we did not believe that this assumption with respect to the equality of the model for all countries is correct. So, the next step was to test for misspecifications in the model for the different countries because we expected some parameters, indicating the reliability and validity to be different in different countries. For the detection of the misspecifications in the first model, we have used the program JRule (Van der Veld et al. 2008) based on the work of Saris et al. (2009b). Using this program, the model is corrected, introducing free parameters in the different countries till the differences between the estimated values of the parameters were so small from one run of the program to the next that one could conclude that it made no sense to continue with the adjustments of the models. We used as a criterion that the differences in estimated values should be in general smaller than .02. We thought that such a difference is not of substantial importance and gives sufficient precision with respect to the estimation in the meta-analysis discussed later.

The results of such a sequential process of model corrections can depend quite heavily on the first steps made in the process. Therefore, we have decided that each data set for each experiment has to be analyzed in the aforementioned way independently by two researchers. Because the two researchers can come to different results, the last step is that they compare the differences and decide together which corrections have to be introduced in the model in order to get a jointly accepted result.

It turned out that this approach worked for all experiments available in the first three rounds of the ESS except for the media data. In the latter case, this approach did not work because the correlations between the traits use of TV, radio, and newspaper are so close to 0 that the model is not empirically identified even with close to 40,000 cases. However, for all other topics, this procedure worked satisfactorily in the sense that the analyses converged to a jointly accepted solution that is difficult to improve and that shows rather small standard errors for the quality estimates.

11.4.2   The Efficiency of the Different Designs

The second issue to be discussed is the efficiency of the different designs. This is a relevant issue because the reduction of the response burden might be gained at the expense of the efficiency of the methods. The efficiency of the different designs has been studied on the basis of the standard errors of the estimates of reliability and validity by Saris et al. (2004). This study showed that for very small method variances, the total sample of a two-group design has to be very large. A much smaller total sample is needed for the three-group design. However, one should realize that the standard error for very small method variances is also minimal unless the variance is equal to 0 as discussed previously. This study also shows that the efficiency of the two- and three-group designs becomes quite similar if the method variance becomes larger. The researcher needs to keep in mind that for both designs the total sample sizes need to be considerably larger than 300, which is the chosen sample size for the one-group design.

   With respect to reliability, the study shows the same pattern, specifically, that the efficiency of the two- and three-group designs becomes nearly the same when the error variance becomes larger. The inefficiency of the two designs for very small error variances compared with the one-group design also becomes apparent. Fortunately or unfortunately, these very small error variances do not occur in survey research.

11.5   SUMMARY AND DISCUSSION

We have pointed out that the classical MTMM design has its disadvantages because respondents have to answer approximately the same requests three times. As a solution, we have suggested an alternative known as the SB-MTMM design. This chapter has shown that the split-ballot MTMM experiment reduces the response burden, by decreasing the number of items to be asked in a questionnaire, without loss of information with regard to reliability and validity measures. Requests concerning the same trait need to be answered only twice and not three times as is required in the classical MTMM approach. The advantage is that it reduces the response burden effects. However, the effects of repeating requests concerning the same concept cannot be eliminated completely. Repeating the requests about the same concepts in different forms is necessary for estimating the reliability and validity of the measures. However, it was shown that two- or three-group designs with repeated observations of exactly the same measures can be used to estimate these effects. We have also seen that the use of two-group design leads to much more problems in estimation than the three-group design. Therefore, it is advisable to avoid the two-group design if that can be done without too large a loss of cases in the analysis.

   For the time being, we suggest analyzing the data of these multiple-group designs using the options available for MGSEM in standard software. With some programs, this may be a bit more complicated task using the missing data approach.

   Concerning the efficiency of the different designs, it has been found that the three-group design is far more efficient than the two-group design at least for the small method variances and error variances. Thus, the total sample sizes can be reduced by using three groups instead of two groups if the errors are rather small. If the errors become larger, the designs become equally efficient, although, for all practical purposes, the three-group design remains more efficient than the two-group design.

Whatever the design may be, it will be clear that MTMM studies will cost extra time and effort and provide information only about a limited number of questions. So, these experiments are not sufficient to cope with the problem of measurement errors in survey research. Therefore, in Chapters 13 and 14, we will discuss a more general solution to this problem.

EXERCISES

1. Specify an MTMM design with one group for three requests of your own questionnaire:
   a. Specify the different requests.
   b. Specify the required sample size.
   c. Specify the correlation matrix that will be obtained.

2. Check how this design changes if you would use a two-group SB-MTMM design:
   a. Specify the different requests.
   b. Specify the required sample size.
   c. Specify the correlation matrices that will be obtained.

3. Check how this design changes if you would use a three-group SB-MTMM design:
   a. Specify the different requests.
   b. Specify the required sample size.
   c. Specify the correlation matrices that will be obtained.

4. Which of the three designs discussed would you prefer to collect information about the quality of requests?

APPENDIX 11.1   THE LISREL INPUT FOR THE THREE-GROUP SB-MTMM EXAMPLE

Analysis of the British satisfaction data with three-group SB-MTMM model group 1

Data ng=3 ni=9 no=140 ma=cm
km
*
1.000
0.469   1.000

```
0.250    0.415    1.000
0.000    0.000    0.0000   1.000
0.0000   0.0000   0.0000   0.0000   1.000
0.0000   0.0000   0.0000   0.0000   0.0000   1.000
−.524    −.322    −.212    0.0000   0.0000   0.0000 1.000
−.313    −.523    −.273    0.0000   0.0000   0.0000 0.509 1.000
−.244    −.313    −.517    0.0000   0.0000   0.0000 0.442 0.461  1.000
me
*
2.39 2.69 2.41  0.0 0.0 0.0  2.09 1.77 2.02
sd
*
.70 .71. 78  1.0 1.0 1.0  .71 .68 .73

model ny=9 ne=9 nk=6 ly=fu,fi te=di,fi ps=di,fi be=fu,fi ga=fu,fi ph=sy,fi
value −1 ly 1 1 ly 2 2 ly 3 3
value 1 ly 7 7 ly 8 8 ly 9 9
value 0 ly 4 4 ly 5 5 ly 6 6
free te 1 1 te 2 2  te 3 3 te 7 7 te 8 8 te 9 9
value 1 te 4 4 te 5 5 te 6 6

free ga 1 1 ga 4 1 ga 7 1 ga 2 2 ga 5 2 ga 8 2 ga 3 3 ga 6 3 ga 9 3
value 1 ga 1 4 ga 2 4 ga 3 4 ga 4 5 ga 5 5 ga 6 5 ga 7 6 ga 8 6 ga 9 6
free ph 2 1 ph 3 1 ph 3 2 ph 4 4 ph 5 5 ph 6 6

out iter=200 adm=off sc

Analysis of British satisfaction group 2
Data ni=9 no=150 ma=cm
Km
*
1.000
0.0    1.000
0.0    0.0    1.000
0.0    0.0    0.0    1.000
0.0    0.0    0.0    0.598   1.000
0.0    0.0    0.0    0.601   0.694   1.000
0.0    0.0    0.0    0.588   0.398   0.517 1.000
0.0    0.0    0.0    0.395   0.690   0.504 0.547   1.000
0.0    0.0    0.0    0.397   0.462   0.571 0.545   0.564   1.000
me
*
.0 .0 .0   5.22 4.30 4.98   1.91 1.69 2.00
sd
*
```

1.0 1.0 1.0 2.27 2.51 2.47 .69 .65 .71

model ny=9 ne=9 nk=6 ly=fu,fi te=di,fi ps=in be=in ga=in ph=in
value 0 ly 1 1 ly 2 2 ly 3 3
free te 4 4 te 5 5 te 6 6
equal te 1 7 7 te 7 7
equal te 1 8 8 te 8 8
equal te 1 9 9 te 9 9
value 1 te 1 1 te 2 2 te 3 3
value 1 ly 4 4 ly 5 5 ly 6 6 ly 7 7 ly 8 8 ly 9 9

out iter=200 adm=off sc

Analysis of the British satisfaction group 3
Data ni=9 no=150 ma=cm
Km
*
1.000
0.469 1.000
0.393 0.605 1.000
−.669 − .454 − .489   1.000
−.512 − .669 − .564   .707   1.000
−.495 − .508 − .742   .693   .729   1.000
0.0   0.0   0.0   .000   .000   0.000 1.000
0.0   0.0   0.0   .000   .000   0.000 0.000 1.000
0.0   0.0   0.0   0.0000   0.0000 0.000 0.000 0.000 1.000
me
*
2.41 2.65 2.50 5.18 4.32 4.99 .0 .0 .0
sd
*
.78 .77 .90 2.39 2.39 2.53 1.00 1.00 1.00

model ny=9 ne=9 nk=6 ly=fu,fi te=di,fi ps=in be=in ga=in ph=in
value 0 ly 7 7 ly 8 8 ly 9 9
equal te 1 1 1 te 1 1
equal te 1 2 2 te 2 2
equal te 1 3 3 te 3 3
equal te 2 4 4 te 4 4
equal te 2 5 5 te 5 5
equal te 2 6 6 te 6 6
value 1 te 7 7 te 8 8 te 9 9
value −1 ly 1 1 ly 2 2 ly 3 3
value 1 ly 4 4 ly 5 5 ly 6 6
out iter=200 adm=off sc

12

MTMM EXPERIMENTS AND THE QUALITY OF SURVEY QUESTIONS

In Chapter 10, the classical MTMM design suggested by Campbell and Fiske (1959) and the analysis of the data obtained by such an experiment using the true score MTMM model have been discussed. This approach has been used in many experiments in the period between 1979 and 1997 by Andrews (1984) and Rodgers et al. (1992) in the United States, Költringer (1995) in Austria, Scherpenzeel and Saris (1997) in the Netherlands, and Billiet and Waege (1989, 1997) in Flanders (Belgium). Scherpenzeel and Saris also used frequently a variant of this design using a panel. In Chapter 11, we have described the SB-MTMM design and analysis developed by Saris et al. (2004). This approach has been used in the European Social Survey (ESS) in the period between 2001 and 2012.

In the earlier studies, the emphasis was much more on variation in the form of the questions, while in the ESS, the emphasis was more on the comparability of the questions across countries and the differences due to the use of different languages and cultures.

In this chapter, an overview is given on the MTMM experiments that have been done from 1979 till 2007. Close to 4000 questions have been evaluated in this way. This research has delivered a lot of information. We will provide some information about these questions. However, in order to correct for measurement error in survey research the obtained information is very little. At this moment, already more than 60,000 questions have been asked in the context of the ESS only, and for proper analysis, one has to correct for the errors in all questions, not only those involved in MTMM experiments. We realized at the start of the ESS that one cannot get quality

Design, Evaluation, and Analysis of Questionnaires for Survey Research, Second Edition.
Willem E. Saris and Irmtraud N. Gallhofer.
© 2014 John Wiley & Sons, Inc. Published 2014 by John Wiley & Sons, Inc.

estimates for all questions. Therefore, a plan was formulated (Saris and Gallhofer 2007a) to code the characteristics of the questions and with this knowledge to develop an algorithm, which will predict the quality of the questions. If that prediction would be successful, the same algorithm could be used to predict the quality of any other question of the ESS as well. In this chapter, the coding procedure will be introduced and the procedure to predict the quality of questions on the basis of the coded characteristics of questions. Because this procedure was rather successful, in the next chapter, the program based on this approach will be introduced, and in the last part of the book, applications of this approach will be discussed.

12.1   THE DATA FROM THE MTMM EXPERIMENTS

Over the whole period, the selection of the questions for the experiments was chosen in the same way. The experiments were incorporated in ongoing research. That means that the questions were chosen from questions that were asked in standard surveys. Some questions from these surveys were chosen to be repeated using a different method at the end of the substantive study. These experiments were directed to evaluate single questions and not composite scores that have been evaluated more frequently in the psychological and marketing literature (Bagozzi and Yi 1991; Lance et al. 2010).

The alternative formulations of the questions were chosen in such a way that they would give information about forms of questions that were also rather common in survey research.

All experiments have been performed using samples from minimal regional populations like the Detroit area or national populations like countries in Europe.

In the first period, 87 MTMM studies have been performed containing 1023 survey items in three languages: English, German, and Dutch. A description of the data from these experiments has been provided in Saris and Gallhofer (2007b). The variation in methods used in these experiments has been summarized in Table 12.1A.

After 1997 the variation in characteristics presented in Table 12.1B was introduced. The table shows that a large variety of standard methods have been used in these experiments. In 87 experiments, it is impossible to systematically vary all these characteristics. To do so, one would have needed many more experiments. In general, one or two characteristics of the requests for answers and/or the data collection method were varied. However, besides that, many other characteristics varied across the questions, because these experiments were the initiative of different researchers independent of each other. Therefore, the topics and the variation of the form of the questions, the context, and the data collection methods were not coordinated. Later, the characteristics of these questions have been coded so that the complete picture of the variation of the characteristics could be observed. In the next section, we will discuss this coding of the questions. For a more elaborate description of these data, we refer to Scherpenzeel (1995) and Saris and Gallhofer (2007b).

The data collection in the second period was coordinated by the Central Coordinating Team (CCT) of the ESS. This organization has included from the very start in 2001 next to the main questionnaire a supplementary questionnaire for methodological purposes in all countries. In the supplementary questionnaire,

TABLE 12.1A   The characteristics of the questions that varied in the experiments between 1979 and 1997

	Group	Specific characteristic
Group 1	The trait	Domain
		Concept
Group 2	Associated to the trait	Social desirability
		Centrality of the topic
		Time specification
Group 3	Formulation of the request for an answer	Trait requested indirectly, direct or no request, and presence of stimulus (battery)
		WH word and what type of WH word
		Type of the request (interrogative, imperative question–instruction, declarative, or none (batteries))
		Gradation
		Balance of request or not
		Encouragement to answer
		Emphasis on subjective opinion
		Information about the opinion of other people
		Absolute or a comparative judgment
Group 4	Characteristics of the response scale	Categories, yes/no answer scale, frequencies, magnitude estimation, line production and more steps procedures
		Amount or the number of categories
		Full or partial labels
		Labels with long or short text
		Order of labels
		Correspondence between labels and numbers
		Theoretical range of scales (bipolar or unipolar)
		Range of scales used
		Fixed reference points
		Don't know option
Group 5	Instructions	Respondent instructions
		Interviewer instructions
Group 6	Additional information about the topic	Additional definitions, information, or motivation
Group 7	Introduction	Introduction and if request is in the introduction
Group 8	Linguistic complexity	Number of sentences
		Number of subordinated clauses
		Number of words
		Number of nouns
		Number of abstract nouns
		Number of syllables
Group 9	Method of data collection	
Group 10	Language of the survey	

TABLE 12.1B   Question characteristics added after 1997

	Group	Specific characteristic
Group 11	Characteristics of the show cards	Categories in horizontal or vertical layout
		Text is clearly connected to categories or if there is overlap
		Numbers or letters shown before answer categories
		Numbers in boxes
		Start of the response sentence shown on the show card
		Question on the show card
		Picture provided

alternative forms of some questions of the main questionnaire were presented to the respondents in order to evaluate the quality of these questions using the SB-MTMM design with two subgroups. This means that all respondents got the chosen question forms in the main questionnaire, while in the supplementary questionnaire, two alternative forms were presented to randomly assigned subgroups of the sample. The purpose of these experiments was to determine the quality of the questions in the different countries. As differences across countries were expected, this information was necessary to correct for measurement errors in the analysis. This means that an effort was made that in all countries the same experiments had to be done.

This aim has not been realized because there was no way at that time that the CCT could control the formulation of all questions in all requests for answers just because of the lack of knowledge of all the languages spoken in the different countries. Now we know, thanks to the coding of the questions, that there are considerable differences in the translations (Saris et al. 2011). This makes it even more important to know the quality of the questions in the different countries so that correction for the differences in quality is possible and comparison across languages can be done.

In the first 3 rounds of the ESS, 4–6 experiments per round were done in 20–32 countries. Given that each experiment is based on at least 9 questions, the total number of questions evaluated in different languages is around 3000.

For details of all the experiments done, we refer to Saris et al. (2011). The following variations in the methods were introduced in these experiments:

- Open questions asking frequencies or amounts versus category scales
- 2-, 5-,7-, and 11-point categorical agree/disagree (AD) scales
- Use of batteries versus separate questions
- 7- and 11-point item-specific (IS) scales
- Use of show cards or not
- 4- and 11-point bipolar versus unipolar scales
- Variation of the position of the statement on the underlying scale
- Variation of the length of the labels of the response categories
- Fixed reference points versus nonfixed reference points

The topics chosen for these experiments can be found in Saris et al. (2011).

12.2 THE CODING OF THE CHARACTERISTICS OF THE MTMM QUESTIONS

In the past, meta-analyses of sets of MTMM experiments have been performed by several scholars (Andrews 1984; Költringer 1995; Scherpenzeel 1995). Looking at the coding systems used in the different countries, Scherpenzeel (1995) came to the conclusion that the results of these studies could not be compared due to the lack of comparability of the coding systems used. Therefore, all questions of these studies have been coded again, using a common coding system. The choice of the characteristics of the questions to code has been derived from the choices that are made in question design described in Parts I and II of this book. We will not repeat this overview here but concentrate on the way we have tried to control the quality of the coding of the translated questions.

In order to code the characteristics of all the questions that were present in the MTMM experiments in all the participating countries, an old version of the program SQP has been used.[1] In the program, more than 60 formal characteristics of survey questions, the answer categories, the show cards, the data collection method, the survey characteristics, etc. are coded. These characteristics evaluated agree to a large extent to the characteristics summarized in Table 12.1.

Because in many of the ESS countries different languages are spoken and used in the questionnaires, coders had to be found who were native speakers for all the languages. Fortunately, it was possible to find sufficient native speakers for all languages in Barcelona.

In order to check the quality of the coding, the first questions of some MTMM experiments were coded by two coders in order to see whether the agreement of the codes was sufficient to rely on a single coder. It turns out that coders often make errors, mostly by mistake. If the two coders spoke about the differences in their coding, it was in general easy to come to a consensus about the correct code.

Given this experience, we have decided that the first two coders should make a consensus coding of the source questionnaire of each round. Consequently, the coding of the MTMM questions of each questionnaire for a specific language and country was done by a native coder. The codes of these questions were compared with the consensus coding of the source questionnaire.[2] If a difference was detected, the coordinator of the coding process spoke with the specific coder about the reasons for the difference.

It could be that a mistake was made in the coding. However, it was also possible that there was an unexpected difference in the coding in the question text in the specific country. In the former case, the code was adjusted; in the later case, the code remained as it was so that now it can be seen that a question on a specific characteristic was different from the characteristic in the source questionnaire. All the codes have been stored in the question data file included with the question text in the different

---

[1] This program has been developed by Daniel Oberski and Thomas Gruner and is now a part of the new SQP program.
[2] For this purpose, again a special program "Compare" was made by Daniel Oberski.

languages. In the next section, we will give some results with respect to the differences that have been found using this procedure.

Using this coding system with more than 60 characteristics for each question, we created a database of questions specifying the characteristics of the questions, the response options, introductions, as well as characteristics of the questionnaire and the data collection method.

12.3   THE DATABASE AND SOME RESULTS

From the MTMM experiments, we obtained quality measures for the questions: the reliability and validity coefficients. Each experiment was analyzed by two researchers. In this way, we obtained two sets of quality estimates for each question. From these two estimates, we derived the point estimates and the standard errors. These were used to construct point estimates and 95% confidence intervals for the original reliability and validity estimates. Histograms of the reliability and validity coefficients are shown in Figure 12.1.

The size of these confidence intervals (difference between upper and lower bounds) ranged between .0009 and .1363 for the reliability coefficients and between .0009 and .2793 for the validity coefficients. The overall average reliability coefficient estimate was .841, and the overall average validity coefficient was .923.

The estimates of the reliabilities and validities of the coded questions were added to the question database with all the codes. So we obtained a database in which question characteristics were joined with the reliability and validity estimates. After deletion of questions that were either not coded or not analyzed, a data set with 3483 questions was obtained.

It is also interesting to mention the quality as the product of the reliability and validity of each question. The mean quality is not so bad (.64) and higher than the average result reported by Alwin (2007) with respect to studies in the United States

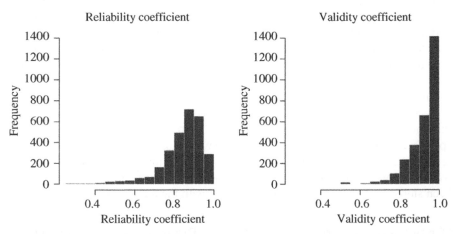

FIGURE 12.1   Reliability and validity coefficient estimates obtained from the MTMM experiments, without transformation.

using a different model. On average, 64% of the variances of the observed variables is explained by the latent variables of interest. In the United States, this was 50%. We also see that there are questions with very low quality (<.40). One may wonder whether these questions are good enough to be used.

12.3.1 Differences in Quality across Countries

Of importance is whether the quality of the questions is not too different across countries. If that is the case, one cannot compare relationships between variables across countries. In Table 12.2, we present the average quality of the questions for all countries that participated.

The mean quality for all countries is .64, and the standard deviation is .175. The country with the highest mean quality is Greece (.69), and the country with the lowest mean value is France (.58). The difference is not that large. On the other hand, we see that the variation in quality within all countries is rather large. This suggests that probably large differences can be seen if we look at different topics within each country. But that could mean that the averages are relatively comparable, but the quality of specific questions can be very different across countries.

So, we looked next at the differences in quality for some questions across the countries and concentrate on the questions in the main questionnaire. The question

TABLE 12.2   The mean quality of the questions in the countries

Country	Mean total quality	Standard deviation	Minimum	Maximum
Austria	.645	.158	.10	.98
Belgium	.603	.172	.02	.93
Czech Republic	.632	.162	.03	.87
Denmark	.621	.188	.02	.92
Estonia	.604	.183	.02	.89
Finland	.624	.178	.03	.96
France	.578	.188	.02	.96
Germany	.637	.156	.03	.95
Greece	.691	.163	.12	.96
Ireland	.594	.180	.03	.95
Netherlands	.672	.183	.01	.98
Norway	.647	.143	.24	.94
Poland	.631	.180	.03	.96
Portugal	.662	.169	.26	.97
Slovenia	.616	.168	.03	.92
Slovakia	.585	.190	.04	.92
Spain	.622	.172	.11	.96
Sweden	.670	.123	.39	.88
Switzerland	.651	.181	.03	.97
United Kingdom	.630	.160	.03	.92
Ukraine	.606	.167	.18	.96
Total	*.639*	*.175*	*.01*	*.98*

with the least variation in the quality across countries is a question about immigration that is asked in the third round as part of the core questionnaire.

B37   still card 14 *How about people from the <u>poorer</u> <u>countries outside Europe?</u> Use the same card.*

Allow many to come and live her	1
Allow some	2
Allow a few	3
Allow none	4
(Don't know)	8

The results for the question with the least variation in quality are presented in Table 12.3. This table shows that the variation is indeed very minimal. The lowest value is in the Ukraine (.686) and the highest in Switzerland (.847).

For a correlation of .6 between two latent variables in both countries with these qualities, this difference in quality would already mean that in the Ukraine, the observed correlation[3] would be $.6 \times .686^2 = .412$ and the correlation in Switzerland would be $.6 \times .847^2 = .501$. This possible difference in observed correlation would not be a substantial difference. This difference would completely be a consequence of the

TABLE 12.3   Quality across countries of the item with smallest variation

Country	Mean quality
Austria	.823
Belgium	.717
Denmark	.826
Estonia	.717
Finland	.748
France	.787
Germany	.796
Ireland	.741
Netherlands	.732
Norway	.781
Poland	.773
Portugal	.814
Slovakia	.717
Slovenia	.717
Spain	.797
Switzerland	.847
Ukraine	.686
United Kingdom	.748
Overall mean	.765
Standard deviation	.046

[3] This result is based on equation 2.1 on the fact that $q^2 = (r \cdot v)^2$ assuming that the method effect is minimal.

difference in data quality as found in the ESS. Note that in both cases, the correlation between the observed variables would be considerably lower than the true correlation between these latent variables (.6).

The largest variation has been found for the question of the core about government intervention. The question is formulated as follows:

> **Card 16** *Using this card, please say to what extent you agree or disagree with each of the following statements.* **Read out each statement and code in grid.**
>
> **B43** *The less that government intervenes in the economy, the better it is for [country]:*
> 1. *Strongly agree*
> 2. *Agree*
> 3. *Agree neither disagree*
> 4. *Disagree*
> 5. *Strongly disagree*

The qualities of this question across the different countries are presented in Table 12.4. In this case, the differences are indeed much larger. The lowest value is .337 in Finland and the highest .943 in Portugal. Fortunately, not all differences are so large because this would lead to very large differences in observed correlations even though the correlation between the latent variables of interest

TABLE 12.4 Quality across countries of item with largest variation

Country	Mean quality
Austria	.638
Belgium	.360
Czech Republic	.590
Denmark	.359
Germany	.383
Finland	.337
France	.365
Greece	.687
Ireland	.352
Netherlands	.400
Norway	.362
Poland	.601
Portugal	.943
Slovenia	.364
Spain	.746
Sweden	.434
Switzerland	.380
United Kingdom	.408
Overall mean	.484
Standard deviation	.175

would be the same. To illustrate this with a correlation of .6 between the latent variables, this would mean that in Finland the correlation would be .6 × .337² = .20 and in Portugal .6 × .943² = .56. This is just a consequence of differences in the size of the measurement errors.

Because the size of the measurement errors has such a big effect on correlations and other measures for relationships, these quality estimates have been estimated. They can be used to correct for measurement errors as we will show in Chapter 15.

12.3.2 Differences in Quality for Domains and Concepts

Saris and Gallhofer (2007a) have shown that there are significant differences in the validity and the reliability—and as consequence in the quality—for items from different domains, concepts, and other associated characteristics. Table 12.5 shows differences in quality depending on the domain of the questions in the MTMM experiments.[4] Items asking about "health" have the lowest quality, .473. Questions about politics vary depending on the specific topic; items on national institutions and national government reported the overall highest quality, .760 and .705, respectively, while in items about "economic or financial matters," the quality was much lower, .595.

TABLE 12.5 Mean total quality of questions for different domains

Domain	Mean total quality	Standard deviation	Minimum	Minimum
Health	.473	.151	.06	.89
Living conditions and back ground variables	.586	.114	.21	.81
Work	.661	.138	.31	.97
Personal relations	.587	.144	.09	.87
Leisure activities	.533	.118	.28	.78
National politics: national government	.705	.120	.24	.96
National politics: national institutions	.760	.081	.07	.93
National politics: economic/ financial matters	.595	.181	.14	.96
National politics: other	.685	.136	.38	.96
Total	*.640*	*.154*	*.06*	*.97*

[4] It should be said that the differences in quality between the different categories of an explanatory variable can also come from other characteristics that are related with this variable. Therefore, a multivariate analysis would give a better indication of the effect of the variable domain. This picture will be given in the next chapter.

TABLE 12.6   Mean total quality of questions for different concepts

Concept	Mean total quality	Standard deviation	Minimum	Maximum
Norm	.723	.110	.43	.96
Policy	.703	.151	.39	.96
Action tendency	.691	.090	.38	.91
Feeling	.676	.126	.07	.96
Evaluative belief	.606	.171	.06	.97
Facts, background, or behavior	.527	.100	.34	.78
Evaluation	.527	.129	.18	.89
Total	*.640*	*.154*	*.06*	*.97*

Table 12.6 shows that there are differences in the quality depending on the concept measured. Norms and policies have the highest quality, while items about facts, background, or behavior and evaluations reported the lowest quality.

12.3.3   Effect of the Question Formulation on the Quality

It can also be expected that question formulation has an effect on the quality of the questions. In an earlier paper, it was reported by Saris et al. (2009a) that AD batteries had a very negative effect on the quality. Without repeating the full report here, we can illustrate this here with two examples. In round 3, experiments were done to study these effects. These experiments have been summarized in Table 12.7.

In the experiment called "imsmetn," the response scales are varied. Three alternatives have been used in the supplementary questionnaire in randomly assigned subgroups of the samples in each country. This means that the data with these three methods have been collected at the same point in time and under the same conditions in randomly assigned groups. The results of this experiment are presented in Table 12.8.

In Table 12.8, we have presented the results with respect to the average quality across the participating countries comparing an IS scale with two AD scales, all three measured in randomized subgroups of the total sample in the supplementary questionnaire.

The table indicates the enormous difference in quality between the different scale types where the IS scale turns out to have much higher quality than the two AD scales. This result is in agreement with an earlier publication of Saris et al. (2009a).

Another result that has been found before (Revilla et al. forthcoming) is that the 5-point AD scale has a higher quality than a 7-point and an 11-point scale as can be seen in Table 12.9 based in experiment with respect to the consequences of immigration (imbgeco).

These examples show very clearly the effect the choice of the method can have on the quality of the questions.

TABLE 12.7   Round 3: the SB-MTMM experiments

			All	Group A	Group B	Group C
Exp	Variables	Meaning	M1	M2	M3	M4
imsmetn	*imsmet*	• [Country] should allow more people of the same race or ethnic group as most [country's] people to come and live here				
	imdfctn	• [Country] should allow more people of a different race or ethnic group as most [country's] people to come and live here	4IS	5AD	4IS	7AD
	impcntr	• [Country] should allow more people from the poorer countries outside Europe to come and live here				
imbgeco	*imbgeco*	• It is generally bad for [country's] economy that people come to live here from other countries				
	imueclt	• [Country's] cultural life is generally undermined by people coming to live here from other countries	11IS	5AD	11AD	7AD
	imwbcnt	• [Country] is made a worse place to live by people coming to live here from other countries				

Note: IS = item-specific scale and AD = agree/disagree scale.

TABLE 12.8   The quality of the questions concerning immigration (imbgeco)

	Question 1	Question 2	Question 3
IS 4	.905	.914	.908
AD 5	.568	.629	.607
AD 7	.525	.597	.562

TABLE 12.9   The quality of the questions concerning consequences of immigration (imbgeco)

	Question 1	Question 2	Question 3
AD 5	.576	.649	.649
AD 7	.352	.462	.490
AD 11	.267	.413	.452

12.4 PREDICTION OF THE QUALITY OF QUESTIONS NOT INCLUDED IN THE MTMM EXPERIMENTS[5]

As we have said before, there are many more questions that cannot all be studied with respect to quality using MTMM experiments. So, in order to make it possible to estimate the quality of questions that were not included in the MTMM experiments, we have studied the possibility to predict the quality of the questions on the basis of the characteristics of the questions. This was done for the questions that were included in the MTMM experiments. If we could find for these questions an algorithm that would give a good prediction of the quality of these questions, this algorithm could also be used for the prediction of the quality of new questions.

The first step to do so was that the estimates (reliabilities and validities) were logit-transformed and the point estimates and standard errors of the logit-transformed reliability and validity coefficients of the coded questions were added to the question database with all the codes. Histograms of logit transformations of the reliability and validity estimates are shown in Figure 12.2.

We fitted separate prediction models for the logit-transformed estimates of the reliability and validity coefficients (r and v). Predictors were obtained using Breiman's (2001) random forests of regression trees, as implemented in the *R 2.13.1* package *randomForest* (Liaw and Wiener 2002; R Development Core Team 2011). A random forest is an ensemble predictor, that is, a collection of many individual predictors whose individual predictions are combined to form the final prediction. In random forests, the individual predictors are regression trees grown with the CART algorithm.

Each of the regression trees using the CART algorithm is grown in the following manner. The data set is split into two groups ("nodes") based on that split on a question characteristic that yields the smallest possible mean squared prediction error for the logit(r) or logit(v). For each new group, the same procedure is repeated until the resulting group would have five or fewer observations or no improvement in mean square prediction error can be found.

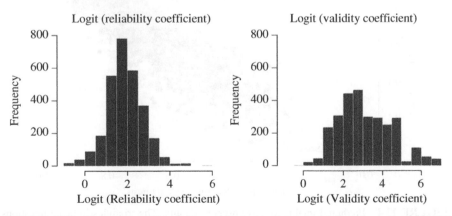

FIGURE 12.2 Reliability and validity coefficient estimates obtained from the MTMM experiments after logit transformation.

[5] This section has been based on the work done by Daniel Oberski; see Saris et al. (2011).

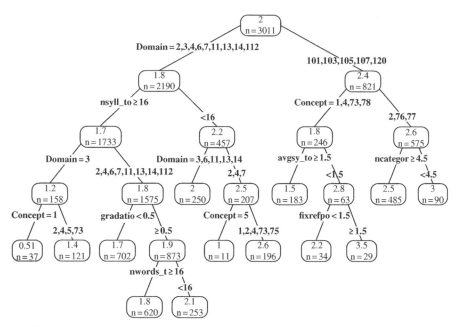

FIGURE 12.3 Example CART tree for prediction of the logit-transformed reliability coefficient.

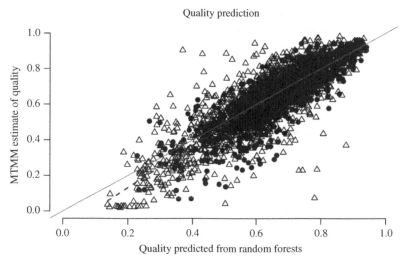

FIGURE 12.4 Predicted quality versus observed quality. The triangles indicate questions from the ESS and the dots questions from the old experiments.

An example of a single regression tree is given in Figure 12.3. The tree in Figure 12.4 gives a prediction of the logit of the reliability coefficient for a question with given characteristics. For example, suppose the question "Do you think the government does a good job?" is asked, with answers ranging from "the worst job" to "quite a good job."

The top node splits off depending on the domain, in this case national government (domain = 101). Afterwards, we follow the split on concept to the left-hand node, because the question asks an "evaluative belief" (concept = 1). The average number of syllables per word is then the next relevant variable, which in this case is 1.22, less than 1.5. Finally, there is only one fixed reference point, so that the prediction of the logit value ends up being 2.2. This prediction was based on 34 observations. This logit value can be transformed in a value of a reliability coefficient by taking the inverse of this value or invlogit $(2.2) = .90$. So according to this tree, the reliability would be predicted to be .90.

In practice, CART trees may suffer from overfitting problems. Their predictive power can be limited, and this has led to pruning techniques, whereby the lower nodes of the tree are removed from the predictor so as to prevent overfitting. In the random forest algorithm, a different approach is taken. Instead of growing just one regression tree, many trees—in our case 1500—are grown without pruning but based on a double randomization of observations and variables used in the prediction. This deals with the overfitting problem by subsuming randomness due to overfitting in the between-tree variance, and automatically using those features that are commonly selected in all bootstrap samples to determine the average, and final prediction.

The tree given in Figure 12.3 is only given as an example output of the CART procedure and does not necessarily correspond to any tree used in the final prediction. In fact, 1500 of these trees are created, and the overall prediction is then taken as an average over all 1500 trees in the "forest." Given the amount of information available in all these trees, a distribution of the predictions is obtained, and this information can be used to determine the means and specify prediction intervals and standard deviations.

A key parameter in the random forest algorithm is the number, m, of variables selected at random for each tree. On the one hand, growing trees with more features gives more predictive power to each tree, which will reduce the mean squared prediction error. On the other hand, increasing the number of features will increase the correlations between tree predictions, reducing the mean squared prediction error or requiring more trees to obtain predictions of the same accuracy. It was found that the mean squared error is not reduced much further after 20 features, which is the default chosen by the *randomForest software*.

12.4.1 Suggestions for Improvement of Questions

Looking once more at the tree in Figure 12.3, we can see that a much higher prediction of invlogit $(3.5) = .97$ would have been given if there had been more fixed reference points, for example, if the final category had not been "quite" but "the very best." It can be seen by looking at the terminal nodes that a large range of different predictions can be obtained depending on the characteristics of the question.

Given the available ensemble predictor, one can, for each predicting variable, vary the code and see what the effect would be on the predicted quality. In this "what-if" analysis, one can get a mean prediction for each possible code of the variable, keeping all other codes the same. Some of these predicted values may be lower than the predicted value of the real question, but others may be higher. In this way, one can get an impression of what improvement in the prediction is possible by changing this characteristic of the question while keeping all other characteristics the same. However, we speak purposely of "impression" because in general one characteristic of a question cannot be changed without also changing other characteristics of the question. For example, increasing the number of categories will change the number of words and syllables and possibly also the instruction or even the labeling of the scale, etc. So, one has to be careful with these suggestions. A more adequate procedure is to reformulate the question and check the prediction of the new question. In addition, it should be kept in mind that the current model is only a prediction model and not a causal model. Therefore, there is no guarantee that actually changing this characteristic will have the predicted effect on the quality.

So far, we have spoken of only one prediction variable. Looking for the possible improvements can already be very tedious for one variable if this variable has many categories, which all have to be checked separately. This can be a rather lengthy process.

12.4.2   Evaluation of the Quality of the Prediction Models

For each tree, the mean squared error of the predictions from the tree is calculated using only questions that are not included in that tree. After growing the entire forest, the prediction error for the overall forest is calculated by combining the prediction errors estimates for the not included questions. Thus, the mean square prediction error estimate is automatically based on cross-validation samples.

From these mean squared error estimates, one can calculate an R^2 measure of the predictive power of the forest as a whole. The R^2 was .84 and .65 for the validity (v) and reliability coefficient (r) logits, respectively. The squared correlations between predicted and observed coefficients on the original scales were .72 for the validity coefficients and .69 for the reliability coefficients.

Another prediction of interest is the prediction of "quality" of the question, which is the product of the reliability and the validity. The scatter plot of predicted versus observed values for the quality is shown in Figure 12.4.

The prediction is rather good as can be expected based on the high R^2 measures for validity and reliability coefficients. However, for the few questions with low-quality parameters, the predictions are systematically too high, as shown by the deviation of the dotted line from the gray 45 degree line.

Overall, we believe that the predictor does a reasonably good job of providing information about the expected quality of a question, with the caveat that the prediction worked less well for the few questions with a very low quality (more than 60%

measurement error). When employing the predictor to obtain quality predictions of questions that were not in this study, it should also be remembered that the questions in the data set cover only a certain range of application.

12.5 SUMMARY

In this chapter, we have given an overview of the variation that existed between the questions involved in the MTMM experiments over the complete period from 1979 to 2012. We have emphasized that in all experiments some question characteristics were varied by purpose but that the variation takes place in a context of a specific topic, a data collection method, a language, a country, etc. These characteristics determine also the resulting quality of a question. Therefore, one may wonder whether the information of a specific question in a specific context is a good indication of the same form of the question in another context. Nevertheless, this information of the quality of many questions is necessary to make more general statements about the quality of questions. Therefore, a lot of experiments have been done and the quality of these questions estimated.

Over the full set of MTMM experiments, we have seen that the differences in quality across countries for some questions can be rather large. As a consequence, one cannot compare relationships between variables across countries without correction for the quality of the questions. It is for this reason that the ESS has decided to include MTMM experiments in the standard operations of the ESS.

Because there is a tremendous variation possible of forms of the questions and this variation leads to differences in quality, it is necessary to know the quality of any type of question. Otherwise, correction for measurement is not possible. MTMM experiments alone cannot provide this information. Therefore, we have looked for the possibility to predict the quality of any question on the basis of the characteristics of the questions themselves. In SQP 1.0, a procedure for this purpose was developed. However, the new procedure for prediction of the quality of questions is considerably different from the previous procedure used to make the predictions in SQP 1.0. In that case, the model was a regression equation based on the absolute values of the reliability and the validity. The reason for the change is that the new procedure gives better predictions and avoids the problem of inacceptable predictions, larger than 1 or smaller than 0. Another advantage of this new procedure is that we do not only provide a point estimate but also a prediction interval. Finally, the new program is based on a much larger range of topics, languages, and countries.

In the next chapter, we will present the program SQP 2.0. This program allows the user to code the question characteristics in a user-friendly interface and provides predictions of the reliability and validity estimates based on the random forest predictors. The program also allows for a direct comparison of the results of the predictions with the results of the MTMM experiments that are available in the database. The results of the present procedure are indeed much better than using SQP1.0. The explained variances for reliability and validity were in the past, respectively, .47 and .61; with the new prediction procedure, the explained variance increased to, respectively, .60 and .85.

This is a considerable improvement. It should be said that the predictions will never be perfect because some questions may be so different that the database does not contain sufficient similar questions. This holds at this moment especially for questions about facts, frequencies, and events.

For these variables, it is difficult to specify MTMM experiments. Therefore, these types of questions are underrepresented here. A procedure that allows the evaluation of the quality of these questions is the quasi-simplex model used by Alwin (2007) to evaluate these types of questions. We recommend the user to look at this publication for the quality of background variables and questions about factual information.

EXERCISES

1. In Table 12.7, the question imsmetn and imdfcm are presented. These questions are supposed to measure the latent variables "tolerance to people of the same race" and "tolerance to people of a different race." Imagine that these latent variables correlate .6. For this situation, calculate what the observed correlation will be on the basis of the information in Table 12.8 of:

   a. Use of a 4-point item-specific scale

   b. Use of a 5-point agree/disagree scale

   c. Use of a 7-point agree/disagree scale

2. Table 12.4 indicates that the question B43 of the ESS has a much lower quality in Germany than in Austria while in these countries they speak the same language. Can the relationships between this variable and other variables be compared across these countries?

3. Imagine that we want to see which variable has more influence on the likelihood to go to vote and we think about an explanation by a norm and an evaluation. An example of a norm is: In how far do you agree or disagree that people should vote? An evaluation may be: In how far do you agree or disagree that voting is useless? The qualities of these two questions are indicated in Table 12.6. If that is so, do you think that the correlations of these two variables with the likelihood to vote can be compared?

4. What has to be done in order to be able to compare the relationships discussed in the previous exercise?

PART IV

APPLICATIONS IN SOCIAL SCIENCE RESEARCH

In Chapter 12, we saw that the quality of the questions can be rather different across methods and across countries. This means that the effect of variables measured with different methods or in different countries cannot be compared without correction for measurement error. We also saw that knowledge of the quality of questions involved in MTMM experiments is only very limited, and therefore, we looked at the possibility of developing an algorithm to predict the quality of any question in any possible language. In the last chapter, we showed that such an algorithm has been developed on the basis of the knowledge we have collected with the MTMM experiments. Surely, the present algorithm is not as general as covering all possible methods and languages, but it is much more general and better than the earlier algorithm incorporated in the program SQP 1.0. Now, the algorithm is covering more methods and many more languages.

This algorithm has been incorporated in a new program, SQP 2.0, which will be introduced in Chapter 13. In this chapter, the program and the possibilities for improvement of questions will be discussed.

In Chapter 14, we will illustrate how the information available in SQP 2.0 can be used to estimate the quality of concepts-by-postulation.

In Chapter 15, the fundamental issue of correction for measurement errors in survey analysis will be discussed. In this case, we will pay attention to the correction for measurement error in models with concepts-by-intuition and models with concepts-by-postulation.

Finally, in Chapter 16, we will illustrate how the information available in SQP can be used to improve cross-cultural research.

Design, Evaluation, and Analysis of Questionnaires for Survey Research, Second Edition.
Willem E. Saris and Irmtraud N. Gallhofer.
© 2014 John Wiley & Sons, Inc. Published 2014 by John Wiley & Sons, Inc.

13

THE SQP 2.0 PROGRAM FOR PREDICTION OF QUALITY AND IMPROVEMENT OF MEASURES

In this chapter, the SQP 2.0 program will be introduced. We will give a stepwise introduction to the program providing screenshots for the results of the different decisions. It should be mentioned that we cannot guarantee that all screenshots as presented in this chapter will appear in exactly the same way when the user applies the steps. This is so because the program is constantly being developed through the introduction of new information and the process of correcting bugs detected in the system. However, users can expect more or less the same information to appear if the same choices are made as those indicated in this chapter, with the exception of some minor details that may differ.

There are, in principle, three different ways in which the new version of the SQP program can be used. The first option is directed at questions that were involved in MTMM experiments. In Chapter 12, we showed that the question database contains, at this moment, all of the questions involved in the MTMM experiments of rounds 1–3, referred to as "ESS round + a number," for all countries that participated in the ESS as well as the questions that were studied in the past (Saris and Gallhofer 2007b) referred to as "old + a number." For all these questions, the quality is available in the database. Soon, the set of questions will be extended with the questions from round 4 to 6. So far, the database contains close to 4000 questions. With the three extra rounds, this number will nearly double. The SQP program can be used to obtain the quality estimates for these questions.

The second option is directed to questions that have not been involved in MTMM experiments. There are more than 60,000 questions in the database. It will be clear

Design, Evaluation, and Analysis of Questionnaires for Survey Research, Second Edition.
Willem E. Saris and Irmtraud N. Gallhofer.
© 2014 John Wiley & Sons, Inc. Published 2014 by John Wiley & Sons, Inc.

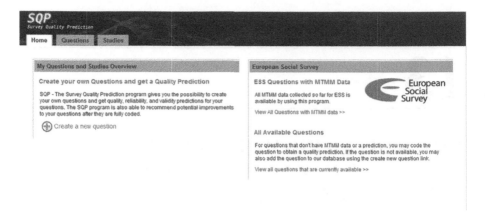

FIGURE 13.1 The page of the SQP program where one can make the basic choice of what one wants to do.

that for these questions, no quality estimates are available. Therefore, in order to obtain these estimates, the user of the program has to code the characteristics of the question, and the program provides the estimates of the quality of the questions.

The third option is directed at questions that are formulated for new studies. In this case, the user first has to introduce the question in the system before the coding can be started, and the program can provide the prediction of the quality and suggestions for improvement of the question.

In the next pages, we will discuss these different options in the sequence indicated previously. However, before we discuss the different options, we will first introduce some basic steps to start up the program. If one goes to the Internet, preferably Mozilla Firefox, and selects SQP.UPF.EDU, then one gets to see the home page of SQP. By clicking on START, it opens the first page of the program. On that page, the program asks you to register as a user. If this is your first time using the program, click on "Register Now" and answer the questions that follow. If you remember your user name and password for the next time, you will be able to bypass this step by entering your user name and password and clicking on "Login." By doing so, you will end up on the screen presented in Figure 13.1.

As can be seen here, this page allows you to make the choices that we have suggested previously. In the next sections, we will illustrate what can and should be done if one makes each one of the specified choices. We start with the choice of the MTMM questions.

13.1 THE QUALITY OF QUESTIONS INVOLVED IN THE MTMM EXPERIMENTS

13.1.1 The Quality of Specific Questions

If we select "View all questions with MTMM data," by clicking on this text, we end up in the next screen presented in Figure 13.2.

FIGURE 13.2 The first page if the MTMM questions have been selected.

On this screen, the user can make a selection for specific questions. The study, the language, and the country of the questions can be specified by making a selection in the boxes at the top left. One can even ask for a specific question by typing the number or name of the question. Imagine that we want to look at questions of round 3 of the ESS, asked in Ireland. We can do so by the specifications presented in Figure 13.3.

Let us say that we want to see the results for a specific one, B38, for example. One can type the number in the text box, but one can also directly click on the question in the screen. By doing so, the screen presented in Figure 13.4 appears.

The pop-up screen presents how the question was formulated, and it indicates what information is available for this question. First of all, we see that the quality of the question estimated in the MTMM experiment is given, which is .557. This means that close to 56% of the variance in the observed variable comes from the variable that it should measure. It also means that close to 44% of the variance is error.

Sometimes, there are also other estimates of the quality of the question available, especially predictions based on the coding of the question and the prediction program discussed in the last chapter. MTMM questions are normally also coded, and therefore, a prediction of the quality by the program can also be obtained. This is also true in this case. The so-called "authorized" prediction is .601. This prediction is called "authorized" because it is based on the coding of this question that has been checked on correctness by our research team at RECSM.

We see that, in this case, the predicted values are not very different from the value obtained by the MTMM experiment. One can get more information about the quality of the question in particular, splitting the quality up into reliability and validity. This can be done for the MTMM results by clicking on "View MTMM Results," but one can also click on "View prediction details." In this case, one gets the details of the

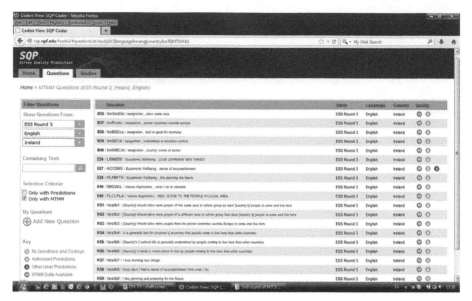

FIGURE 13.3   The first 20 questions asked in Ireland and involved in MTMM experiments.

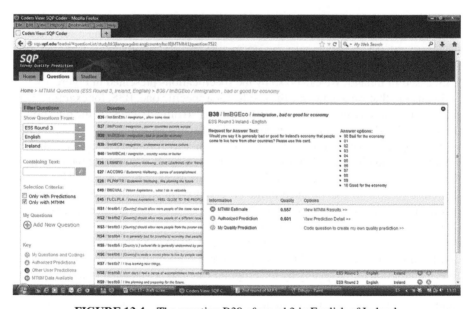

FIGURE 13.4   The question B38 of round 3 in English of Ireland.

MTMM and the SQP prediction results together. Choosing the latter option, the screen presented in Figure 13.5 appears.

We see that in this case, the estimates by MTMM and the predictions by SQP 2.0 are rather similar. In general, this can be expected given the high correlation between

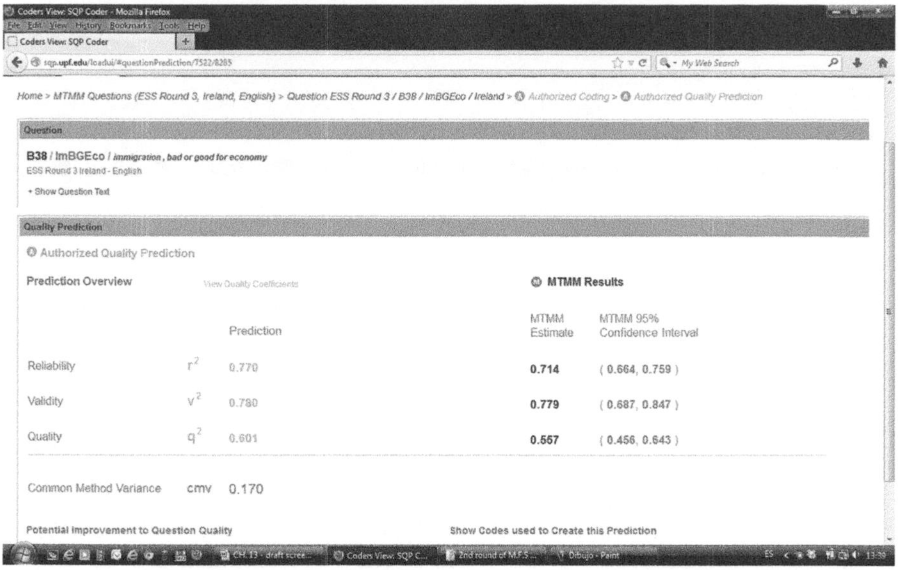

FIGURE 13.5   Detailed information about the quality indicators of question B38 from Ireland.

these two estimates reported in the last chapter, but there are exceptions because, occasionally, questions can deviate from the most common questions or because the analysis has led to a rather deviant result.

The common method variance (CMV) is also presented on the screen. The CMV is an estimate of the correlation that the method would produce between variables that are measured with the same method and have the same quality. In this case, one can say that due to the method used, the correlations would be .170 too high. In Chapter 16, we will discuss how this information with respect to the data quality can be used in data analysis in order to correct for measurement errors.

To get a different picture of the quality, one can also ask for the quality coefficients by clicking on "View quality coefficients." By doing so, the screen image presented in Figure 13.6 appears.

The quality coefficients are comparable with standardized factor loadings. In the MTMM experiments, these coefficients have been estimated. They are the square root of the estimates for reliability, validity, and quality. In this screenshot, the uncertainty ranges are also presented for all three estimates. It is clear that for a large part, they overlap. This is what you expect if the two estimates give approximately the same result. It should, however, be clear that these two sets of estimates are based on very different information. One is based on the MTMM data and the other on the codes of the question characteristics and the prediction procedure described in the previous chapter.

In order to see the codes, one should click on "View prediction codes." In doing so, the screenshot presented in Figure 13.7 appears.

FIGURE 13.6   The comparison of the quality coefficients.

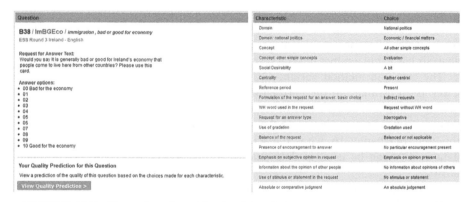

FIGURE 13.7   The codes selected for the different characteristics of the question B38.

At the right-hand side, we can see the codes for all the characteristics of the question. On the left side is the text of the question. These are the authorized codes of the characteristics approved by the team of RECSM.

13.1.2   Looking for Optimal Measures for a Concept

Typical for the MTMM experiments is that alternative forms of questions for a concept have been tested, that the quality of the questions is evaluated in the experiments, and that a prediction of the quality is also available if the codes of the questions were completed.

For the concept discussed previously concerning good or bad consequences of immigration for the economy, this is typically the case. We can get this information by first going back to the screen presented in Figure 13.3 and asking the program to

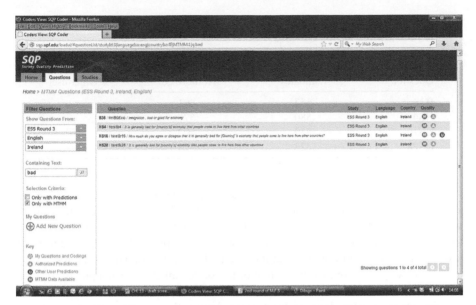

FIGURE 13.8 The result of the search for questions with the typical word "bad."

search for questions that are characterized by a keyword. In this case, the keyword "bad" can be used because that is typical for these questions. There may be other questions that are also characterized by this word. Nevertheless, we will surely end up with all questions characterized by the word "bad." If we search with this word, we get the screen presented in Figure 13.8.

It happens that in this case, all four questions do indeed measure bad consequences for the economy due to immigration. It seems that no other question contains the same word. In round 3 of the ESS, an MTMM experiment was carried out involving four questions for the same concept-by-intuition. In the main questionnaire, question B38 was asked according to an item-specific scale with 11 categories going from "bad for the economy" to "good for the economy."

In the supplementary questionnaire, all three questions were of the agree/disagree type. In the first group, question HS4 was asked with the use of a fully labeled five-point scale going from "agree strongly" to "disagree strongly." In the second group, HS16 was presented with an 11-point scale with only the endpoints labeled from "strongly disagree" to "strongly agree." In the third group, question HS28 used an eight-point scale going from "disagree strongly" to "agree strongly."

Figure 13.8 shows that all questions were evaluated in the MTMM experiment (m) and completely coded and checked by RECSM, which is called an authorized coding (a), while one question was also completely coded by another user that was not controlled by RECSM (u). Clicking on a question, one can get the quality estimates from each of the sources and more detailed information as we have shown previously. For illustrative purposes, we have summarized the estimates provided for the different questions in Table 13.1.

For the first two questions, the MTMM estimates of the quality and the predictions based on the authorized codings are rather comparable. This is not the case for

TABLE 13.1 The quality estimates of four questions about consequences of immigration for the economy

Question	MTMM Estimates	SQP predictions	
		Authorized	Other user
B38	.557	.601	—
HS4	.580	.579	—
HS16	.250	.503	.503
HS28	.322	.523	—

the last two questions, that is, the agree/disagree questions with 8- and 11-point scales. These two questions are typical cases described in the previous chapter where the MTMM estimate of the quality is very low and the prediction based on the question characteristics is more moderate. This is probably so because it is not always the case in our data set that the 8- and 11-point scale have such a negative effect on the quality. In fact, the length of the scale can even have a positive effect, but not in the case of the agree/disagree scales as we have seen in Chapter 12. It may be, therefore, that the prediction overestimates the quality of these questions.

It seems that a user did not trust the predicted low quality of question HS16 and coded the question again. Based on these codes, the quality prediction was, however, exactly the same. This does not have to happen. One can compare these codes, and then, one can see if the codes were indeed exactly the same.

We give this example because it illustrates that SQP can be used to see which type of question is better for the concept-by-intuition one would like to measure. This information helps to make a choice for a question in a new study. In this specific case, the last two types of agree/disagree questions should not be chosen. The first two types are better. The two questions have very different forms, but they are quite similar in quality. However, this is not the end of the story because we will show in Section 13.3 that the SQP program can also give suggestions for the improvement of the questions. In that respect, the first type of question can be more improved than the second.

This example illustrates how one can use the information from the MTMM experiments provided by SQP for selection of higher-quality questions for new research. However, it should be mentioned that this information does not actually exist for many questions. One cannot do MTMM experiments for all questions given the costs and the amount of work associated with them and because different forms of the same questions cannot always be formulated. This is, for example, true for factual and background questions.

13.2 THE QUALITY OF NON-MTMM QUESTIONS IN THE DATABASE

Moving to the second option of the SQP program, we have to go back to the first page by clicking on the word "Home," and then, we click on "View all questions that are currently available." If we select on this screen as study round 1, English as language, and as country Ireland, then we get the screen presented in Figure 13.9. Here,

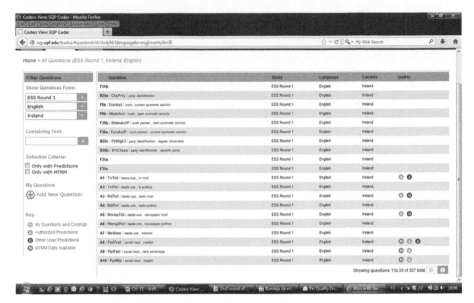

FIGURE 13.9 Overview of the questions of round 1 from Ireland.

there are two possibilities: either the questions have already been coded or they have not. Looking at Figure 13.9, we see both examples; question A3, for example, has been coded, while question A4 has not been coded. A1 was already coded, but A2 was not coded. In reality, the latter is now coded because we used this question as an example.

If we select question A1, we once again get the pop-up screen for this question as before, presenting the quality prediction by SQP based on the approved coding. We can also find the details of the quality estimates and the specification of the codes. So far, it goes the same as before.[1]

If we select A2, however, the process is different because A2 has not been coded so far. In selecting this question, we get the screen presented in Figure 13.10.

In order to get a prediction of the quality of this question, the first step is to code the question. If you click on "Code question to create my own prediction," the screen appears that is presented in Figure 13.11.

Selecting "Begin coding" leads one to the screen presented in Figure 13.12.

On the lower left-hand side, the question and answer categories are presented. On the top left side, the first characteristic that we should code is presented. This is the domain of the question. The possible categories have been indicated. At the side, some information about this characteristic is indicated. If one selects a category, the choice is presented on the right side of the screen, and the next characteristics to be coded appear at the left side. This characteristic is coded in the same way, and this process goes on until all characteristics are coded.

---

[1] Except that in this case, one coder did not finish the task and therefore no prediction was generated (−99).

FIGURE 13.10 Screenshot of question A2 of round 1 in Ireland.

FIGURE 13.11 The screen with the question and the option to begin coding.

For some codes, you will need to refer to the questionnaire, given that in the SQP program, only the basic information is provided about the introduction, the request, and the answer categories. Information whether instructions were given, a show card was available, or a don't know answer was possible or registered are not provided in SQP.

FIGURE 13.12   The first screen of the coding procedure.

FIGURE 13.13   The screen after the coding has been completed.

Sometimes, the program makes a suggestion for a possible answer. For example, it suggests how many sentences and words there are in the questions. In that case, you can accept the suggestion by clicking on "next," or you can correct the number and click on "next" to go to the next characteristic. When the coding is done for all characteristics, the screen captured in Figure 13.13 appears.

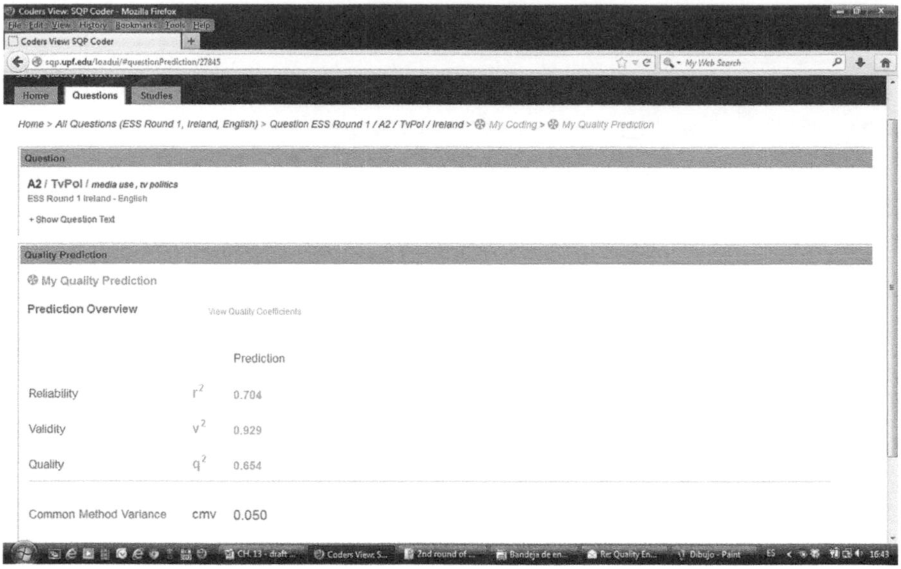

FIGURE 13.14   The quality prediction for question A2 in Ireland.

At this stage, you can ask for a prediction of the quality of the question by clicking on the text "Get Quality Prediction." If predictions are requested, the screen captured in Figure 13.14 will appear.

In this case, only the prediction of SQP is presented because no quality estimate was obtained for this question in an MTMM experiment. If one would like the predictions of the quality coefficients that are the square root of the quality predictions, one has to click on the button at the right saying "View quality coefficients." In that case, one also gets the prediction intervals.

It will be clear that by coding this question, the first screen presented in Figure 13.9 is no longer the same because question A2 has now also been coded. In order to follow this process, the user can repeat the coding for the same question or, even better, code another question that has not been coded so far.

13.3   PREDICTING THE QUALITY OF NEW QUESTIONS

For the third option of the program, we have to go back to the home page of SQP and select the option "Create a new question." If one chooses to introduce a new question, a screen appears that asks for information about the particular study and question. In this case, let us specify that we are doing a study called "immigration" in English. The name in the questionnaire is A1 and the concept is "equality." This information is also presented in the screen in Figure 13.15.

The next step is to introduce the question itself. In this case, we have chosen to introduce the question about the value of equal opportunities in the Schwartz

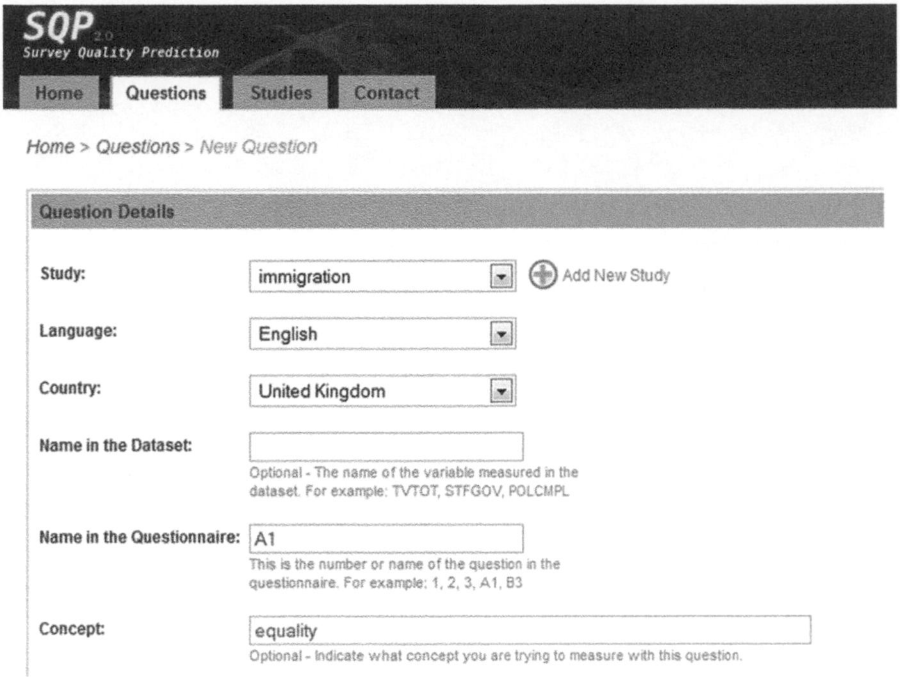

FIGURE 13.15   Input of the basic information about the question.

Human Value scale. This question has an introduction, a question with stimuli, and six response categories. Figure 13.16 is the continuation of the page captured in Figure 13.15.

After having saved the question, the screen presented in Figure 13.17 appears.

The next thing one has to do is code the question. This process begins with clicking on "Begin coding." The coding process was already described previously. If this process is finished, one should ask the program to perform the quality prediction. In this case, we obtained the result presented in Figure 13.18.

It will be clear that this question is not very good. The quality is .55, and so 45% of the variance in the observed scores is error. We can therefore ask the program for suggestions for improvements. In this case, the results after evaluation of all characteristics are the suggestions presented in Figure 13.19.

This analysis shows that several improvements can be made. We see that choosing another country would help. Of course, this is an impossible option. Possible alternatives are presented by the characteristics "avgwrd_total," "stimulus," "visual," etc. One should realize that this table gives the improvement for one question characteristic keeping all the other characteristics the same as they are. This means that by combining several of these characteristics, one may be able to improve the question even more. The program gives suggestions for this, but one would have to test the new version again. We cannot just add the different

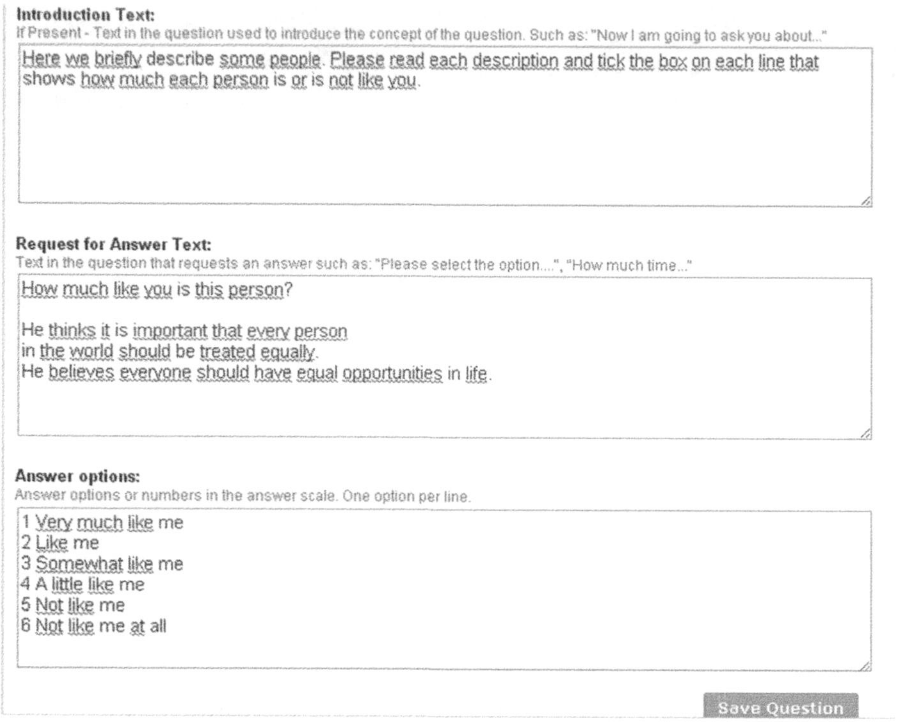

Introduction Text:
If Present - Text in the question used to introduce the concept of the question. Such as: "Now I am going to ask you about..."
Here we briefly describe some people. Please read each description and tick the box on each line that shows how much each person is or is not like you.
Request for Answer Text:
Text in the question that requests an answer such as: "Please select the option....", "How much time..."
How much like you is this person?
He thinks it is important that every person
in the world should be treated equally.
He believes everyone should have equal opportunities in life.
Answer options:
Answer options or numbers in the answer scale. One option per line.
1 Very much like me
2 Like me
3 Somewhat like me
4 A little like me
5 Not like me
6 Not like me at all
Save Question

FIGURE 13.16 The form in which to specify the question.

SQP
Survey Quality Prediction
Home
Questions
Studies
Home > All Questions (ESS Round 2, Ireland, English) > Question Immigration / equality / equality / United Kingdom > My Coding
Question
Question Coding
There are no codings yet for this question.
Begin Coding
Introduction Text:
Here we briefly describe some people. Please read each description and tick the box on each line that shows how much each person is or is not like you.
Request for Answer Text:
How much like you is this person? He thinks it is important that every person in the world should be treated equally. He believes everyone should have equal opportunities in life.
Answer options:
1 Very much like me
2 Like me
3 Somewhat like me
4 A little like me
5 Not like me
6 Not like me at all
Edit Question

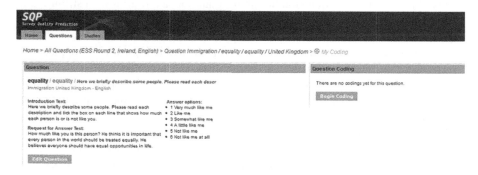

FIGURE 13.17 The question with the button "Begin coding."

improvements together. We have seen in Figure 13.19 that SQP suggests several improvements, especially with respect to the length of the text, the use of stimuli, the data collection, the concept, etc. Let us start with the latter issue. The question asked about the similarity of the respondent to the person described in the stimulus, while there was a mixture of two concepts presented in the stimulus: a value statement and a norm. This is what Saris and Gallhofer (2007a) have called a

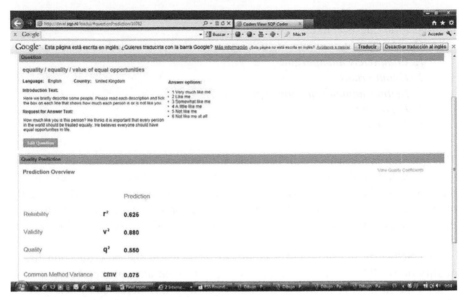

FIGURE 13.18   The prediction of SQP 2.0 of the quality of the question "equality."

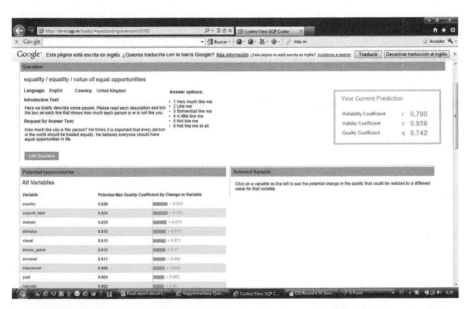

FIGURE 13.19   Several suggestions for improvement of the question "equality."

complex concept because the item asks a similarity about other concepts. Besides that, two different concepts have been combined in the stimulus. This could lead to a great deal of confusion for the respondent. We have also seen that the use of batteries of statements has a negative effect; therefore, following the suggestions of Saris and Gallhofer (2007b), we would advise to measure the value with an item-specific question like:

> *How important or unimportant is it for you that all people be treated equally?*
> 1. *Completely unimportant*
> 2. *Unimportant*
> 3. *Neither unimportant nor important*
> 4. *Important*
> 5. *Extremely important*

This question is much shorter, it has a bipolar item-specific scale, and no statement is used. Note that the change of some aspects of the request also changes a lot of other aspects. Therefore, it makes sense to evaluate once again how good the quality of this question is according to the SQP program. In order to check this, we introduce this new question in the program, code the question, and ask for the quality prediction. The result is presented in Figure 13.20.

The Schwartz question had a quality of .55, while the new question has a quality of .64. This would mean that the explained variance of the observed variable by the variable of interest—the value equality—has increased by nearly 10%. One can also look at further possible improvements, but the explained variance will never be perfect. This means that measurement errors will still remain. Therefore, correction for measurement error is also important as we will see in the next chapters.

Note that the improvement in quality was mainly obtained by the increase in the validity. This means that by using this formulation, the systematic effect of the

FIGURE 13.20   The quality of the reformulated question.

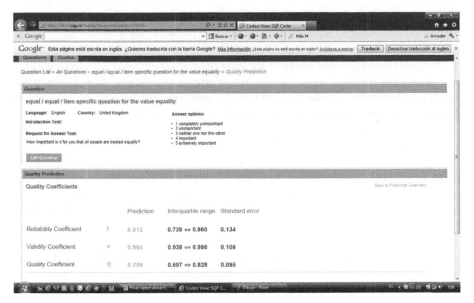

FIGURE 13.21   The quality coefficients, interquartile range, and standard error.

method, that is, the complement of the validity, has been reduced. This can also be seen in the reduction in the CMV, which is now rather small.

A more detailed picture of the quality can be obtained by clicking on the text at the right "View quality coefficients." If we do so, we arrive at the screen presented in Figure 13.21.

The quality coefficients are the square root of the quality indicators themselves. These are the coefficients that are estimated in the MTMM experiments. In this screenshot, we can see the uncertainty that exists in the estimates presented in the interquartile range and the standard error. It will be clear that a considerable range of uncertainty remains.

Nevertheless, the attractiveness of this approach lies in the fact that we can get these estimates before the data have even been collected. While the MTMM experiments are time consuming and expensive, the quality estimates using SQP 2.0 are obtained with minimal efforts and allow researchers to improve their data collection before they spend a lot of money on it. It is not possible to take into account 60 question characteristics while formulating a question. SQP makes it possible to evaluate the questions and suggests improvements. This is the major advantage of this procedure.

13.4   SUMMARY

In this chapter, we have shown that the SQP 2.0 program can be used to obtain (1) the quality estimates that were obtained by MTMM experiments, (2) the quality predictions of questions that are in our database but not part of an MTMM experiment, and (3) the quality predictions of new questions that a researcher would like to evaluate. We have also shown that the program provides suggestions for improving questions in a simple way.

The most important advantage of the new SQP 2.0 program above version 1.0 is that it provides predictions of the quality of questions in more than 22 countries based on a database of more than 3000 extra questions that have been evaluated in MTMM experiments to determine the quality of the questions.

Another very important advantage of the new program compared to the earlier version is that the earlier program had to be downloaded and used on one's own PC. The new version is an Internet program with a connected database of survey questions, which now contains all questions used in both the old and new experiments, as well as all questions asked so far in the ESS. This means that there are already more than 60,000 questions in all languages used in the ESS that appear in the SQP database. The number of questions will grow in three ways: first, by way of the new studies carried out by the ESS, which, in each round, adds another 280 questions across all languages used; second, through the new studies that are added to the database by other large-scale cross-national surveys; and, third, thanks to the introduction of new questions on the part of researchers, using the program to evaluate the quality of their questions.

Therefore, the SQP program is a continuously growing database of survey questions in most European languages with information about the quality of the questions and the possibility to evaluate the quality of the questions that have not been evaluated so far. In this way, the program will be a permanently growing source of information about survey questions and their quality. To date, there is no other program in the world that offers such possibilities.

EXERCISES

1. As we have said, the relationships between variables cannot be compared across countries if the quality of the questions is not comparable. Check with SQP whether relations of questions measuring political trust can be compared with each other. For which countries can comparisons be made, and for which countries can they not be compared?

2. Social trust is measured with three indicators:
   a. Check which indicator in your language is the best for social trust.
   b. Are the MTMM estimates and the predictions of SQP comparable?
   c. Is the best indicator good enough, or does this question need improvement?
   d. How could you improve this question?
   e. Specify the new question.
   f. Test the new question by introducing the question in SQP in the study called "Test" and evaluate if the new question is better than the ESS question.

14

THE QUALITY OF MEASURES FOR CONCEPTS-BY-POSTULATION

In this chapter, we pick up the discussion of the first chapter about concepts-by-postulation (CP) and concepts-by-intuition. This is important because often, the concepts people want to study are not so simple that they can be operationalized by concepts-by-intuition. Several concepts-by-intuition are also combined into one CP in order to obtain a measure of the concept of interest with better reliability and/or validity. To date, we have become familiar with the quality of measures for concepts-by-intuition. In this chapter, we want to show how this information can be used to say something about the quality of measures for CP. This is possible because a measure of a CP is an aggregate of several measures of concepts-by-intuition.

First, we will introduce the possible structures of CP. The logic is that the measures of CP are based on concepts-by-intuition. In order to determine the score of the CP, the relationships between the concepts-by-intuition and the CP need to be known. In some cases, one can control whether the expected relationships indeed exist. Depending on the hypothesized relationships, different tests for the structure of the measures are performed. In this chapter, we will discuss these different structures and indicate how the quality of the measures for the CP can be determined on the basis of the estimated quality of the measures for the concepts-by-intuition.

Design, Evaluation, and Analysis of Questionnaires for Survey Research, Second Edition.
Willem E. Saris and Irmtraud N. Gallhofer.
© 2014 John Wiley & Sons, Inc. Published 2014 by John Wiley & Sons, Inc.

14.1 THE STRUCTURES OF CONCEPTS-BY-POSTULATION

The structures of CP and their tests have a strong research tradition. In fact, the entire depth of this topic is beyond the scope of this chapter. For more elaborate discussions of this topic, we refer the reader to the following authors: Bollen (1989), Cronbach (1951), Guttman (1954), Hambleton and Swaminathan (1985), Messick (1989), and Nunnally and Bernstein (1994). We will proceed with the two most commonly used structures.

The first common structure type assumes that the CP is the variable that causes the correlations between the measures of the concepts-by-intuition. This is, for example, the basic model for one of the definitions for attitude. Fishbein and Ajzen (1975) suggested that an attitude is a learned predisposition, where one consistently reacts in a positive or negative manner to an object. Therefore, a *factor analysis* or a more general *latent variable model* should apply, where the indicators are called *reflective* because they reflect the score of the latent variable.

The second most common structure is that the CP is an aggregate of several measures of different concepts-by-intuition. For example, the concept socioeconomic status (SES) is defined as a resultant of income, education, and occupational status. Since these indicators determine the definition of the concept, they are called *formative*. The initial formulation of this type of measurement model can be traced back to Blalock (1964). Other relevant sources are Bollen and Lennox (1991) and Edwards and Bagozzi (2000).

Many other measurement models have been developed for dichotomous or ordinal responses (see Guttman 1954; Rasch 1960; Mokken 1971). For a more general discussion of item response theory or IRT models, we can refer the reader to Hambleton and Swaminathan (1985). Other scales are developed for preference judgments, as, for example, the unfolding scale (Coombs 1964; Van Schuur 1988; Münnich 2004). These different scales do not fit very well the context of the models discussed here. So for these measurement models, we recommend the literature. In the following text, we will concentrate on the first two models.

14.2 THE QUALITY OF MEASURES OF CONCEPTS-BY-POSTULATION WITH REFLECTIVE INDICATORS

If the CP is defined as a causal latent variable that affects a number of observable measures of concepts-by-intuition, then the latent variable models can describe the structure of the CP. This is a common assumption on which our example of the definition for attitude by Fishbein and Ajzen (1975) mentioned earlier rests. The frequently used concept of "political efficacy" has also been operationalized using this assumption. In the pilot study of the ESS, the set of five questions presented in Table 14.1 is used to measure this concept.

Although some researchers speak of "political efficacy" as if it were one concept, most researchers assume that there are two CP behind these items. The first three items are supposed to measure "subjective competence" or "internal efficacy," while the last two are supposed to measure "perceived system responsiveness" or "external efficacy" (Thomassen 2002).

TABLE 14.1   Survey items for "political efficacy" in the first wave of the ESS *Card C1:*
Using this card, how much do you agree or disagree with each of the following
statements? Firstly…READ OUT.

	Agree strongly	Agree	Neither agree nor disagree	Disagree	Disagree strongly	(Don't know)	
"Sometimes, politics and government seem so complicated that I connot really understand what is going on."	1	2	3	4	5	8	357
"I think I can take an active role in a group that is focused on political issues."	1	2	3	4	5	8	358
"I understand and judge important political questions very well."	1	2	3	4	5	8	359
"Politicians do not care much about what people like me think."	1	2	3	4	5	8	360
"Politicians are only interested in people's votes but not in their opinions."	1	2	3	4	5	8	361

It is assumed in this case that the more "subjective competence" people have, the more likely it is that they will score lower on the first and higher on the second and third items. Therefore, it is also assumed that "subjective competence" explains the correlations between these three items. For the last two items, it is believed that people who think that the political system (via the politicians) facilitates them to influence it will score higher on them. Also, the correlation between the last two items is assumed to be explained by a general opinion about the responsiveness of the system. It is not clear if it can be expected that people with a higher "subjective competence" score also perceive more "system responsiveness." If that were the case, there would be a strong indication that these two CP are correlated, which is an important consideration for further analysis. Therefore, it is necessary to test the factor model before further computations of the quality of the measures are made. Consequently, we will proceed in the following manner. We will first test this type of model; following this, we will discuss the way the measure for the CP is estimated and conclude with an estimation of the quality of the measures.

14.2.1   Testing the Models

Figure 14.1 represents the model mentioned previously, where the unknown correlation (?) between the latent traits that represent the two CP ("subjective competence" and "perceived system responsiveness") is ready for testing. In this figure,

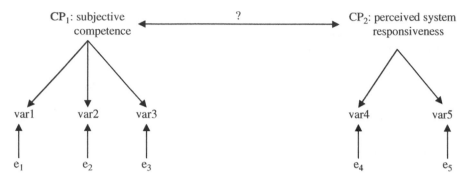

FIGURE 14.1 The two-factor model for political efficacy as suggested in the literature.

var1...var5 represent the responses to the requests 1–5, while $e_1...e_5$ represent random errors contained in these responses. Arrows represent direction of influence.

On one hand, if it is assumed that the two latent traits correlate perfectly (with correlation equal to 1), the model reduces to a one-factor model, and then the factor can be called "political efficacy," and it does not make sense to speak about two separate concepts. On the other hand, if the correlation between the latent traits is 0, it means that knowing something about a respondent's "subjective competence" does not indicate anything about his/her "perception of the system responsiveness." Both possibilities are theoretically acceptable; however, they lead to quite different measurement instruments. There is also the third possibility that there is a correlation between 1 and 0.

In the Dutch pilot study of the ESS, the correlations presented in Table 14.2 have been obtained for the five variables from Figure 14.1. The first three variables, which supposedly measure "subjective competence," correlate higher with each other than with the last two variables, which, in turn, correlate quite strongly with each other. However, there are also correlations different from 0 between the two sets of variables. Using the procedures discussed in Chapter 10, different models can be estimated and tested. The model assuming that there is no correlation between the latent variables is rejected because this model cannot explain the correlations between the two sets of variables.

Also, the model assuming that there is only one factor, "political efficacy," behind these observed variables is rejected because of the size of the residuals. However, the model allowing for the correlation between these two latent variables fits the data; this model is presented in Figure 14.2.

Although the model in Figure 14.2 with the estimated values of the parameters neatly reproduces the correlation matrix, this does not mean that it necessarily is the correct model. In fact, after studying the errors in the data and finding that the batteries of agree/disagree (A/D) items can produce quite large random and systematic measurement errors due to the method used, we think that the model is not correct. The systematic method effects also explain the correlations between these two sets of variables.

TABLE 14.2   Correlations between the five "political efficacy" variables in the Dutch ESS pilot study with a sample size of 230

	var1	var2	var3	var4	var5
var1	1.00				
var2	−0.33	1.00			
var3	−0.52	0.43	1.00		
var4	0.16	−0.13	−0.13	1.00	
var5	0.12	−0.18	−0.13	0.68	1.00

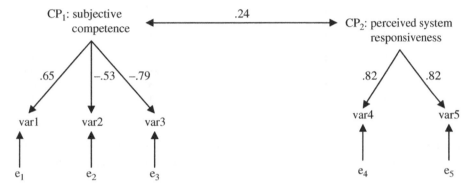

FIGURE 14.2   The two-factor model for political efficacy as estimated on the basis of data from Table 14.1 assuming correlation between the two factors.

The alternative model for our data taking into account reliability, validity coefficients, and method effects is presented in Figure 14.3. In this model, the lower part of the figure presents the result obtained by the MTMM experiment done in the Dutch pilot study of the ESS using three different methods. For our present purpose, only the results for the A/D method are included. This part of the model is consistent with the model specified in Chapter 9, where a distinction was made between the observed variables (var), the true scores (T), the variables of interest (F), and the "A/D method factor." New to this model is that above the variables of interest, another level of variables appears that represents the CP. Until now, our analysis stopped at the level of the variables of interest, which represented concepts-by-intuition. Now, we go further by looking at the CP that explains the correlations between the different concepts-by-intuition (corrected for measurement error). We assume that each of these observed variables has a unique component "u." This would not be the case if all the items only measured the same variable. Given that we know the reliability, validity, and method effects from earlier studies, the coefficients for the relationships between the highest-order latent variables (representing the CP) and the latent variables "F" (representing the concepts-by-intuition) can also be estimated.

If these effects are estimated, a very good fit is obtained where the effect of "subjective competence" on the lower level latent variables is respectively − .66 for

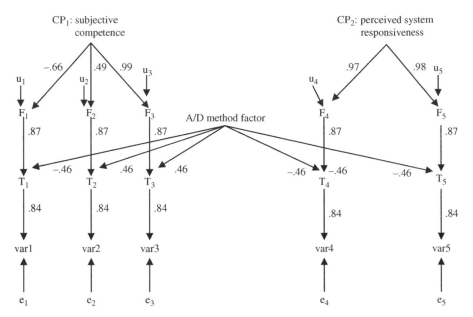

FIGURE 14.3 An alternative factor model for political efficacy combining the two-factor model with information about the quality of the measurement of the concepts-by-intuition.

F_1, .49 for F_2, and .99 for F_3, while the effect of "perceived system responsiveness" is .97 for F_4 and .98 for F_5. In this model, it turns out that the correlation between the two highest latent variables is not significantly different from 0 (.04), and therefore, it is reasonable to assume that the two latent variables vary independently from each other.

This model demonstrates that a different explanation for the correlations between the two sets of variables is possible. This is because in this particular model it is assumed that a part of the correlation between the observed variables is spurious, due to a factor that has no substantive meaning (in this case a method factor). In the model of Figure 14.2, this was not assumed, and the correlation between the latent variables was seen as substantive correlation. It is important to note that this difference between the models causes the estimates of the quality of the instruments to differ. Therefore, it is absolutely necessary to test the model and to ensure that it is correct before starting to estimate the quality of the measures.

14.2.2 Estimation of the Composite Scores

Only after testing the latent variable model is it advisable to move to the next phase of constructing the measure of the CP. Equation 14.1 demonstrates a possibility for estimating the scores of the respondents on the latent variables by using the weighted average of observed variables (S) as a measure for the CP (CP_i):

$$S = \sum_{i=1}^{k} w_i \, \text{vari} \tag{14.1}$$

In this equation, w_i is the weight for the ith observed variable. However, the question of choosing the weights is still left up to the researcher. Most of the time, this problem is avoided by choosing the value 1 for the weights, which leads to an unweighted sum score. However, this approach is rather inefficient if the different variables differ in quality. Therefore, procedures have been developed to estimate weights that are optimal for some specific applications. Well-known criteria are as follows:

1. The sum of the squared differences in scores between the variable of interest and the sum score should be minimal. The weights derived using this criterion are known as *regression weights*. They are the most appropriate when trying to obtain scores for individual persons.
2. The sum score should be an unbiased estimate of the variable of interest and should satisfy criterion 1. The weights derived using these two criteria are known as the *Bartlett weights*. This would be the best procedure for comparison of means for different groups.
3. The relationships between the sum scores should be the same as the relationships between the different factors, and criterion 1 needs to be satisfied. The weights derived using these two criteria are known as the *Anderson and Rubin weights*. This is the preferred method when attempting to estimate the relationships between the latent variables in more detail.[1]

For further reading, we recommend Lawley and Maxwell (1971) and Saris et al. (1978). In general, it can be observed that the different methods generate only slightly different results if the different observed variables are approximately equally good. However, it will be shown later that unequal weights have important advantages if this condition is not satisfied.

If the observed variables are expressed in deviation or standardized scores and, therefore, have a mean of 0, then the sum score will also have a mean of 0. The variance of the sum can be calculated as follows:

$$\text{var}(S) = \sum_{i=1}^{k} w_i^2 \, \text{var}(\text{vari}) + 2 \sum_{i,j} w_i w_j \, \text{cov}(\text{vari},\text{varj}) \tag{14.2}$$

Here, "var(S)" stands for the variance of "S," while "cov(vari,varj)" is the covariance between the variables "i" and "j." By taking the square root of the var(S), the standard deviation of S can be obtained denoted by "sd(S)."

---

[1] In the next chapter, we will show that this approach is not necessary because it is possible to directly estimate the relationships between the latent variables.

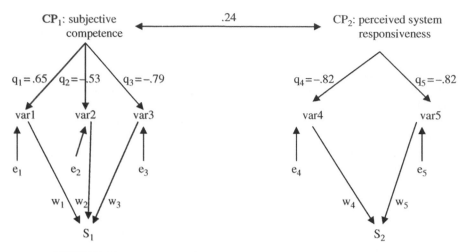

FIGURE 14.4   The model for evaluation of the quality of the sum scores.

14.2.3   The Quality of Measures for Concepts-by-Postulation

In the previous chapters, we have discussed the quality of only single items as indicators for concepts-by-intuition. Now, we want to introduce the quality of the sum scores of concepts-by-intuition for the CP. Until now, quality was always defined as the correlation between the theoretical variable of interest and the observed variable. This definition still holds true for our current case. We will also define method effects as the complement of validity and random error variance as the complement of unreliability.

If we combine the specification for generating a sum score with the knowledge we have about the model describing the relationship between the latent variables (which represent the CP and the observed variables), we get, for the model presented in Figure 14.2, the results presented in Figure 14.4.

In this figure, we see the relationships between the latent variables representing the CP and the sum scores (S) for these variables. The *quality of these sum scores as indicators for these latent concepts can be expressed in the correlation squared between these two variables.*

In Chapter 9, we mentioned that the correlation between two standardized variables is the sum of direct and indirect effects, spurious relationships, and joint effects. In this case, there are only indirect effects, and therefore, it follows that the correlation is equal to the sum on the indirect effects of "subjective competence" (CP_1) and S_1 and between "perceived system responsiveness" (CP_2) and S_2. This leads to the following general result for k observed variables[2]:

---

[2] Here, it is assumed that all variables are standardized as is done in the whole text except the new variable S_i. For a more general formulation, we mention Bollen (1989: 209–222).

$$\rho(CP_1, S_1) = \sum_{i=1}^{k} \frac{q_i w_i}{sd(S_1)} = \left\{ \frac{1}{sd(S_1)} \right\} \sum_{i=1}^{k} q_i w_i \qquad (14.3)$$

The regression weights that minimize the sum of squared differences between the CP and the sum score are computed: for var1, it is .29; for var2, it is .19; and for var3, it is .55, while the weight for var4 and var5 turned out to be negligible. Using these weights, the variance of the sum is calculated at .715, and the standard deviation is .845. Therefore, the correlation between CP_1 and S_1 becomes

$$\rho(CP_1, S_1) = \left\{ \frac{1}{.845} \right\} [(.29 \times .65) + (-.19 \times -.53) + (-.55 \times -.79)] = .857$$

The strength of the relationship between this weighted sum score and CP_1 is this correlation squared, which equals .73.

In practice, often the unweighted sum of the observed variables is often used. In that case, the weights are all equal to $1/sd(S)$. Hence, formula 14.3 can be simplified to[3]

$$\rho(CP_1, S_1) = \sum_{i=1}^{k} \frac{q_i}{sd(S_1)} = \left\{ \frac{1}{sd(S_1)} \right\} \sum_{i=1}^{k} q_i \qquad (14.4)$$

If the unweighted sum[4] of the observed variables var1 through var3 is used, then the variance of this sum is 7.0, and the standard deviation $sd(S_1)$ is equal to 2.646. From this, it follows that

$$\rho(CP_1, S_1) = \left\{ \frac{1}{2.646} \right\} [(1 \times .65) + (-1 \times -.53) + (-1 \times -.79)] = .745$$

This means that the strength of the relationship between "subjective competence" (CP_1) and S_1 is

$$\rho(CP_1, S_1)^2 = .56$$

This correlation indicating the quality of the unweighted sum score is considerably lower than the quality for the weighted sum score, meaning that in this case the weighted sum is considerably better than the unweighted sum score.

In fact, it can be proved that the regression method provides weights that are optimal in the sense that it produces the highest possible correlation between the factor and the sum score (Lawley and Maxwell 1971). *It produces the sum score with the best quality.* The major advantage of the regression method is that the quality of the sum score can never be lower than the quality of the best indicator.

---

[3] See also Heise and Bohrnstedt (1970) and Raykov (1997, 2001).
[4] In this calculation, a correction for the direction of the scale is necessary. Therefore, the second and third weights are −1.

However, we have demonstrated with this example that this is not necessarily true for the unweighted procedure because variable 3 has better quality than the unweighted sum score.

In the literature, people often use the so-called Cronbach's α as a measure of the quality of the sum score. This quality index is calculated in many ways that are equal if the unweighted sum is used and all indicators have equal quality (q^2). Under this assumption, it follows from Equation 14.4 that

$$\rho(CP_1,S_1)^2 = \alpha = \left\{\frac{1}{var(S)}\right\} k^2 q^2 \qquad (14.5)$$

If all loadings are equal to q, then the correlations between all observed variables are q^2; therefore, it could also be said (see Bollen 1989) that

$$\rho(CP_1,S_1) = \alpha = \left\{\frac{1}{var(S)}\right\} k^2 r \qquad (14.6)$$

Since the correlations are normally not exactly equal, it is customary to take the mean of the correlations as an estimate for r. From our example, we derive an estimate of the quality of the sum score of .56, which is much lower than the estimated reliability for the weighted sum using formula 14.3. It is known from the literature (Raykov 1997) that the Cronbach's α is only the lower bound of the reliability. Only if the indicators satisfy the condition that they are all equally good indicators for the CP of interest and the unweighted sum score is used is this estimate equal to the more general estimate presented in Equation 14.3. We have seen previously that this assumption is not necessarily true and that the quality of the sum scores can easily be calculated with the latter formula.

So far, we have been talking about the estimation of the quality of the sum score as an indicator for the CP; now, we would like to illustrate the consequences of the selected model on the quality of the sum score. We will show what happens if we follow the same procedure with the model of Figure 14.2 as we did with the model of Figure 14.3. In order to proceed, we simplify the model. In Figure 14.3, we showed that the effect of CP_1 and the method factor (A/D) on each observed variable is indirect. It can also be proved that the sizes of these indirect effects are equal to the products of the coefficients along the path from the causal variable to the effect variable (see Chapter 9). This leads to Figure 14.5.

This model differs quite a bit from the model in Figure 14.2 because in this case there is no correlation between the variables CP_1 and CP_2. The correlations between the observed variables are explained by the systematic effect of the method factor. We also see that by introducing this factor all measures for the quality of the different variables are lower than in Figure 14.2. These differences also affect the evaluation of the quality of the sum scores based on the observed variables. In Figure 14.6, we introduce the model for the estimation of the quality of the sum scores. The estimation of the quality of the sum score S_1 for CP_1 can be calculated by formula 14.3; however, the fundamental difference

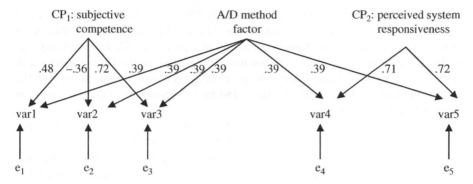

FIGURE 14.5   The simplified model derived from Figure 14.3.

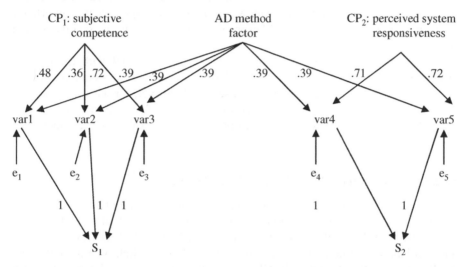

FIGURE 14.6   Model for calculation of the quality of the sum scores derived for the model in Figure 14.5.

is that a second factor influences the sum score S_1, which is the A/D method factor. The effect of this A/D method factor can be computed with the same formula but now substituting the quality coefficients (q_i) by the method effect coefficients (m_i):

$$\rho(\mathbf{AD}, S_1) = \sum_{i=1}^{k} \frac{m_i w_i}{sd(S_1)} = \left\{ \frac{1}{sd(S_1)} \right\} \sum_{i=1}^{k} m_i w_i \qquad (14.7)$$

This correlation also indicates the invalidity coefficient caused by the method in the sum score. If the unweighted procedure for estimation of the sum scores is used, the results of the estimation are for the quality of the sum score .35 and for invalidity .20. These results indicate that quite a large part of the systematic variance in the sum score S_1 is due to method effect (20%) and only 35% is due to the variable it should represent (CP_1), while 45% is random error. These values differ significantly from the estimates derived on the basis of the model in Figure 14.2. There, the quality was higher (58.3%) and invalidity was 0, because there was no method effect assumed. In principle, the error variance of the sum score should remain the same along with the sum of the valid and the invalid variance, but minor deviations in the estimates can cause small differences.

The regression method improves the outcome: CP_1 explains 53.5% of the sum score and the method explains only 5.3% while the random error is 41.2%.

This example has illustrated how dependent estimates are on the specification of the model and on how the sum scores are computed. Our example showed that it is always safer to use one of the weighted procedures because they will give a sum score with better quality. Also, our examples clearly illustrated the effect that the model has on the estimation. Again we would like to emphasize the importance of testing the measurement model before starting to evaluate the quality of sum scores.

After demonstrating the superiority of the second model with the method effects, we can state that the sum scores derived from the first model are biased. They overestimate the quality of the sum scores. However, for the second model, we have to conclude that a sum score of a quality of .535 is not sufficiently high. Therefore, we think that indicators for the CP "subjective competence" need to be improved.

14.2.4   Improvement of the Quality of the Measure

There are three reasons for the lack of quality of the sum score of "subjective competence" that we were discussing in Section 14.2.3: (1) the lack of reliability of the three items; (2) the invalidity due to the method effect; and (3) at least two of the three variables have large unique components, making the link with the CP rather small (respectively, $.66^2$ and $.49^2$). All three points can be improved. Suggestions for the improvement of the reliability and validity can be obtained from analysis of the individual survey items using SQP. The program suggested that the results would be much better if the first three items would be reformulated into requests with trait-specific scales as shown in Table 14.3.

In this case, the resulting reliability and validity coefficients for the "subjective competence" questions are presented in Figure 14.7.

Applying the same simplification method as before, we get the result presented in Figure 14.8, where the derived effects are equal to the indirect effects of the model in Figure 14.7.

Comparing this result with the result presented in Figure 14.5, we see that the strength of the relationships between CP_1 and all observed variables is stronger than in the A/D format. This improvement is due to higher reliability and validity derived from the changes we made. Given the stronger relationships, the sum score of these variables should also be a better indicator for "subjective competence." In this case, the regression weights are estimated at .55 for var1, .16 for var2, and .38 for var3.

TABLE 14.3   The question format for "subjective competence" used in the first wave of the ESS

Var1: *How often seem politics and government so complicated that you cannot really understand what is going on?*

Never	*1*
Seldom	*2*
Occasionally	*3*
Regularly	*4*
Frequently	*5*
(Don't know)	*8*

Var2: *Do you think that you could take an active role in a group that is focused on a political issue?*

Definitely not	*1*
Probably not	*2*
Not sure either way	*3*
Probably	*4*
Definitely	*5*
(Don't know)	*8*

Var3: *How good are you at understanding and judging political questions?*

Very bad	*1*
Bad	*2*
Neither good nor bad	*3*
Good	*4*
Very good	*5*
(Don't know)	*8*

The variance of the sum score derived, including the weights, is .832 and the $sd(S) = .911$. While applying the procedure presented in Equation 14.3, we see that the correlation coefficient between the CP and the sum score is .912, the quality of the sum score as an index for "subjective competence" is .83, while the method effect turns out to be .08. Our results indicate little allowance for random errors. It seems that this new sum score contains minor invalidity and only minimal random errors (9%) and that the quality of the measure is high with an 83% explained variance of the sum score by the factor of interest. This is because the IS format generates a higher data quality than does the A/D format. Our example also illustrates how the SQP predictions can be used to indicate the direction in which the quality of the measures for the concepts-by-intuition can be improved, thereby also improving the measures for the CP.

In addition, the aforementioned illustration shows that the unweighted procedure for estimating the sum scores should not be automatically used. It can lead to a significant decrease in the quality of the measure of the CP compared with the measure obtained by the regression method. We have also shown that the Cronbach's α is not always the best index by which to estimate the quality of a sum score, certainly not if using weighted sums. We recommend the easy-to-use alternative discussed in this section, which results in better estimates.

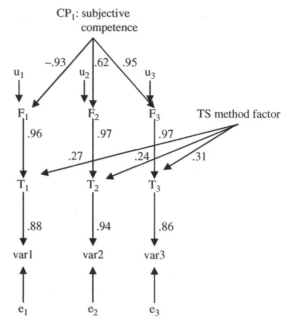

FIGURE 14.7   The alternative factor model for "subjective competence" with information about measurement error.

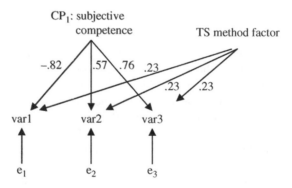

FIGURE 14.8   The simplified model of Figure 14.7.

14.3   THE QUALITY OF MEASURES FOR CONCEPTS-BY-POSTULATION WITH FORMATIVE INDICATORS

A different model should be used if the variable of interest is the effect of several other variables such as when indicators, known as *formative*, determine or define the CP. The example in our introduction was the relationship between SES and the causal variables income, education, and occupation. The fundamental

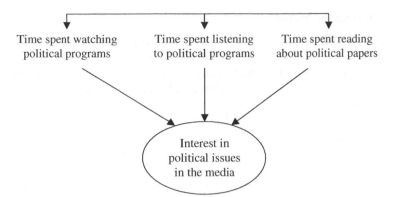

FIGURE 14.9   The effect variable as CP.

difference between the concept SES and the previous type of concept is that the observed variables are now the causal variables and not the effect variables. This has very different consequences. For example, in the previous section, the model indicates that the correlations between the observed variables are spurious relationships, due to the unobserved causal variable. In the present case, the unobserved variable has no effect on the observed variables. Whether there are correlations between the observed variables is not explained by the variable of interest. The model suggests only that there is an effect of each of the observed variables on the unobserved effect variable. Let us give two other examples of CP with formative indicators.

Our first example, in Figure 14.9, is the measurement of "interest in political issues in the media" that is based of the time spent on politics in the media. Time spent watching TV programs or radio broadcastings or reading articles in the newspapers is observable. The "interest in political issues in the media" can be operationalized as the total time spent on political issues in the media and can logically be the sum of time spent on these three observable variables.

A second example is the measure of "social contacts" that is a key variable for research related to social capital and its effects. The measure includes "informal contacts" and "formal contacts," and an obvious measure for the CP is the sum of these two observable variables. In this case, the causal structure for this concept is the same as indicated in Figure 14.9. Typical for both examples is that the observed causal variables do not have to correlate with each other. TV watching and reading newspapers can be done either in isolation or in combination.

A consequence of this model is, first, that it is difficult to test, because the effect variable is not measured. Second, the weights of the different variables are left to the arbitrary choice of the researcher. A third issue is that the quality of the measure for the sum score as the correlation between the latent variable and the sum score cannot be determined. Therefore, different approaches have to be specified.

In the following section, the solutions to the issues we raised will be discussed in the same sequence as was done for the CP with reflective indicators.

14.3.1   Testing the Models

A solution for testing this type of model and determining the weights can be to add extra variables to the models that are a consequence of the CP. In this way, it becomes clear whether the effect really comes from the CP or the concepts-by-intuition. Also, it becomes possible to estimate the effects of the different components on the CP.

We will illustrate this procedure for our two examples. For the concept "social contacts," we add an effect on a latent variable "happiness" that has been measured by two variables: a direct question concerning "satisfaction" and a direct question concerning "happiness." It has been mentioned in the literature that socially active people are happier than socially inactive people. The theory does not state that this is due more to informal or formal interaction. Therefore, we assume that it is a consequence of the contacts in general.

For the measurement of "interest in political issues in the media," we can add the effect of "political interest in general" that will be operationalized by a direct question about political interest and by a measure of "knowledge of politics."[5] There is no doubt that these two variables are caused by the variable "political interest in general" and not by the different variables measuring "time spent on political issues in the media." Hence, we do not expect direct effects of these observed variables on the "political interest" indicators.

Taking into account that there is a difference between the concepts-by-intuition and the observed variables due to measurement error, we have created models for our two examples in Figure 14.10 and Figure 14.11. These figures indicate the information that has been collected previously with respect to the quality of the requests. The information came from another source because the quality of the measures cannot be estimated by this type of model. The two sources for this information that have been discussed are the SQP program and the MTMM experiments (which both can estimate the quality of single items). We know that the contact variables were asked only once in the first round of the ESS; therefore, the quality was estimated by the SQP program. As it turns out, the quality coefficients are relatively good: .79 for informal contact and .68 for formal contacts. Since the measures are so different from each other, no correlation due to method effects is expected. Given the quality coefficients, the error variances can also be calculated as $1 - .79^2 = .38$ for "informal contact" and $1 - .68^2 = .54$ for "formal contact."

The measures about "time spent on programs in the media" were included in an MTMM experiment.[6] It turned out that the quality coefficients are .52 for TV, .73 for radio, and .48 for newspapers. Because these items had the same format and were

---

[5] This measure is based on the number of times the respondents answer "don't know" on questions concerning political issues in the ESS. Direct questions about political knowledge were not asked in the first round of the ESS.

[6] The MTMM experiments were conducted with the general questions about "media use" in the pilot study of the first round of the ESS. We assumed that the results for the questions about "political issues" will have the same quality characteristics since they share the same format.

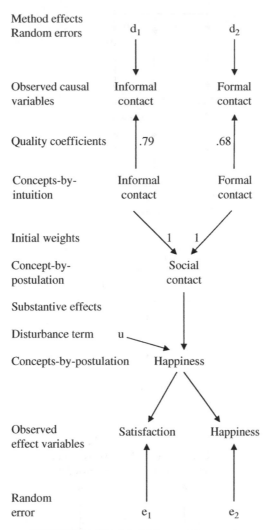

FIGURE 14.10   Model for social contact.

presented in a battery, a method effect of .09 was also found. Therefore, given these low-quality coefficients, large error variances were found: .73 for TV, .47 for radio, and .77 for newspapers. In our models, the method effect is included as the correlated errors between the measurement error variables. This is a possible approach if the method factor itself is not specified.

These models are very different from the models for CP with reflective indicators. Moreover, the measurement approach with formative indicators is more common in research than one would think. For example, in Likert scales, different items are introduced to measure aspects or dimensions of a CP, and therefore, there is no reason to expect correlations between the separate items (although such

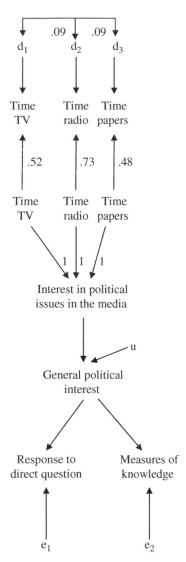

FIGURE 14.11   Model for interest in political issues in the media.

correlations cannot be ruled out). The quality of this type of model cannot be evaluated in the same way as we have elaborated in the previous section, but the models specified for the two examples in Figure 14.10 and Figure 14.11 can be used.

These two models have been estimated from round 1 data of the ESS. The LISREL input for the analysis is complex and available in Appendices 14.1 and 14.2. We will turn our attention to whether our analysis determines if the CP are plausible. If the analysis shows that effects have to be introduced from the observed causal variables directly on the effect variables, it suggests that the CP are not needed. Then the separate variables should be worked with as concepts-by-intuitions that have direct effects on other variables. On the other hand, if the effects are not needed, then the CP are plausible because all effects go through them to other effect variables.

In our specific examples, no direct effects were needed. It is a very convincing result because in both cases the effective sample size was 1500 cases (ESS 2002) making the power of such tests very high, meaning that even small effects would already lead to strong indications of misspecifications in the models and therefore to rejection of the models. Therefore, we can conclude that in our examples the CP play the role that has been specified for them in their respective models.

14.3.2 Estimation of the Composite Score

In the earlier analyses, the weights for the composite score were chosen to be equal to 1, making the composite score a simple sum of the different concepts-by-intuition. However, this is not necessarily the most accurate method. For example, it may be that "informal contacts" contribute more to the "happiness" of a person than do "formal contacts" or vice versa. The same is true for the media attention. It may be that reading about political issues in the newspaper is a much better indicator of interest in politics than passively listening to radio or TV news.

Structural equation model (SEM) programs show if the weights should be different from 1 in the expected parameter change (EPC) indices. These indices indicate the extent to which fixed coefficients will change if they are freely estimated. If these changes are substantively relevant, it would be wise to consider it.

In both our previous examples, the EPCs for the weights were substantial. For the "social contact" variables, the program suggested that the "informal contact" weighting would decrease by .87. After asking the program to freely estimate the coefficients, the weights became .14 for "informal contacts" and .94 for "formal contacts." These differences are sufficiently large to be considered as substantively relevant. The results indicate that the social contact variable, with a much higher weight for "formal" than for "informal contacts," is a better causal variable for happiness than a "social contact" variable with equal weights.

For the CP "interest in political issues in the media," we see a similar phenomenon. Allowing a greater weighting for "reading about political issues in the newspapers" than for "radio" and "TV," the composite score better predicts "general political interest" than does a model with equal weights. The weights turned out to be .31 for TV, .1 for radio, and .8 for newspapers. Here, the differences in prediction quality also are substantively relevant.

Our analysis suggests that unequal weights should be used to estimate the scores of the CP in both of our examples. The formula is the same as for calculating the concepts with reflective indicators (Equation 14.1). However, in the following text, we will demonstrate that the evaluation of the quality of the composite scores is quite different in this particular case.

14.3.3 The Estimation of the Quality of the Composite Scores

So far, we have evaluated measurement instruments by estimating the squared correlation between the observed variable and the latent variable of interest. There is,
however, another equivalent way to evaluate measurement instruments. If the latent
variable is called "F" and the observed variable "x" and the error variable "e," it has
been shown by several authors (Bollen 1989) that

$$\text{Quality of } x = \rho_{Fx}^2 = \frac{\text{var}(F)}{\text{var}(x)} = 1 - \frac{\text{var}(e)}{\text{var}(x)} \tag{14.8}$$

In this situation, we cannot use the squared correlation as a measure for the quality
of the composite scores for the CP with formative indicators, but we can use the last
form. The quality of the sum score S can thus be defined as[7]

$$\text{Quality of } S = 1 - \left[\frac{\text{var}(e_S)}{\text{var}(S)} \right] \tag{14.9}$$

where var(e_S) is the variance of the errors in S and var(S) is the variance of the
sum score S. The variance of the composite score can be obtained directly after
calculating the composite score by asking for the variance of it.

If for the different observed variables the weights (w), the error variances (var(e_i)),
and covariances (cov($e_i e_j$)) are known, we can estimate the error variance of the
composite score "var(e_S)" as follows:

$$\text{Var}(e_S) = \sum w_i^2 \, \text{var}(e_i) + 2 \sum w_i w_j \, \text{cov}(e_i e_j) \tag{14.10}$$

Formula 14.10 can be simplified to the first term if the error terms are not correlated
(no method effects) and further reduced to the sum of the error variances if all weights
for the components are equal to 1.

For the concept "interest in political issues in the media," we employ the complex
formula because the error terms are correlated. On the other hand, for the concept
"social contact," the second term can be ignored because the correlated error terms
are equal to 0.

The results presented in the last two sections indicate that the variance of the
errors for the concept "interest in political issues in the media" is

$$\text{Var}(e_S) = .31^2 \times .7 + .1^2 \times .42 + .81^2 \times .75 + .31 \times .1 \times .09 + .31 \times .81 \times .09$$
$$+ .1 \times .81 \times .09 = .60$$

The weights were estimated in such a way that the variance of the composite score is
equal to 1. Hence, the quality of the composite score as an indicator for the concept
"interest in political issues in the media" is

---

[7] This is true if it can be assumed that the CP is exactly defined as a weighted sum of the concepts-
by-intuition. If that is not the case and a disturbance term is specified, the result becomes more
complex (Bollen and Lennox 1991).

$$\text{Quality} = 1 - (.6/1) = .4$$

It will be clear that a quality score of .4 is not a very good result.

For the concept "social contact," the calculation simplifies because the correlations between the errors are 0, and we have to evaluate only the first term:

$$\text{Var}(e_S) = .14^2 \times .384 + .92^2 \times .535 = .46$$

The weights were estimated in such a way that the composite score had a variance of 1, and the quality for this concept resulted as follows:

$$\text{Quality} = 1 - (.46/1) = .54$$

This composite score of .54 is of better quality than the previous one (.4), but it still is not very good. Both examples indicate that composite scores, as measures for CP, can have considerable errors that should not be ignored. For both examples, we recommend that researchers consider improving these measures before moving on with substantive research.

14.4 SUMMARY

This chapter showed that there are several different models for representing the relationships between measures for concepts-by-intuition and CP. In fact, the definition is the model. The testing of such models is highly recommended. It is simpler if the model is a factor model. It becomes more difficult if the CP is the effect of a set of measures for concepts-by-intuition. In this chapter, we have shown how these tests can be performed.

Since the CP are defined as a function of the measures of the concepts-by-intuition, the quality of the composite scores can be derived directly from the information about the quality of the measures for the concepts-by-intuition. Therefore, evaluating the quality of concepts-by-intuition is very important, and we have focused on this issue in this book.

We have also demonstrated that the composite scores (as measures of CP) can contain considerable errors that can cause further substantive analysis to be biased. Therefore, the next chapter will show how to take these errors into account during the substantive analysis. In this context, calculating the composite scores is highly advisable because we have seen that the models can become rather complex if substantive and measurement models need to be combined. Using composite scores simplifies the models. However, this should not be an excuse to ignore the measurement errors in the composite scores because they introduce considerable biases into the analyses.

EXERCISES

1. Choose the ESS data of one country for the following exercises:
   a. Compute the correlation matrix, means, and standard deviations for the indicators of the model of Figure 14.1.
   b. Estimate the parameters of the model on the basis of the estimated correlations.
   c. How high is the correlation between the factors?
   d. What do you conclude—can we speak of a variable "political efficacy" or should we make a distinction between two different variables?

2. For the same data set, perform the following tasks:
   a. Estimate the regression weights for the indicators for the concepts found.
   b. Estimate the individual composite scores.

3. Evaluate the quality of the composite scores.
   a. Find the strength of the relationship between CP_1 and S_1 (the weighted sum score).
   b. Find the strength of the relationship between CP_2 and S_2 (the weighted sum score).
   c. Find the Cronbach's α for the two relationships: $CP_1 - S_1$ and $CP_2 - S_2$.

4. From the ESS data of the same country, select the indicators for "formal" and "informal contact" and answer the following questions:
   a. Why are the indicators for "social contact" not reflective but formative indicators?
   b. Use the SQP program to determine the quality of the indicators.
   c. How large is the measurement error variance of these two variables?
   d. Now, compute the unweighted composite score for "social contact."
   e. What is the variance of this variable?
   f. Calculate the quality of this composite score.
   g. Is the quality of the composite score good enough to use the composite score as an indicator for "social contact?"

APPENDIX 14.1 LISREL INPUT FOR FINAL ANALYSIS OF THE EFFECT OF "SOCIAL CONTACT" ON "HAPPINESS"

mimic partcip - satisfaction in The Netherlands
data ni=4 no=2330 ma=km
km
1.00
.660 1.00

.121 .134 1.00
.178 .181 .274 1.00
sd
1.647 1.416 1.356 .952
me
7.62 7.79 5.28 2.78
labels
satif happy infpart formpart
model ny=2 nx=2 ne=2 nk=2 ly=fu,fi te=di,fr lx=fu,fi td=di,fi ga=fu,fi be=fu,fi
ps=sy,fi ph=sy,fi
value 1.0 ly 1 2
free ly 2 2
value .785 lx 1 1
value .682 lx 2 2
value .384 td 1 1
value .535 td 2 2
value 1 ga 1 2
free ga 1 1
free be 2 1
free ps 2 2
value 1 ph 1 1 ph 2 2
free ph 2 1

start .5 all

out mi sc adm=of ns

APPENDIX 14.2   LISREL INPUT FOR FINAL ANALYSIS OF THE EFFECT OF "INTEREST IN POLITICAL ISSUES IN THE MEDIA" ON "POLITICAL INTEREST IN GENERAL"

Political interest in The Netherlands
data ni=5 no=2330 ma=km
km
1.00
.215   1.00
−.056 −.262 1.00
−.046 −.126 .151 1.00
−.073 −.327 .247 .164 1.00
sd
1.249 .797 1.1356 1.56158 .93348
me
.401 2.28 2.28 1.366 1.054
labels
knowl polint tvtime radiotime paptime

```
select
1 2 3 4 5/
model ny=2 nx=3 ne=2 nk=3 ly=fu,fi te=di,fr lx=fu,fi td=sy,fi ga=fu,fi be=fu,fi
ps=sy,fi ph=sy,fi
value 1 ly 1 2
free ly 2 2
free be 2 1 ps 2 2

value .52 lx 1 1
value .73 lx 2 2
value .48 lx 3 3
value .64 td 1 1
value .38 td 2 2
value .69 td 3 3
value .09 td 2 1 td 3 1 td 3 2

value 1 ga 1 1
free ga 1 3 ga 1 2

value 1  ph 1 1 ph 2 2 ph 3 3
free ph 2 1
free ph 3 1 ph 3 2

start .5 all

out mi sc adm=of ns
```

15

CORRECTION FOR MEASUREMENT ERRORS

Measurement errors will remain, no matter how much we do our best to improve the questions. That means that the estimates of the relationships between the variables will be affected by these errors. Therefore, it is necessary to correct for these errors. Thus, the most important application of our research on the quality of questions and composite scores is the use of the quality estimates for correction for measurement errors in the analysis between variables. In the past, complex procedures have been discussed for the correction of measurement errors using multiple indicators. In this chapter, we will present a very simple procedure for the correction of measurement errors that is not much more difficult than regression analysis or causal modeling without latent variables.

15.1 CORRECTION FOR MEASUREMENT ERRORS IN MODELS WITH ONLY CONCEPTS-BY-INTUITION

In this section, we want to show by a simple example how this can be done. The example we want to use is a model to explain opinions about immigration. Variables to explain this opinion have been collected in the third round of the ESS. Some of the questions have already been discussed. For this example, we suggest the model presented in Figure 15.1.

Design, Evaluation, and Analysis of Questionnaires for Survey Research, Second Edition.
Willem E. Saris and Irmtraud N. Gallhofer.
© 2014 John Wiley & Sons, Inc. Published 2014 by John Wiley & Sons, Inc.

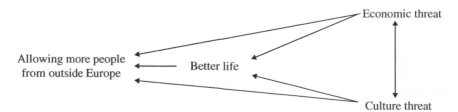

FIGURE 15.1 A simple model for explaining opinions about immigration.

The concepts presented in this model can be seen as concepts-by-intuition. These concepts can therefore be measured by direct questions. The English version of the question used for the dependent variable, which we will call "allow," was used in round 3 of the ESS and was measured by the second question mentioned in the following text:

B35 CARD 14 Now, using this card, to what extent do you think [country] should allow people of the same race or ethnic group as most [country's] people to come and live here?

Allow many to come and live here	1
Allow some	2
Allow a few	3
Allow none	4
(Don't know)	8

B37 STILL CARD 14 How about people from the poorer countries outside Europe? Use the same card.

Allow many to come and live here	1
Allow some	2
Allow a few	3
Allow none	4
(Don't know)	8

The questions used for the other variables are mentioned in the following text with the name we have given to the different variables.

"Economic threat"
B38 CARD 15 Would you say it is generally bad or good for [country]'s economy that people come to live here from other countries? Please use this card.

Bad for the economy								**Good for the economy**			**(Don't know)**
00	01	02	03	04	05	06	07	08	09	10	88

"Cultural threat"

B39 CARD 16 And, using this card, would you say that [country]'s cultural life is generally undermined or enriched by people coming to live here from other countries?

Cultural life undermined								**Cultural life enriched**			**(Don't know)**
00	01	02	03	04	05	06	07	08	09	10	88

"Better life"

B40 CARD 17 Is [country] made a worse or a better place to live by people coming to live here from other countries? Please use this card.

Worse place to live								**Better place to live**			**(Don't know)**
00	01	02	03	04	05	06	07	08	09	10	88

Note that the explanatory variables are measured using the same scale. That may cause common method variance (cmv). The dependent variable is measured in a rather different way, and therefore, we do not expect cmv between this variable and the other three variables.

The correlations between these variables obtained in Ireland in round 3 of the ESS were as indicated in Table 15.1.

The quality of the questions has been determined in two different ways. First of all, these questions were involved in an MTMM experiment. The quality of the questions was therefore determined by an experiment. Secondly, these questions were also coded in order to be included in the meta-analysis over all evaluated questions. In that case, the SQP 2.0 program can be used to predict the quality of these questions. Here, we will illustrate the procedure using the SQP predictions because this is the more general procedure. Using the MTMM estimates, the procedure is of course the same.

In Table 15.2, we present the quality predictions obtained using the authorized codes of the questions presented previously.

On the basis of the data in Table 15.1 and the quality information in Table 15.2, the effects presented in Figure 15.1 can be estimated with and without correction for measurement errors. Normally, the analysis is done without correction for measurement errors. In that case, the estimation is done on the basis of the correlation matrix of Table 15.1 without any adjustment.

TABLE 15.1 The correlations between these variables obtained in Ireland (n = 1700)

Allow	1.00			
Better life	−.470	1.00		
Economic threat	−.423	.662	1.00	
Culture threat	−.447	.718	.704	1.00

TABLE 15.2   The quality of the questions predicted by SQP

Variable	Method	r^2	v^2	m^2	q^2
Allow	SQP2.0	.826	.906	.094	.747
Economic threat	SQP2.0	.770	.780	.220	.601
Culture threat	SQP2.0	.761	.705	.295	.537
Better life	SQP2.0	.748	.725	.275	.543

TABLE 15.3   The correlations adjusted for cmv below the diagonal with the quality estimates on the diagonal and the cmv's above the diagonal

Allow	.747	.00	.00	.00
Better life	−.470	.543	.186	.215
Economic threat	−.423	.476	.601	.195
Culture threat	−.447	.503	.509	.537

If one wants to correct for measurement error, one has to correct the correlations in the matrix in the following way:

$$\text{Corrected } r_{ij} = \left(\frac{\text{observed } r_{ij} - \text{cmv}}{q_i . q_j} \right) \quad (15.1)$$

$$\text{where the } \text{cmv} = r_i \, m_i \, m_j \, r_j \quad (15.2)$$

This result follows directly from Equation (9.1). The observed correlations are obtained from the collected data (Table 15.1), and the cmv and the quality estimates can in general be obtained from SQP 2.0 (here Table 15.2). In general, then, the corrected correlations can be obtained in this way. The simplest way is to calculate the cmv first and subtract these values from the observed correlations. After that, one can substitute the 1 s on the diagonal of the correlation matrix by the quality of each question. If one then asks a SEM program to analyze the correlation matrix, the program first transforms the provided matrix into a correlation matrix that automatically contains the corrected correlations for all variables. This is so because this transformation is done by dividing all cells by the product of the quality coefficients as suggested in Equation 15.1.

First, the correlation matrix is adjusted by subtracting the cmv for each cell. In this case, we do not expect cmv for the "allow" variable and the other variables because their measurement procedure is different. For the other cells, we have calculated the cmv using Equation 15.2.

The adjusted matrix for this example is presented in Table 15.3. The cmv's for the different cells are presented above the diagonal. We see that quite a large part of the correlations is due to cmv. As a consequence, the cells below the diagonal indicating the correlations corrected for the cmv are much smaller than the observed correlations presented in Table 15.1.

On the diagonal, the qualities of the different variables have been presented as predicted by SQP. Note that the cmv's are calculated on the basis of the reliability coefficients and the method effects while on the diagonal the qualities are presented.

TABLE 15.4 The correlations corrected for measurement error

Allow	1.00			
Better life	−.738	1.00		
Economic threat	−.631	.833	1.00	
Culture threat	−.706	.931	.896	1.00

TABLE 15.5 The estimates of the effects of the explanatory variables on the variables allow and better life

	Without correction	With correction for errors
Effects on allow from		
Better life	−.265[a]	−.609[a]
Economic threat	−.133[a]	.001
Cultural threat	−.154[a]	−.140[a]
Total explained (R²)	*.254*	*.547*
Effects on better life from		
Economic threat	−.310[a]	−.007
Cultural threat	.500[a]	.938[a]
Total explained (R²)	*.564*	*.868*

[a]Means significantly different from 0.

The lower triangular matrix is not a correlation matrix anymore; however, by transforming this covariance matrix into a correlation matrix, one will get the correlations between these variables corrected for measurement error. The result is presented in Table 15.4.

In the first column, the correlations have only been corrected for random measurement errors. The other cells were also corrected for systematic errors. Comparing this table with Table 15.1, we see that all correlations have been increased by this correction for measurement error even though the systematic errors (cmv) were quite large. Given the changes in the correlations, we should also expect that the estimates of the effects will be different. The effects have been estimated with the ML estimator of LISREL. The inputs for these analyses are presented in Appendix 15.1. The results without and with correction for measurement error are presented in Table 15.5.

The results change considerably by correction for measurement errors. A systematic effect is that the effect of the variable economic threat is reduced and the effect of the better life and cultural threat have considerably increased. In particular, the conclusions with respect to the effects on the variable "allow" change considerably. The effect of the variable "better life" gets much larger. While without correction for measurement error, the effects of the economic and cultural threat are approximately equal. After the correction, the effect of the cultural threat is approximately the same, while the effect of the economic threat is only minimally different from 0. A similar effect can be seen for the other equation.

Besides that, we see that in both equations, the unexplained variance is reduced considerably. After correction for measurement errors, it is clear that there is still reason for looking for other explanatory variables because the lack of explanatory power cannot come from measurement errors. This conclusion would not have been possible if one had not corrected for measurement errors.

We give this example in order to show that taking into account the quality of the questions (i.e., correction for measurement errors) can have a considerable effect on the results of the analysis of the relationships between variables. Therefore, we are of the opinion that the information about the quality of questions is essential for the analysis of survey data and even more so in comparative research.

We hope that we have demonstrated that the procedure for taking into account the quality of questions is very simple. There is therefore no reason not to correct for measurement errors.

15.2 CORRECTION FOR MEASUREMENT ERRORS IN MODELS WITH CONCEPTS-BY-POSTULATION

In the estimation of models with concepts-by-postulation, one normally computes composite scores for these concepts and evaluates the quality of the composite scores using Cronbach's α, and after that, one continues with the analysis as if there were no measurement errors, while a Cronbach's α below 1 indicates that there are measurement errors. In doing so, one will get biased estimates of the effects one is interested in. This is not necessary. Therefore, we will illustrate here how one can get unbiased estimates of effects of models containing concepts-by-postulation. We will use for this purpose a topic that has been very popular recently.

During the last 15 years, a lot of attention has been given to the theory of "social capital" (Coleman 1988; Putnam 2000; Newton 1999, 2001; Halpern 2005). This theory suggests that investment in "social contact" functions for people as an asset that results in trust in other people and in the political system. We take these hypotheses as the starting point for our model and add more variables to it because we think that not only "social contact" influences "social trust" and "political trust." We enrich the model by adding the variables "experience of discrimination" and "political interest" for explanation of "social trust" and "political trust." To explain "political trust," the variables "political efficacy" and "political interest" are added. Figure 15.2 incorporates these variables into a simple substantive model.

In this model, it is assumed that the variables "social contact," "experience of discrimination," and "political interest" cause a spurious correlation between "social trust" and "political trust" and that these two latter variables, possibly, also have a reciprocal causal relationship. The reciprocal effect is included because it is plausible and to date has not been falsified.

15.2.1 Operationalization of the Concepts

In Table 15.6, we give an overview of the operationalization of the different concepts defining the chosen approach of the ESS in the first round. Most concepts are concepts-by-postulation with several reflective indicators.

"Social contact" is a concept-by-postulation with two formative indicators as has been discussed in Chapter 15. "Social trust," "political trust," and "political efficacy"

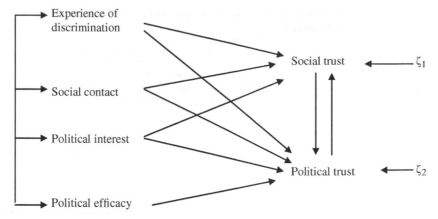

FIGURE 15.2 A structural model of a simple theory about effects of "social contact" and other variables on "social trust" and "political trust."

TABLE 15.6 The operationalization of the concepts in Figure 15.2

Concept name	Concept type	Observed indicator	Characterization of Indicators
Social contact	Postulation	Informal contact	Formative
		Formal contact	Formative
Social trust	Postulation	Can be trusted	Reflective
		Fair	Reflective
		Helpful	Reflective
Political trust	Postulation	Parliament	Reflective
		Legal system	Reflective
		Police	Reflective
Political efficacy	Postulation	Complex	Reflective
		Active role	Reflective
		Understand	Reflective
Discrimination	Intuition	Discriminated	Direct question
Political interest	Intuition	Interested	Direct question

are concepts-by-postulation with reflective indicators. "Experience of discrimination" is a concept-by-intuition measured by a direct question. "Political interest" could have been measured in different ways, but we opted for a direct question as a measure for the concept-by-intuition.

Part III of this book demonstrated how to estimate the size of the errors or the quality of a single question by using MTMM experiments; at least three forms of the same request for an answer are needed. In Chapter 13, we showed that an estimate of the size of the errors can also be obtained through the SQP program. It reduces the number of concepts to be measured to one for each indicator, which is more efficient than the MTMM approach.

Furthermore, in Chapter 14, we have already seen that the quality of a measure for a concept-by-postulation can be derived if the qualities of the measures for

TABLE 15.7 Possible designs of a study with respect to the number of observed variables included in the model

Concept name	Number of observed variables			
	Composite scores	Indicators for each concept	MTMM	In the ESS
Social contact	1	2	6	2
Social trust	1	3	9	9
Political trust	1	3	9	9
Political efficacy	1	3	9	9
Discrimination	1	1	3	1
Political interest	1	1	3	1
No. of observed variables	6	13	39	31

the concepts-by-intuition are known. Therefore, the number of observed variables can be reduced to 1 for each variable in the model. Our overview of the different possibilities to evaluate the quality of the measures in a study leads to designs that differ with respect to the number of observed variables and complexity of the model. Table 15.7 summarizes the possibilities.

This table shows that only 6 observed variables are needed if composite scores for all concepts mentioned in the model are calculated while 13 variables are used if one form of each of the indicators for these concepts is employed. The option to evaluate the data quality through MTMM analysis for this substantive research corresponds to the need for 39 observed variables. Finally, Table 15.7 informs us that there are 31 observed variables in the ESS out of the 39 mentioned.

Our advice is to avoid making models with 31 or 39 variables, because it increases the risk of serious errors in the design and analysis. It calls for a complex model of a combination of MTMM models for each concept and the corresponding substantive model of Figure 15.2. Therefore, in the following discussion, we will concentrate on the use of composite scores (six observed variables).

The following two steps are needed to reduce the design of the analysis while correcting for measurement error:

1. An evaluation of the quality of the measurement instruments
2. An analysis of the substantive model correcting for the detected errors

In the next section, we will give an overview of the data quality of the possible observed variables.

15.2.2 The Quality of the Measures

It is beyond the scope of this chapter to describe in detail how all the questions were evaluated. Some of the results of the studies of the quality of the measurement instruments have been presented previously. The results of these evaluations have been summarized in Table 15.8. The indicators for "social trust," "political trust," and "political

TABLE 15.8   Quality estimates of the 13 indicators from the Dutch study in the ESS

Concept name	Indicator	Coefficient for				Method used
		Reliability	Validity	Quality	Consistency	
Social contact	Informal	.79	1.0	.79	—	SQP
	Formal	.68	1.0	.68	—	SQP
Social trust	Be trusted	.87	1.0	.87	.84	MTMM
	Fair	.83	1.0	.83	.94	MTMM
	Helpful	.84	1.0	.84	.66	MTMM
Political trust	Parliament	.85	.95	.81	.66	MTMM
	Legal system	.90	.96	.86	.99	MTMM
	Police	.94	.96	.90	.66	MTMM
Political efficacy	Complex	.88	.96	.85	.89	MTMM
	Active role	.94	.97	.91	−.57	MTMM
	Understand	.86	.97	.83	−.78	MTMM
Discrimination	Direct request	.72	.72	.52	—	SQP
Political interest	Direct request	.96	.80	.77	—	SQP

efficacy" were evaluated by MTMM experiments,[1] while the other indicators have been evaluated by SQP. In the table, we see that the quality of the indicators[2] evaluated by an MTMM experiment is much better than the quality of the indicators evaluated by the SQP program. Given that the SQP program is based on the MTMM experiments, there is no reason to think that this difference is due to the evaluation method used. The real reason is that MTMM experiments were done in the pilot study, and the best method was selected for the main questionnaire in the definitive research. The results from the study confirm that this procedure is successful. The questions evaluated with the SQP program were not developed in the same way. They were not involved in an MTMM study in the pilot study and therefore were not improved upon.

This table also shows that for "social trust" and "social contact," the method effects are 0 so that the validity coefficient, which is the complement, is equal to 1. For the concepts "political trust" and "political efficacy," this is not true; there, the validity coefficients are not 1.

The low value of the quality of the "experience of discrimination" variable is of concern. The quality of this indicator is low because the explained variance in the

[1] In this chapter, the Dutch data of the first official round of the ESS are analyzed and not the data from the pilot study. As a consequence, the coefficients are slightly different from those presented in Chapter 14. As far as predictions with SQP are made, they are made with SQP 1.0. The predictions are not sufficiently different to repeat the analysis with the new numbers because the procedures are exactly the same and the result will be only minimally different.

[2] The reader is reminded that the quality coefficient is the product of the reliability and the validity coefficient and the quality itself is the quality coefficient squared, which can be interpreted as the percentage explained variance in the observed variable by the concept-by-intuition.

observed score is only 27%. This is due partially to the lack of precision of the scale used, which is a yes/no response scale. Here, a scale with gradation would result in a better quality measure. However, in the context of our illustration, this lack of quality will serve to show just how large the effect of correcting for measurement error can be.

Table 15.8 also shows the size of the consistency coefficients of the different reflective indicators for the concept-by-postulation that they are supposed to measure. We have included these coefficients because they play a role in calculating the measures of the composite scores (Chapter 14). Such relationships do not exist for concepts with formative indicators or concepts-by-intuition.

Finally, we have to mention that we did not specify the method effects because they are the complement of validity (1—validity coefficient squared). These effects are important because the method factors cause correlations between the observed variables, which have nothing to do with the substantial correlations. In this study, such method effects can be found within sets of variables for the same concept, but not across the different concepts of the model, since the methods are too different for the different substantive variables.

Now that we have discussed the quality of the indicators, we can turn to the quality of the composite scores for the different concepts-by-postulation that have been included in Figure 15.2. Chapter 14 discussed the procedures to estimate the quality of the composite scores for the "social contact" and "political efficacy" concepts. The measures for "social trust" and "political trust" are calculated using regression weights, followed by evaluation of the quality of these composite scores, using Equation (14.3). The results for these four concepts-by-postulation have been summarized in Table 15.9.

TABLE 15.9   The quality of the measures for the concepts-by-postulation

Variable name	Indicator	Construct		Composite score	
		Validity coefficient	Regression weights	Quality coefficient	Method effect
Social contact				.74	.00
	Informal	.79	.14		
	Formal	.68	.92		
Social trust				.81	.00
	Can be trusted	.73	.35		
	Fair	.81	.50		
	Helpful	.55	.10		
Political trust				.87	.31
	Parliament	.53	.09		
	Legal system	.86	.74		
	Police	.59	.13		
Political efficacy				.86	.22
	Complex	.76	.53		
	Active role	−.52	−.20		
	Understand	−.66	−.34		

In this table, the construct validity coefficient represents the effect of the concept-by-postulation on the observed indicator. This coefficient is the product of the quality of the indicator and the consistency coefficient, which were presented in Table 15.8.

This table shows that the four concepts-by-postulation differ in quality. In the next section, we will see that these differences play an important role when estimating the effects of the different variables on each other. Two concepts also contain invalidity due to method effects. However, we will not worry about this, because the methods were different across concepts and therefore the method effects could not affect the correlations between the different concepts.

15.2.3   Correction for Measurement Errors in the Analysis

Table 15.8 and Table 15.9 summarize the quality of the variables that can be used as observed variables in the analysis to estimate the effects in the substantive model presented in Figure 15.2. Therefore, we can now start with a discussion about the different possibilities to correct for measurement errors in the analysis. In Saris and Gallhofer (2007b), we discussed the analysis with different numbers of indicators for the different concepts-by-postulation. Here, we will only present the simple approach using the composite scores as the observed variables for the concepts-by-postulation. In that case, we have to estimate the relationships between the six variables presented in Figure 15.2. The correlations between the scores for the six variables of the model in Figure 15.2 are presented in Table 15.10.

Having reached this point, we are in the same situation as in the previous section with variables measured by direct questions. Here, for each concept, we also have one observed variable, and we have estimates of the quality of the observed variable (the composite scores) so we can use the same procedure to correct for measurement errors. This can be done by reducing the variances on the diagonal of the correlation matrix to the quality coefficient squared of the composite scores and to specify in the program that this matrix is a covariance matrix and that one would like to analyze the correlation matrix. The program will then automatically correct all correlations for measurement errors and estimate the values of the parameters corrected for measurement errors. The LISREL input for this approach is given in Appendix 15.2.

TABLE 15.10   Correlation matrix for the composite scores (regression weights)

soctr	1.00					
poltr	.41	1.00				
discr	.15	.12	1.00			
socconr	.11	.10	.06	1.00		
poli	−.15	−.23	.02	−.10	1.00	
poleff	−.13	−.26	.05	−.15	.52	1.00

TABLE 15.11   Estimated values of the standardized parameters for the model presented in Figure 15.2 based on composite scores

Causal relationships	No correction	Using variance reduction
Effects on social trust from		
Political trust	.35	.12
Discrimination	.12	.34
Social contact	.07	.08
Political interest	−.07	−.21
Effects on political trust from		
Social trust	.03	.39
Discrimination	.12	.14
Social contact	.05	−.01
Political interest	−.12	−.05
Political efficacy	−.19	−.25

Table 15.11 presents the final result of the analysis with correction for measurement errors and without correction for measurement errors.

This table shows that the analysis without correcting for measurement errors gives very different results than the analysis correcting for measurement errors. If we do not correct for measurement errors, the effect of social trust on political trust is very small and that of political trust on social trust is rather large. Correcting for measurement errors, the opposite result is obtained. Besides that, we see that also other coefficients have considerably been changed in value especially the effects of discrimination and political efficacy.

15.3   SUMMARY

Using two examples, we have shown in this chapter that correction for measurement error makes a real difference in the final results. In order to be sure about the estimates of the relationships between variables, one should correct for measurement error. This requires knowledge about the error variance or quality of the observed variables. In previous chapters, we have shown that the more complex method used to obtain this information with respect to a single question is the MTMM experiment. A simpler way is to use the predictions that the SQP program provides.

Given the information about the quality of single questions, one can also estimate the quality of composite scores for concepts-by-postulation as we have shown in Chapter 14. In this chapter, we have shown how this information can be used to estimate the effects that the variables have on each other, corrected for measurement errors. In an earlier version of this book (Saris and Gallhofer 2007b), we demonstrated that this can be done in different ways and, if it is properly done, the different approaches should lead to similar results. Therefore, here we recommend using the simplest method that consists of using the calculation of composite scores and their quality and applying the "reduction of the variance method" to correct for measurement errors.

A more commonly used, but wrong, approach is as follows. First, indicators are developed for all theoretical variables, and the best are selected by factor analysis. Next, unweighted sum scores are computed, and the quality of the composite scores are evaluated with Cronbach's α, hoping that the quality coefficients are close to .7 or higher. Then the model is estimated without any correction for measurement error. This approach leads to biased estimates of the relationships because the correction for measurement errors is not applied.

We should mention that the procedure presented in this chapter has also its limitations. We have presented this procedure to make correction for measurement errors accessible for all researchers. A disadvantage of this procedure is that the standard errors of the estimates are underestimated because we assume in the analysis after correction of the correlation matrix for measurement errors that there are no measurement errors at all. If we would use a model that makes the difference between latent and observed variables explicit, then the estimates would be the same but the standard errors would have been bigger. However, as long as we are mainly interested in relatively large effects, this disadvantage is minimal compared with the advantage that the presented procedure provides that we can correct for measurement errors in a simple way.

EXERCISES

1. Estimate the model presented in the following text, which is a simplified version of the model used in this chapter.

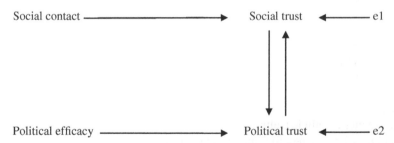

The quality of the composite scores of "social contact" and "political efficacy" has been evaluated in the last chapter. Therefore, your first task is to:

a. Test the measurement models for "political" and "social" trust for the same country that you used in the exercises of Chapter 14.

b. Compute the composite scores and determine the quality of the composite scores as was done in the last chapter.

c. Calculate the correlation matrix, means, and standard deviations for the four composite scores.

d. Estimate the effects of the above model with and without correcting for measurement errors on the basis of the correlation matrix.

2. Given that we have the estimated values of the effects, the following questions can be asked:
   a. Does the model fit the data?
   b. If not, what has to be changed to fit the model?
   c. If so, which coefficients are not significant? Can these parameters be omitted while the model remains acceptable?
   d. What is your interpretation of the final model?

3. The parameters can also be estimated on the basis of the covariance matrix.
   a. How should one correct for measurement errors if one uses the covariance matrix as the data for the estimation?
   b. Estimate the parameters on the basis of the covariance matrix.
   c. Why are the values of the parameters different?
   d. Ask for the completely standardized solution.
   e. If you did a correct analysis, the completely standardized solution should be approximately the same as the one in Exercise 2.

APPENDIX 15.1 LISREL INPUTS TO ESTIMATE THE PARAMETERS OF THE MODEL IN FIGURE 15.1

The LISREL input for the analysis without correction for measurement errors

Immigration Ireland
da ni=4 no=1700 ma=km
km
1.00
−.470 1.00
−.423 .662 1.00
−.447 .718 .704 1.00

labels
Allow better econimie Culture
model ny=2 nx=2 be=fu,fi ga=fu,fi ps=di,fr
fr be 1 2 ga 1 1 ga 1 2 ga 2 1 ga 2 2
out nd=3

The LISREL input for the analysis with correction for measurement errors

Immigration Ireland
da ni=4 no=1700 ma=km
cm
 .747
−.470 .543
−.423 .476 .601
−.447 .503 .509 .537

labels
Allow better econimie Culture
model ny=2 nx=2 be=fu,fi ga=fu,fi ps=di,fr
fr be 1 2 ga 1 1 ga 1 2 ga 2 1 ga 2 2
out nd=3

These two inputs show that the only part that has been changed is the correlation matrix as indicated in Table 15.3.

APPENDIX 15.2   LISREL INPUT FOR ESTIMATION OF THE MODEL WITH CORRECTION FOR MEASUREMENT ERRORS USING VARIANCE REDUCTION BY QUALITY FOR ALL COMPOSITE SCORES

Analysis of composite score using the variance reduction method

```
data ni=6 no=2300 ma=km
cm
.656
.408  .757
.114  .099  .548
−.132 −.260 −.146 .735
−.155 −.227 −.101 .522 .593
.158  .115  .061  .049 .022 .270
labels
soctr poltr socconr poleff poli discr
select
1 2 6 3 5 4 /
model ny=2 nx=4 be=fu,fi ga=fu,fi ps=di,fr

free ga 1 1 ga 1 2 ga 1 3 ga 2 1 ga 2 2 ga 2 3 ga 2 4
free be 2 1 be 1 2
out
```

16

COPING WITH MEASUREMENT ERRORS IN CROSS-CULTURAL RESEARCH

In this chapter, we will introduce the problem of measurement errors in cross-cultural research. In particular, we will focus on comparative research across countries. In the last chapters, we established that measurement errors have strong effects on results of research. Therefore, when the effects of measurement error differ in the individual countries of a study, cross-country comparisons become quite challenging.

Two types of comparisons are most frequently made: comparisons of *means* and comparisons of *relationships* of different variables across countries. Often, comparisons based on single requests or on composite scores of the latent variables are made. In this chapter, we will add to this the comparisons based on latent variables.

The problem of such comparisons is that one can compare the results across different countries only if, in fact, the data are comparable, that is, if the measures used in the different countries have the same meaning. This topic is studied under the heading of functional equivalence or invariance of measures in different countries.

This chapter will concentrate on the procedures to determine the equivalence of measurement instruments. But, before we can introduce this topic, we have to introduce the notation for this topic. So far, we used standardized variables and concentrated on the effects of these variables on each other. However, in cross-cultural research, the means of the variables are frequently compared, and therefore, we need to introduce unstandardized relationships, slopes, intercepts, and means.

Design, Evaluation, and Analysis of Questionnaires for Survey Research, Second Edition.
Willem E. Saris and Irmtraud N. Gallhofer.
© 2014 John Wiley & Sons, Inc. Published 2014 by John Wiley & Sons, Inc.

16.1 NOTATIONS OF RESPONSE MODELS FOR CROSS-CULTURAL COMPARISONS

In order to introduce the notations in this chapter, we will use the already familiar example of "political efficacy" and, to a lesser extent, "political trust." In Chapter 15, we introduced a measurement model for "political efficacy"; here, we will concentrate on "subjective competence." Normally, three requests are used to measure this concept. For the moment, we will use only one request called "understand" with a 5-point response scale (see Appendix 16.1). The operationalization of this concept is illustrated in Figure 16.1.

At the top of the model, the relationship between the concept-by-postulation "subjective competence" (CP_1) and the reaction to a specific request (f_1: "understand") is specified. This relationship represents a brain process that is triggered by the stimulus posed by the request. The output of this process is a variable that represents a possible reaction not yet expressed in the requested form (Van der Veld 2006). Assuming a linear relationship, we can expect the relationship that is presented in Figure 16.2 and Equation (16.1):

$$f_1 = a + cCP_1 + u \qquad (16.1)$$

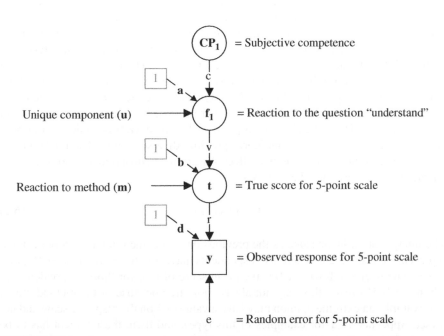

FIGURE 16.1 A measurement model for "subjective competence" using the request "understand" and a 5-point scale with a unique component (u), systematic method factor (m), and random error (e). New in this model is the specification of an effect of a "variable" called "1". This symbol is used to make a distinction between the effects of real variables and the effect of the intercepts.

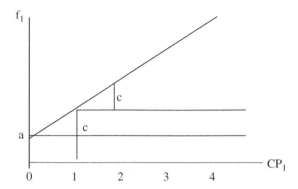

FIGURE 16.2   The linear model presenting the cognitive response process.

Because "subjective competence" has been measured with different requests (see Chapters 14 and 15), u represents the unique component of the specific request (understand). We assume that the scores of this unique component vary around 0 so that the mean score of this unique component over the whole population is 0.

The unit of measurement of these variables is unknown, but what we can say is that "a" represents the mean value of f_1 if $CP_1 = 0$. If $a = 0$, persons who have the impression that their "competence in politics" is 0 will also have a reaction that on average could be represented by 0. Furthermore, it is assumed that whatever the "subjective competence" of the person is, an increase in "subjective competence" of 1 unit will lead to an increase of c in the "reaction" (f_1). Therefore, we can conclude that this relationship is linear, as can be seen in Figure 16.2 and Equation (16.1), where the coefficient "a" is called the *intercept* and the c the *slope* of the function.

In the next step, respondents have to express their "reaction to the request" in a certain format. Here, we choose a 5-point scale; however, this choice is arbitrary, and for different choices, we find corresponding diverse results. Therefore, in this relationship, we introduce a method effect. Assuming again a linear relationship, the equation can be formulated as follows:

$$t = b + vf_1 + m \tag{16.2}$$

The interpretation is the same as the previous one: m is the random component and represents the different reactions respondents have to the method used (5-point scale). The intercept b will be 0 if the mean score of $t = 0$ for those respondents for whom $f_1 = 0$. However, this may not always hold true because of the method effect. For example, an extreme opinion may have a value of 1 on the response scale and not 0 (see Appendix 16.1 for examples of this type), and then, the intercept has to be equal to 1.

The final step is the real response selected on the 5-point scale. Assuming a linear relationship, we get

$$y = d + rt + e \tag{16.3}$$

Here, e represents the disturbance due to random errors with a mean of 0. The coefficient d is the intercept that is 0 if the mean of y is 0 for those respondents for whom t=0. Normally, we can assume that r=1 and d=0 because there is no reason to expect otherwise when the translation goes from a 5-point scale to a 5-point scale.

In the models used in the previous chapters, the variables and coefficients were standardized, and then, the distinction between Equations (16.2) and (16.3) makes sense because v would be the validity coefficient and r the reliability coefficient. However, in unstandardized forms, these coefficients represent the slopes of their corresponding equations. Assuming that d=0 and r=1, we obtain the following:

$$y = t + e \tag{16.4}$$

This is the form commonly used in the classical test theory (Lord and Novick 1968).

Substituting Equation (16.2) into Equation (16.4), we get

$$y = b + vf_1 + m + e \tag{16.5}$$

Equations (16.1) and (16.5) together specify the response process, with the difference now being the use of unstandardized variables (variables expressed in their original units of measurement). Equation (16.1) presents the cognitive process started by the request for an answer and finishing with a preliminary reaction, while Equation (16.5) represents the measurement process going from the preliminary reaction to an observed score. Both processes can differ between countries, and it may also be true that the differences come from only one of the two processes.

Most authors do not make this distinction (Davidov 2008; Grouzet et al. 2006), and they only use the equation that can be obtained by substituting Equation (16.1) into Equation (16.5):

$$y = b + v(a + cCP_1 + u) + m + e$$

or

$$y = b + va + vcCP_1 + vu + m + e$$

This can be simplified to

$$y = \tau + \lambda CP_1 + \zeta \tag{16.6}$$

where

$$\lambda = vc \tag{16.6a}$$

$$\tau = b + va \tag{16.6b}$$

$$\zeta = e + m + vu \tag{16.6c}$$

Equation (16.6) is the equation frequently used for the evaluation of measurement instruments also in cross-cultural research. This equation looks rather simple, but the coefficients are complex because they consist of different components. Unfortunately, these components cannot be derived if Equation (16.6) is used in the estimation of the response process as a whole.

On the other hand, we should say that the original model consisting of Equations (16.1)–(16.3) is too complex to be estimated as it is. It requires more data to estimate all the parameters.

Assuming that the means of the disturbance term (u), the method effect variable (m), and the random error component (e) are equal to 0, then the mean over the observed responses (μ_y) of the respondents can be expressed as a function of the mean of the variable of interest (μ_{CP_1}):

$$\mu_y = \tau + \lambda \mu_{CP_1} \tag{16.7}$$

The result shows that the mean of the responses (μ_y) is not necessarily equal to the mean of the latent variable of interest (μ_{CP_1}). This will normally only be the case if

$$\tau = 0 \quad \text{and} \quad \lambda = 1 \tag{16.8}$$

This requires that there be no systematic effects on both due to the request asked and the method used. It is unlikely that these conditions are fulfilled for all cases. However, the assumptions mentioned in Equation (16.8) are commonly made because, normally, the observed mean is treated as the mean of the variable of interest.

So far, we have formulated the response model in the original units of measurement. In the previous chapters, we have always made use of standardized variables. The relationships between these different formulations can be found in the appendix of Chapter 9.

In cross-cultural research, a translated request can be perceived differently across countries and languages. It is also possible that the use of a 5-point scale can create different reactions in different countries. Such country differences may change not only the correlations between the variables but also the slopes and intercepts of the different equations.

Normally, it has been recommended that researchers compare responses across groups only if the requests can be seen as functionally equivalent. If *functionally equivalent* means that λ and τ in Equation (16.6) are the same across countries, then the means across countries can be compared, even though we are not sure that they represent the mean of the opinions of the sample on the latent variable of interest. But, if these restrictions do not hold true, what can we do? One question is whether we can make cross-cultural comparisons when these conditions are not satisfied. Another question is how we can evaluate whether requests are functionally equivalent. We will address these topics in the next sections.

16.2 TESTING FOR EQUIVALENCE OR INVARIANCE OF INSTRUMENTS

In this section, we will introduce the standard textbook approach. Then, we will indicate the problems of this approach. After that, we will formulate an alternative approach that is in line with the basic ideas in this book.

16.2.1 The Standard Approach to Test for Equivalence

Scholars commonly make a distinction between *configural, metric,* and *scalar invariance* (Horn et al. 1983; Meredith 1993; Steenkamp and Baumgartner 1998). While discussing the different forms of equivalence or invariance, they employ the model specification of Equation (16.6) using several indicators for the same latent variable of interest. This is necessary because with one observed variable, the equivalence of the measures cannot be tested. In fact, it has been discussed in Chapter 10 that three observed variables are needed to estimate the three effect parameters and error variances of a reflective measurement model with one latent variable. In case of the measurement of "subjective competence," there are indeed three indicators and the model can be estimated. The model would consist of three equations, one for each of the indicators. Using the formulation of Equation (16.6), we obtain

$$y_1 = \tau_1 + \lambda_{11}CP_1 + \zeta_1 \qquad (16.9a)$$

$$y_2 = \tau_2 + \lambda_{21}CP_1 + \zeta_2 \qquad (16.9b)$$

$$y_3 = \tau_3 + \lambda_{31}CP_1 + \zeta_3 \qquad (16.9c)$$

Since the scale of the latent variable needs to be fixed in one equation, the τ should be 0 and the λ equal to 1. This means that the latent variable CP_1 will be expressed in the same units as the observed variable and that the observed score is 0 if the latent variable has a score of 0. If these restrictions are not made, the model is not identified. Further assumptions for the standard model are

$$\text{Covariance }(CP_1, \zeta_i) = 0, \quad \text{for all i} \qquad (16.9d)$$

$$\text{Covariance}(\zeta_i, \zeta_j) = 0, \quad \text{for all i} \neq j \qquad (16.9e)$$

This means that the model of Figure 16.3 is used in the test.

The literature says that comparison of means and relationships across cultures requires:

1. *Configural invariance,* meaning that the model of (16.9) holds for all the countries involved
2. *Metric invariance,* meaning that, besides configural invariance, the slopes are the same in all the countries studied

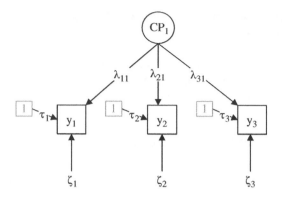

FIGURE 16.3   The measurement model of "subjective competence."

These two requirements are sufficient for comparison of relationships. The comparison of means requires:

3. *Scalar invariance*, meaning that, besides metric invariance, the intercepts are the same across all countries being compared

These hypotheses can be tested using multiple-group analysis in SEM programs.

In this approach, the same model (16.9) is simultaneously estimated in several samples, taking into account the invariance restrictions on the parameters from the different samples. Such restrictions can be tested using an overall test statistic, which is the sum of the test statistics for the different samples. The degrees of freedom of the test are equal to the sum of the degrees of freedom for the different samples, meaning that

$$\chi_k^2 = \chi_{k_1}^2 + \chi_{k_2}^2 + \cdots \qquad (16.10a)$$

where

$$k = k_1 + k_2 + \cdots \qquad (16.10b)$$

If the samples are independent, then χ_k^2 is also χ^2 distributed with df$=$k if the model is correct and distributed noncentral χ^2 if the model is incorrect. Here, standard χ^2 tests can be used to test the hypotheses. The input of the scalar invariance test for the concept-by-postulation of "subjective competence" is presented in Appendix 16.3. The results of the tests of the different invariance restrictions for three countries, the United Kingdom, the Netherlands, and Spain, are presented in Table 16.1.

This table shows that scalar invariance, where all loadings (slopes) and intercepts are equal across countries, must be rejected. The model, assuming that only the loadings are equal (metric invariance), fits much better to the data. However, this model can be improved upon if one loading in Spain is allowed to be different from those in the other countries. Then, the model is acceptable, even though the sample sizes are large. The conclusion would be that the indicators are not scalar invariant, and

TABLE 16.1   Results of the tests of different requirements of invariance for the concept of "subjective competence" based on data from three countries

Invariance restrictions	χ^2	df	Probability
Scalar	84.5	8	.00
Metric	13.3	4	.01
Metric except for λ_{21} in Spain	8.0	3	.39

according to the literature, this means that we cannot compare the means across the different groups. In our case, it is even questionable whether we can compare relationships given that one slope in Spain deviates from those in the other countries and, therefore, metric invariance also does not hold true across these countries.

While this is the standard procedure used to test for equivalence across languages and countries or any other groups, researchers also complain that this procedure leads nearly always to rejection of scalar invariance suggesting that the means cannot be compared across countries and often leads to rejection of metric invariance, which means that relationships cannot be compared. We think that there are good reasons for these problems. These reasons will be indicated in the next section.

16.3   PROBLEMS RELATED WITH THE PROCEDURE

There are three reasons why the standard approach leads to these problems. The first reason is that one does not take the power of the test into account. Particularly, the test of the equality of the intercepts is normally very powerful. This means that the probability of rejecting the assumption of scalar in variance is very high even for very small deviations. Later, we will show how this problem can be solved.

The second reason is that the test is evaluating whether the understanding of the question and the response process are the same. However, it is essential to test if the understanding of the questions is the same, while the response process may be different across countries. We will show later that one can correct for the latter but should require that the meaning of the questions be the same for all people. So, in this respect, the standard test is too strict.

A third issue we would like to mention is that statistical significance is not necessarily the same as substantive relevance. If the difference between the measures is significant but small, one can nevertheless decide to continue with cross-cultural comparison. All three issues will be discussed in the next sections.

16.3.1   Using Information about the Power of the Test

Saris et al. (2009b) have indicated that the chi-square test discussed cannot be trusted because the power of the test is not taken into account. In fact, they suggest that full models cannot be tested at all. They suggest that a model containing one or more *relevant* misspecifications is not a good model. Starting from that principle, Saris et al. (1987) suggested evaluating the quality of a model using the combination of expected parameter change (EPC) and the modification index (MI). They noted that the EPC gives a direct estimate of the size of the misspecification for all fixed parameters, while the MI provides a significance test (with one degree of freedom) for the estimated misspecification (for more details, we refer to the paper by Saris et al. 1987). However, one should realize that the MI has the same problem as the chi-square test statistic, which is that the value of the estimate depends on other characteristics of the model. In addition, the direct EPC misspecification estimates are problematic because sampling fluctuations can be rather large as will be shown later. To tackle this issue, we will introduce the standard error of the EPC and the power of the MI test.

Fortunately, the following simple but fundamental relationship exists between the three statistics mentioned in the preceding text (Saris et al. 1987: 121):

$$MI = \left(\frac{EPC}{\sigma^2} \right) \qquad (16.11)$$

where σ is the standard error of the EPC.

From this relationship, it follows that

$$\sigma = \frac{EPC}{\sqrt{MI}} \qquad (16.12)$$

Thus, σ can be estimated from EPC and MI, statistics that are nowadays provided by most of the SEM software.

Knowing the size of the EPC and the MI also provides a simple way to estimate the power of the test for the size of each misspecification. Consider a specific deviation δ for which one would like to know the power. One could choose a value of δ that would be the minimum size of the misspecification that one would like to be detected by the test with a high likelihood (power). By standard theory, under deviation from the null hypothesis, the asymptotic distribution of the MI is noncentral χ^2 with the noncentrality parameter (ncp) given by

$$ncp = \left(\frac{\delta}{\sigma} \right)^2 \qquad (16.13)$$

By combining (16.12) and (16.13), we obtain

$$ncp = \left(\frac{MI}{EPC^2} \right) \delta^2 \qquad (16.14)$$

This ncp is expressed as a function of statistics provided by the standard software and the user-specified value δ of maximally acceptable misspecification. This ncp can be

TABLE 16.2   The decisions to be made in the different situations defined on size of the MI and the power of the test

	High power	Low power
Significant MI	Inspect EPC (EPC)	Misspecification present (m)
Nonsignificant MI	No misspecification (nm)	Inconclusive (I)

used to determine the power of the test of a misspecification of δ for any value of the significance level α of the test and for all restricted parameters. The power of the test can be obtained from the tables of the noncentral χ^2 distribution (or using any computer-based routine) as

$$\mathrm{Prob}(\chi^2(1,\mathrm{ncp}) > c_\alpha) \qquad (16.15)$$

where c_α is the critical value of an α-level test based on a χ^2 distribution with df = 1 and $\chi^2(1, \mathrm{ncp})$ is the noncentral chi-square distribution with ncp.

Note that this approach requires the specification of the deviation δ. Saris, Satorra and Van der Veld (2009), in the context of the equivalence testing, that for a standardized slope and for a correlated error term, a misspecification of .1 is so big that it should be detected by a test with a high likelihood (power). For intercepts, we suggest that differences of more than 5% of the number of scale points should be detected with high power. These values are merely suggestions, and one could use other values for δ that are more appropriate within a specific situation.

We propose to distinguish four possible situations shown in Table 16.2, which result from combining the significance or not of the MI test and the high/low power of the MI test.

When MI is significant and the power of the MI test is low, we conclude that there is a misspecification because the test is not very sensitive (low power) but a significant value of the MI has nevertheless been obtained. This is the cell in Table 16.2 labeled m for misspecified. The program JRule[1] that can be used to make these decisions indicates this situation with the number 2.

Using the reversed argument, the decision is also simple if the MI is not significant and the power of the MI is high. In that case, the conclusion is that there is no misspecification, so the corresponding cell in Table 16.2 is labeled "nm." In the program, this situation is denoted by the number 1.

The situation is more complex if the MI is significant but the power of the MI test is high. In that case, it may be a serious misspecification, but it may also be that the MI is significant due to a high sensitivity of the test for this misspecification. Therefore, in that situation, the suggestion is to look at the substantive relevance of

---

[1] A computer program, JRule, that produces statistics for all the restricted parameters, based on the output of LISREL, has been developed by Van der Veld et al. (2009). The program can be requested by sending an e-mail to vdveld@telfort.nl putting JRule in the subject line.

the EPC: if the EPC is rather small, one concludes that there is no serious misspecification. This makes sense because, generally, we do not want to adjust our model for a standardized coefficient of .001, even though this coefficient is significant. However, when the EPC is large, for example, larger than .2, it is concluded that there is a relevant misspecification in the model. This cell in Table 16.2 is labeled "EPC," for EPC use. In the program JRule, this situation is indicated by the number 3.

The fourth and last situation is that in which the MI is low and the power of the MI test is also low. In that case, it should be concluded that one lacks sufficient information to make a decision. This is a frequently occurring situation. In this case, the conclusion is that not enough information is available to reach a decision. This case is labeled in the table as inconclusive "I." In JRule, it is indicated by the number 4.

Let us apply this approach on the example we discussed in the last section. First of all, we evaluate the model where we assume scalar invariance across all three countries. We have seen before in Table 16.1 that the chi-square test indicates as usual that this hypothesis has to be rejected. We have suggested that this is mainly due to the fact that the power of the test for the equality of the intercepts is very high.

In the new approach, we are not looking at the fit of the whole model but at each restricted parameter separately and evaluating, taking into account the power of the test, whether the equality of the parameter was misspecified. The results of this test using JRule are presented in Table 16.3. In this table, we see the evaluation of all restricted parameters: TY represents the intercept, LY the slope, and TE the error variances and covariances. MI is the modification index, that is, the chi-square value for this specific restriction; EPC is the expected parameter value if this parameter would be estimated without a restriction; and DELTA is the chosen unstandardized value of the deviation of the parameter that should be detected with high power. This value is transformed from a standardized value into a value for an unstandardized value for the slopes, the error variances, and the covariances. For the intercept, the value is equal to .05 times the number of scale points. In this case, the number of scale points was 5, so the deviation to be detected with high power is .25 or a fourth of a scale point. PWR stands for the power of the test, and JR is the judgment rule that was received where 1=no misspecification, 2=misspecified, 3=look at the EPC, and 4=inconclusive because of lack of power.

It is immediately clear from the table why the standard procedure using the chi-square test for the full model leads to rejection if we test for scalar invariance. Many tests for the intercepts have a power of .99 and very large significant MI. That is denoted in JRule with a three and suggests that one has to look if the expected deviation (EPC) will indeed be so large that the model has to be rejected. If we look at the EPCs, we see that in most cases, the expected value is considerably smaller than the value we have chosen to be detected. In fact, only one intercept in Spain is expected to be larger than the critical value for DELTA we have chosen. This suggests that in most cases, we should see the restriction not as a misspecification because the large chi-square value is due to the high power of the test and not because of the size of the deviation. This is exactly the reason why the new testing procedure, taking into account the power of the test, was developed by Saris et al. (2009a).

TABLE 16.3   JRule results for scalar invariance

FIRST DATA SET FROM THE NETHERLANDS

JUDGMENT RULES FOR TAU-Y

PARAMETER	INTERCEPT OF		MI	EPC	DELTA	PWR	JR	EPC/D
TY 1		VAR 1	41.88	−0.12	0.25	0.99	3	−0.48
TY 2		VAR 2	0.04	0.01	0.25	0.99	1	0.04
TY 3		VAR 3	18.86	−0.07	0.25	0.99	3	−0.28

JUDGMENT RULES FOR LAMBDA-Y

PARAMETER	TO	FROM	MI	EPC	DELTA	PWR	JR	EPC/D
LY 1 1	VAR 1	ETA 1	0.08	−0.03	0.16	0.32	4	−0.19
LY 2 1	VAR 2	ETA 1	0.94	−0.05	0.19	0.96	1	−0.26
LY 3 1	VAR 3	ETA 1	0.54	0.04	0.14	0.75	4	0.29

JUDGMENT RULES FOR THETA EPSILON

PARAMETER	RELATION	BETWEEN	MI	EPC	DELTA	PWR	JR	EPC/D
TE 1 1	VAR 1	VAR 1	−	−	−	−	−	−
TE 2 1	VAR 2	VAR 1	0.54	−0.04	0.14	0.74	4	−0.29
TE 3 1	VAR 3	VAR 1	0.94	0.05	0.11	0.55	4	0.45
TE 2 2	VAR 2	VAR 2	−	−	−	−	−	−
TE 3 2	VAR 3	VAR 2	0.08	0.01	0.13	0.95	1	0.08
TE 3 3	VAR 3	VAR 3	−	−	−	−	−	−

SECOND DATA SET FROM UK

JUDGMENT RULES FOR TAU-Y

PARAMETER	INTERCEPT OF		MI	EPC	DELTA	PWR	JR	EPC/D
TY 1		VAR 1	37.32	0.17	0.25	0.99	3	0.68
TY 2		VAR 2	15.88	0.14	0.25	0.99	3	0.56
TY 3		VAR 3	29.06	0.14	0.25	0.99	3	0.56

PARAMETER	TO	FROM	MI	EPC	DELTA	PWR	JR	EPC/D
LY 1 1	VAR 1	ETA 1	0.00	0.00	0.17	0.99	1	0.00
LY 2 1	VAR 2	ETA 1	1.53	−0.11	0.21	0.64	4	−0.52
LY 3 1	VAR 3	ETA 1	1.65	0.13	0.16	0.36	4	0.81

JUDGMENT RULES FOR THETA EPSILON

PARAMETER	RELATION	BETWEEN	MI	EPC	DELTA	PWR	JR	EPC/D
TE 1 1	VAR 1	VAR 1	−	−	−	−	−	−
TE 2 1	VAR 2	VAR 1	1.61	−0.07	0.15	0.79	4	−0.47
TE 3 1	VAR 3	VAR 1	1.49	0.07	0.12	0.55	4	0.58
TE 2 2	VAR 2	VAR 2	−	−	−	−	−	−
TE 3 2	VAR 3	VAR 2	0.00	0.00	0.14	0.99	1	0.00
TE 3 3	VAR 3	VAR 3	−	−	−	−	−	−

(Continued)

TABLE 16.3   (*Cont'd*)

THIRD DATA SET FROM SPAIN

JUDGMENT RULES FOR TAU-Y

PARAMETER		INTERCEPT OF	MI	EPC	DELTA	PWR	JR	EPC/D
TY	1	VAR 1	1.00	0.06	0.25	0.99	1	0.24
TY	2	VAR 2	28.93	−0.30	0.25	0.99	3	−1.20
TY	3	VAR 3	1.47	−0.07	0.25	0.99	1	−0.28

JUDGMENT RULES FOR LAMBDA-Y

PARAMETER		TO	FROM	MI	EPC	DELTA	PWR	JR	EPC/D
LY	1	1	VAR 1	ETA 1	0.13	0.05	0.14	0.18	4	0.36
LY	2	1	VAR 2	ETA 1	5.34	0.23	0.13	0.24	2	1.77
LY	3	1	VAR 3	ETA 1	4.01	−0.22	0.14	0.25	2	−1.57

PARAMETER		RELATION BETWEEN		MI	EPC	DELTA	PWR	JR	EPC/D
TE	1	1	VAR 1	VAR 1	–	–	–	–	–	–
TE	2	1	VAR 2	VAR 1	4.47	0.18	0.13	0.31	2	1.38
TE	3	1	VAR 3	VAR 1	5.90	−0.23	0.14	0.32	2	−1.64
TE	2	2	VAR 2	VAR 2	–	–	–	–	–	–
TE	3	2	VAR 3	VAR 2	0.13	−0.03	0.12	0.31	4	−0.25
TE	3	3	VAR 3	VAR 3	–	–	–	–	–	–

The table indicates that there is a serious misspecification in only a few cases (2). This occurs only in the Spanish data set for LY(2,1), LY(3,1), TE(2,1), and TE(3,1). In all these cases, the power is very low (<.35) but the EPC is large. This suggests that the misspecification must be large; otherwise, with low power, one cannot get a significant MI. It is remarkable that, again, for these serious misspecifications, the chi-square value is much smaller than for the intercepts, which do not deviate much. This is another illustration of how a chi-square without information about the power can lead to very wrong conclusions. Finally, we would like to say that concerning these results, the misspecifications indicated in the preceding text are related, so that one needs to relax only one parameter in order to solve the problem. A problem is to choose which parameter.

If we do the test for metric invariance, the JRule program indicates only mis-specifications for the LY and TE parameters we have mentioned in the preceding section as well. If we introduce LY(2,1) as a free parameter in Spain, then JRule does not detect any misspecification anymore.

If we test scalar invariance allowing LY(2,1) in Spain to be different from the value in the other countries, then JRule suggests that TY2 is still misspecified.

Although the result of the analysis is quite similar in this case as in the standard procedure, it should be clear that the test for misspecifications has the potential, by taking the power into account, to prevent rejecting scalar invariance with the high frequency of the standard procedure.

16.3.2   An Alternative Test for Equivalence

We are now going to discuss the second reason why we think that invariance is rejected too frequently. It is our opinion[2] that the invariance of the parameters in Equation (16.1), the cognitive process equation, should be evaluated because one cannot compare the answers to questions that are differently understood by people of different groups. On the other hand, the parameters in the measurement equation can be different because these differences can be corrected. Therefore, we suggest that *cognitive equivalence of measurement instruments should be required, that is, invariance after correction for differences in the measurement process.*

Consequently, the model specified in Equations (16.1) and (16.5) should be used and not the derived model presented in Equation (16.6). The model in Figure 16.4 gives us the second-order factor model that can be applied to test equivalence if the coefficients (v) and intercepts (b) of the measurement equation are known. If this information is available, the test of equivalence can be conducted in the same manner as previously indicated. This approach leaves us with the task to estimate the parameters of the measurement equations, and this can be done in a separate study using MTMM experiments or using predictions from SQP.

An alternative would be to simultaneously estimate and test the quality measures and the intercepts assuming equality of the consistency coefficients over the different countries. A model for this approach is shown in Figure 16.5. The model illustrates that for identification reasons the slopes of the second indicator of each concept-by-intuition (f_i) are set to 1. This is possible because the same scale was applied in all

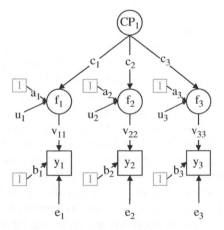

FIGURE 16.4   The alternative measurement model for "subjective competence" to be evaluated in cross-cultural research. For simplicity we have ignored here method effects therefore the error terms are denoted by e and not by e + m as in Equation (16.5).

---

[2] At first glance, it appears as if Little (1997) and Grouzet et al. (2006) make the same observation, but on a closer examination of their approach, we see that they share the opinion of other cited authors on this issue.

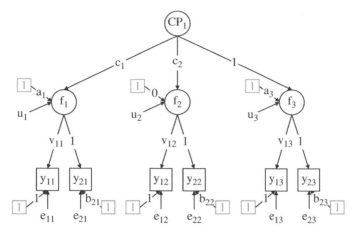

FIGURE 16.5 The measurement model for "subjective competence" to be evaluated for three countries.

three requests, and therefore, we expect the same unit of measurement. Also, the intercept of the first indicator of each concept-by-intuition is set to 1 because the scales have been specified[3] in such a way that an extreme value (fixed reference point) is equal to 1. Hence, we expect that a 0 value on the latent variable goes together with a score of 1 on the observed variable and not 0. Finally, the intercept for the concept "active role" was set to 0 because we expect that participants who have no competence will respond that they cannot play a role in political activities, and c_3 was fixed on 1 for identification of the variance of CP_1. For the LISREL input of this model, we refer the reader to Appendix 16.4.

The result of the analysis, assuming that the slope coefficients (c) and the intercepts (a) are equal across countries, while all other coefficients are not fixed,[4] was that JRule indicated only one possible misspecification in the cognitive part, especially c_2 in Spain that was detected as not being equal to the value in the other two countries. Not constraining c_2 in Spain to the value in the other countries showed that this coefficient was indeed considerably different. These results suggest that the model with one minor adjustment fits the data. The differences between the countries were found mainly in the measurement equations. Therefore, we can conclude that most of the differences in the parameters of the standard model (16.6) come from the measurement part of the model, which we can correct. This can also be seen in Table 16.4 where the parameter values for the different countries are presented. This table shows that the slopes of the measurement equations are quite similar in the three countries and that the intercepts are different. Furthermore, the coefficient values for the

---

[3] This, however, is not clear for the "understand" requests. In that case, the extreme category cannot be called a fixed reference point. This may be the explanation for having to remove the restriction for the Spanish data set.

[4] To avoid a nonsignificant negative error variance, we fixed the variance of the unique component for "understand" in Spain on 0 to avoid an improper solution with this coefficient equal to –.02.

TABLE 16.4   Estimates of parameters of the model presented in Figure 16.5 for three countries[a]

Parameter	United Kingdom	NL	Spain
Slopes of the measurement equations			
v_{11}	1.15	1.06	.98
v_{12}	1.15	1.02	1.00
v_{13}	1.02	.93	1.13
Intercepts of the measurement equations			
b_{21}	1.28	1.00	.81
b_{12}	1.00	1.00	1.28
b_{22}	1.22	1.09	1.46
b_{23}	1.06	.87	1.37
Slopes of the cognitive process equations			
c_1	−.94	−.94	−.94
c_2	.91	.91	.60
c_3	1.00	1.00	1.00
Intercepts of the cognitive process equations			
a_1	3.09	3.09	3.09
a_2	1.00	1.00	1.00
a_3	−.84	−.84	-.84

[a] We present only the estimated values of relevant parameters.

relationships between the variable of interest, "subjective competence," and its three indicators are very similar, except for one deviation in Spain on the second indicator.

This result shows that the scale invariance test that is normally used to test for equivalence can lead to a rejection of the model, while the cause of the problem is not that the indicators have a different interpretation across countries, but that instead the respondents in the different countries employ the scales distinctively. However, if the differences due to the measurement procedure are corrected, rather good comparable indicators across countries may be obtained. Therefore, we prefer the less restrictive requirements of *cognitive equivalence* over the standard requirements.

16.3.3   The Difference between Significance and Relevance

A final point we would like to make is that a detected invariance not automatically should mean that we cannot compare the results across groups. One reason for this statement is that in case of reflective indicators, one may accept that one measure out of the three is incomparable across countries, as in our example, and nevertheless compare latent means and relationships between latent variables across countries. In that case, one speaks of partial equivalence (Byrne et al. 1989). Allowing the parameters for the deviating measure to be different, one can still get consistent estimates of the latent means and the relationships between latent variables.

In case of the use of composite scores, this is in principle not true. However, if the deviations are rather small and one realizes that each measure will be weighted by

1/m where m is the number of measures, then one can imagine that in case of very small deviations, the systematic error in the composite score will be so small that it will not harm the comparison across countries.

In this section, we suggested a different approach to invariance testing that takes into account the power of the test with respect to the test of the cognitive invariance, allowing for differences in the response process and accepting single measures to be deviant under certain conditions. Using this alternative procedure, we expect that invariance much less frequently will be rejected and so more often comparative research is possible. Examples of this approach where invariance was less frequently rejected can be found in the paper of Coromina et al. (2008).

16.4 COMPARISON OF MEANS AND RELATIONSHIPS ACROSS GROUPS

The most common applications of comparative research are comparisons of means and relationships across groups. This can be done using single questions, composite scores, and latent variables. In the next sections, we will discuss the limitations and possibilities in the different situations.

16.4.1 Comparison of Means and Relationships between Single Requests for Answers

There is a strong research tradition that concentrates on differences in means (see, e.g., Torcal et al. 2005) and relationships (see, e.g., Newton 1997) between responses to single requests across countries. This, however, is a very questionable activity if the requests are not previously checked for equivalence. As a consequence, we do not know the source of the response differences; it could be due to differences in measurement errors, cognitive processes, or substantive differences between countries or a combination thereof.

Previously, we established that at least three observed variables for the same concept are needed in order to evaluate the quality of its measures. For separating measurement and cognitive processes, repeated observations are needed. However, most studies lack this type of information, and therefore, their comparisons may be incorrect.

The magnitude of this problem can be illustrated with the items for "subjective competence": "complex," "active," and "understand." It was found in the previous section that the second item was not scalar invariant for the countries the United Kingdom, the Netherlands, and Spain. Given the estimated values of the parameters of the response process, one can expect, assuming that the mean of the latent variable of interest is 3, that the mean for y_{12} in the United Kingdom is equal to 4.13, in the Netherlands it is 3.78, and in Spain it is 3.14. Without having collected the information about the differences in the response processes of the different countries, we could not have known that these differences have no substantive meaning, because in all three countries the mean on the latent variable is equal to 3.

For relationships between variables, the same problems occur. The correlation between the variables "complex" and "understand" is $-.44$ in Greece and $-.514$ in the

Czech Republic, but after correction for measurement error, the difference is much larger. In Greece, the correlation becomes .59, and in the Czech Republic, it is .77. In this case, the differences would have been underestimated. However, the opposite can also occur. If we compare the correlation for the same variables in the Czech Republic and Slovenia, we get for the observed correlations, respectively, $-.514$ and $-.449$. But after correction for measurement errors, these correlations are exactly equal and have the value .77. So, in this case, one could think that there are substantive differences, while the differences are due merely to differences in data quality.

This overview demonstrated that the means and relationships of single requests in cross-cultural research cannot be compared unless the measurement instruments are equivalent or the differences in the response process are corrected. Given the usual lack of information, we advise to proceed with caution when attempting to compare the results based on single requests.

16.4.2 Comparison of Means and Relationships Based on Composite Scores

In Chapter 14, we demonstrated how composite scores can be computed and evaluated. In Chapter 15, it was indicated how relationships between composite scores can be estimated. In this chapter, we deal with the comparison of relationships across countries, which requires equivalence of the measurement instruments. Normally, it is suggested that for comparison of relationships, it is sufficient if metric invariance requirements are met, while for comparison of means, scalar invariance has been required. The equality of slopes (λ) of factor models like the one in Figure 16.3 across the different countries would be required for comparison of relationships, and the equality of slopes and intercepts (τ) would be required for comparison of means. We suggested that these requirements are too strict and have proposed that these restrictions should be required for the cognitive part of the model in Figure 16.4. This means that the slopes (c) in the model of Figure 16.4 should be invariant across countries for comparison of relationships, while for comparison of the means, the coefficients c and a should be identical across countries. The argument is that we can correct for measurement errors and that it can be shown that the slopes and intercepts of the cognitive process should be invariant.

After correction for differences in the measurement equations, the composite score for CP_1 can be computed as an unweighted sum (C_1) of the three latent variables:

$$C_1 = (f_1 + f_2 + f_3) \tag{16.16a}$$

or

$$C_1 = (a_1 + c_1\, CP_1 + u_1) + (a_2 + c_2\, CP_1 + u_2) + (a_3 + c_3\, CP_1 + u_3) \tag{16.16b}$$

or

$$C_1 = (a_1 + a_2 + a_3) + (c_1 + c_2 + c_3)\, CP_1 + (u_1 + u_2 + u_3) \tag{16.16c}$$

This shows that whenever one or more intercepts and/or slopes are different across countries, the computed composite score is most likely different for the different countries.[5] Even if two countries would have the same mean value on the variable of interest (CP_1), the means of the composite scores will usually be different if not all coefficients in Equation (16.16c) are equal (scalar invariance) across countries. Hence, without scalar invariance, the means of the computed composite scores cannot be used as indicators to compare the means across countries. If scalar invariance holds true, the comparison can be made even though these means are not equal to the means of the variables of interest. In fact, the latter applies only if all the intercepts (a) are equal to 0 and the sum of the slopes (c) is equal to 1, which is quite rare.

A similar argument can be made for the comparison of relationships based on composite scores. The covariance between the variables of interest CP_1 and CP_2 is denoted by "$\sigma_{CP_1 CP_2}$" and the composite scores for CP_1 and CP_2 are simple unweighted sums for the reaction variables (f). This means that

$$C_1 = f_{11} + f_{21} + f_{31} \quad \text{and} \quad C_2 = f_{12} + f_{22} + f_{32} \tag{16.17a}$$

where f_{ij} represents the reaction to the ith request, which is an indicator for CP_j. Substituting the relationship with the concept-by-postulation, we get (see Appendix 16.6)

$$\sigma_{C_1,C_2} = (c_{11} + c_{21} + c_{31})(c_{12} + c_{22} + c_{32})\sigma_{CP_1 CP_2} \tag{16.17b}$$

This result shows that where composite scores are calculated for several different populations, the covariances across the countries cannot be compared if not all slope coefficients are invariant (see note 5). This is because differences between the covariances can stem from two sources: from the differences in slopes or from substantive differences in covariances between the latent variables of interest. Therefore, our derivation shows that the minimum requirement for comparing relationships based on composite scores is that the slopes after correction for measurement errors are invariant.

In an earlier section, we showed how we think that one can test for scalar and metric invariance. In that analysis, we have found that the indicators for "subjective competence" are not metric invariant and therefore also not scalar invariant. One of the items generated a different reaction in Spain than in the other two countries. Now, we are left with the question of what to do. It is possible to leave one country out of the analysis. Another possibility is to omit one item and to reduce the number of indicators to two; however, in doing so, the concept-by-postulation will change. For our example, this option will not critically affect the analysis, because all indicators may not be necessary for the definition of the concept-by-postulation in a measurement model consisting of reflective indicators. In case of formative indicators, one cannot omit an indicator without changing the concept. However, if the deviation is rather

[5] Except for the unlikely case that some deviations cancel each other out.

small, one may consider if it really harms the comparability as we have suggested in Section 16.3.3.

We will not continue these computations here because in the two previous chapters, we have already shown how to compute and evaluate composite scores and how to estimate the relationship between them.

16.4.3 Comparison of Means and Relationships between Latent Variables

Composite scores are frequently used to compare the means of latent variables of interest. However, it is much easier and safer to use the estimated means of the latent variables for the comparison. In the estimation of the model of Figure 16.5 presented in Appendix 16.4, the means (ka) of the latent variable are also estimated. Even though the second indicator was not equivalent according to the popular definition of equivalence, the estimates of the means were correct because two indicators are sufficient to identify the means.

An interesting advantage from this approach is that by specifying the restriction that the means in the different countries are identical, we can test if the means are the same. Specifying this restriction, we get a $\chi^2 = 53.8$ with df $= 27$ where Pr $= .002$. The difference with the model without the equality constraint was equal to 20.1 with df $= 2$. So we can conclude that the means are significantly different from each other. When allowing the estimation of the means to be different, they were calculated at 1.25 for the United Kingdom, 1.27 for the Netherlands, and .62 for Spain. This result indicates that for all three countries the mean of "subjective competence" is rather low when applying a 5-point scale. But the results also show that the "subjective competence" in Spain is significantly lower than in the United Kingdom or in the Netherlands.

This approach is much easier than the previous one and less prone to computation errors. Moreover, the fact that one of the indicators is not equivalent does not harm the estimates of the means and does not require any additional effort. Scholars have described this approach as partial equivalence, stating that under this condition means of latent variables can be compared. Although this is statistically true, it should also be said that we changed the operationalization to determine the mean through two equivalent indicators, while the third indicator was treated as another effect variable of the latent variable. It is a theoretical question whether this correction of the interpretation of the measurement model is acceptable.

If the model is correctly specified, it is also possible to simultaneously estimate the relationships between the latent variables with the quality of the measurement instruments in the different countries with a minor extension of the input for the program we have used before.

However, the problem is that the unique components of the variables for the different constructs may be correlated. These correlations will not be detected if composite scores for each concept-by-postulation are calculated separately, but they will be discovered when the two measurement models are combined in order to estimate a relationship between them.

Let us illustrate our point with the relationship between the variables "subjective competence" and "political trust." Both of them are defined as a concept with three reflective indicators. Appendices 16.1 and 16.2 present the requests for these two concepts. The equivalence of the measures for "subjective competence" has already been tested. The equivalence of the measures of "political trust" has been tested in the same way.

After correction for measurement error and assuming that the slopes are identical across countries, we get a χ^2 of 60.7 with 18 degrees of freedom. This is not a good fit. The program suggests that the slope for the second item in Spain should not be constrained to be equal to the same coefficients in the other countries. Repeating the analysis with this correction, the fit improves to 48.5 with 16 degrees of freedom. Without making any further improvements, we are satisfied with the results for the cognitive process part of the model. The final result is presented in Table 16.5.

Again, there is one item in Spain that is not invariant across countries. Previously, this meant that we had to make the choice of omitting one country or one item. However, the attractive characteristic of directly estimating the relationships between latent variables is that we no longer have to make this choice. This is because we can view the "not invariant" item as just another consequence of the latent variable defined by the other two equivalent items. In other words, allowing for a free estimation of the not invariant parameters will produce consistent estimates for the relationships between the latent variables.

Now, we can specify the input for estimating the relationships between the two concepts-by-postulation, allowing for two deviations from metric equivalence after correction for measurement error. The model estimated is a combination of two measurement models like the one presented in Figure 16.3: one for "subjective competence" and the other for "political trust." The null model estimated is presented in Figure 16.6. Appendix 16.5 lists the LISREL input for this analysis.

The estimates of the parameters (v_{ij}) for the measurement equations have been inputted as fixed parameters. The results of this analysis are presented in the first row of Table 16.6, and they indicate that the fit of this model is not very good. The table shows that the covariances between "subjective competence" and "political trust" for the three countries could also be estimated in this analysis. Applying this approach, we detect that the model contains misspecifications. Therefore, before proceeding

TABLE 16.5   The unstandardized loading of the factor model for "political trust" for three countries

Indicator	UK slopes	NL slopes	Spain slopes
Parliament	1.00[a]	1.00[a]	1.00[a]
Legal system	1.53	1.53	1.09
Police	1.06	1.06	1.06

[a] These parameters have been fixed on 1 for identification.

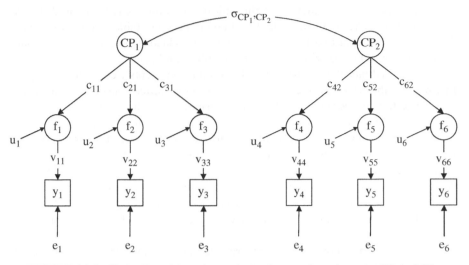

FIGURE 16.6   The null model used to estimate the covariance between CP_1 and CP_2.

TABLE 16.6   **The fit of the model and the estimated relationships between "subjective competence" and "political trust" for three countries**

| Model | χ^2 | df | prob | Covariances in | | |
				United Kingdom	NL	Spain
Null model	138.9	30	.000	.08	.25	.21
$+c_{41}$	117.3	27	.000			
$+c_{12}$	73.9	24	.000			
$+c_{51}$	42.3	21	.004	−.11	.01	.14

c_{ij}, covariance (e_i, e_j); df, degrees of freedom; prob, probability.

further, we need to check if the estimates are correct and to search for the misspecifications. It could be that correcting for misspecifications will change the values of the estimates as well.

Using the program JRule, we concluded that several coefficients have not been introduced in the model that should be there. So, we introduced the following cross loadings in sequence for all three countries: c_{41}, c_{12}, and c_{51}. These misspecifications could not have been detected without combining the two concepts into one model, because they represented the effects from one concept on indicators of another concept. The specified effects suggest that:

- An increase in "subjective competence" will also increase the "trust in the government" (c_{41}) and the "legal system" (c_{51}).
- An increase in "political trust" will also increase the frequency with which people think that politics is too complex (c_{12}).

These effects suggest reduction of the unique components (u) for the different indicators that are related with variables of other concepts. Table 16.6 shows us how these corrections lead to a better fit, and after introducing these coefficients, only minor errors remain that will not affect the results significantly.

For this final model, the covariances between the concepts were estimated again. In row 4 of the table, we see that these covariances differ considerably from the previous estimates, because a part of the covariances was due to misspecifications. The newly introduced effects absorb a large part of the relationships between the observed correlations, so that very little covariance is left between the two concepts. It turns out that these estimates are not significantly different from 0.

This exercise illustrates several important points. Most importantly, this approach detects misspecifications within a model that cannot be detected using composite scores. This is a critical point because a misspecified model can generate quite different estimates of the parameters than a correctly specified model as we have shown in the preceding text.

Another important point is that we were able to perform the analysis even though in each operationalization one item in one country was not invariant, even after correction for measurement errors. We did not remove this one item. In fact, if we would have done so, the last model could not have been identified anymore. Therefore, this approach is much more flexible than using composite scores.

We have also seen that we learned more about the measurement instrument with this approach than otherwise, because now we know more about the unique components of the different indicators.

Altogether, we hope to have shown that directly estimating the relationships between latent variables is a more efficient and therefore a better way of comparing relationships across countries than using composite scores. But, the composite scores can also be used if the model becomes too complex and one has some assurance that the model specified is correct.

16.5 SUMMARY

In this chapter, we have shown that cross-cultural comparisons are also affected by measurement errors. They require different kinds of equivalences for measurement instruments that are sometimes not present or unknown. We have also insisted that comparing results across countries requires that the data are corrected for measurement errors. Without the corrections, we run the risk of giving explanations for differences between countries on substantive grounds that could be due to differences in measurement quality of the instruments. Even though we agree with the requirements of metric and scalar equivalence, after correction for measurement errors, we think that the commonly used requirements for equivalence are too strict.

We suggested a different approach to invariance testing that takes into account the power of the test with respect to the test of the cognitive invariance, allowing for differences in the response process and accepting single measures to be deviant under certain conditions. Using this alternative procedure, we expect that invariance much less frequently will be rejected and so more often comparative research is possible. Examples of this approach where invariance was less frequently rejected can be found in the paper of Coromina et al. (2008).

A problem with this approach is that the needed information is seldom available. Especially in cases of comparing means, correlations using a single request, or concept-by-intuition, we find that the information about the quality of the requests is missing. In such instances, the information about the quality can be derived from external sources such as MTMM experiments or SQP predictions.

In case of the use of composite scores for concepts-by-postulation, the comparison across countries requires perfect metric invariance for comparison of relationships and perfect scalar invariance for comparison of means. These requirements are very strict and will rarely be satisfied. We have indicated that there are some ways to be a bit less strict in cases where scalar invariance has been rejected.

An alternative we have shown is comparing means and relationships between latent variables across countries because in that case one does not have to require perfect invariance. Consistent estimates of the means and relationships are also possible with partial equivalence. Therefore, using latent variables is a more flexible approach than employing composite scores. The only disadvantage is that the models become more complex.

Overall, the conclusion is that cross-cultural comparisons are not as simple as they seem and that the comparison of means is even more difficult than the comparison of relationships between variables.

EXERCISES

1. In these exercises, we continue with the exercises of the previous chapter, and we add at least one more country to the analysis:
   a. Calculate the means for the variables "social trust" and "political trust."
   b. Are the means the same? What is your conclusion?
   c. Calculate the correlations between the indicators of "social trust" and "political trust" for both countries.
   d. Are these correlations similar or different?
   e. What can we say on the basis of these results about the relationship between "social trust" and "political trust"?

2. Let us now turn to measurement models for all four variables:
   a. Test the measurement models for all four variables on the basis of the data in the new country.
   b. Compare using multiple-group analysis whether there is some level of invariance across the countries.

    c. Given the results, can you make comparisons across countries using these variables?

    d. If so, which comparisons can be made, and what is the result?

3. Let us now consider the use of latent variables:

    a. Given the invariance of the measurement instruments, can you make comparisons across countries using the latent variable means and/or the relationships between the latent variables?

    b. If comparisons are possible, make these comparisons and state your interpretation of the results.

APPENDIX 16.1   THE TWO SETS OF REQUESTS CONCERNING "SUBJECTIVE COMPETENCE"

The set of requests in the supplementary questionnaire

    C5   **CARD C4** *How often do politics and government seem so complicated that you cannot really understand what is going on?*

Never	*1*
Seldom	*2*
Occasionally	*3*
Regularly	*4*
Frequently	*5*
(Don't know)	*8*

    C6   **CARD C5** *Do you think that you could take an active role in a group that is focused on political issues?*

Definitely not	*1*
Probably not	*2*
Not sure either way	*3*
Probably	*4*
Definitely	*5*
(Don't know)	*8*

    C7   **CARD C6** *How good are you at understanding and judging political questions?*

Very bad	*1*
Bad	*2*
Neither good nor bad	*3*
Good	*4*
Very good	*5*
(Don't know)	*8*

The set of requests in the supplementary questionnaire

How far do you agree or disagree with the following statements?

L4 "Sometimes politics and government seem so complicated that I cannot really understand what is going on."
Please tick one box
Strongly disagree ☐
Disagree ☐
Neither disagree nor agree ☐
Agree ☐
Strongly agree ☐

L5 "I think I could take an active role in a group involved with political issues."
Strongly disagree ☐
Disagree ☐
Neither disagree nor agree ☐
Agree ☐
Strongly agree ☐

L6 "I am good at making my mind up about political issues."
Strongly disagree ☐
Disagree ☐
Neither disagree nor agree ☐
Agree ☐
Strongly agree ☐

APPENDIX 16.2   ESS REQUESTS CONCERNING "POLITICAL TRUST"

CARD C8: Using this card, please tell me on a score of 0 to 10 how much you personally trust each of the institutions. I read out: 0 means you do not trust them at all, and 10 means you have complete trust. Firstly…**READ OUT**.

	No trust at all									Complete trust		(Don't know)
C10 …the British government?	00	01	02	03	04	05	06	07	08	09	10	88
C11 …the legal system?	00	01	02	03	04	05	06	07	08	09	10	88
C12 …the police?	00	01	02	03	04	05	06	07	08	09	10	88
C13 …the politicians?	00	01	02	03	04	05	06	07	08	09	10	88
C14 …the European Parliament?	00	01	02	03	04	05	06	07	08	09	10	88
C15 …the United Nations?	00	01	02	03	04	05	06	07	08	09	10	88

APPENDIX 16.3   THE STANDARD TEST OF EQUIVALENCE FOR "SUBJECTIVE COMPETENCE"

Factor model test for Ned Spain and UK; Netherlands
data ng=3 ni=3 no=1190 ma=cm
km
1.00
–.334 1.00
–.465 .389 1.00
me
3.00 2.18 2.96
sd
1.09 1.30 .986
model ny=3 ne=1 ly=fu,fi te=di,fr ps=fu,fr ty=fu,fi al=fu,fr
free ly 2 1 ly 3 1
free ty 2 ty 3
value 1 ly 1 1
out sc

Spain
data ni=3 no=280 ma=cm
km
1.00
–.306 1.00
–.508 .403 1.00
me
3.44 1.65 2.64
sd
1.20 1.044 1.175
model ny=3 ne=1 ly=in ty=in ps=sp al=sp te=di,fr
out sc

UK
data ni=3 no=885 ma=cm
km
1.00
–.317 1.00
–.381 .301 1.00

me
3.21 2.32 3.13
sd
1.122 1.355 1.069
model ny=3 ne=1 ly=in ty=in ps=sp al=sp te=di,fr
out sc

APPENDIX 16.4   THE ALTERNATIVE EQUIVALENCE TEST FOR "SUBJECTIVE COMPETENCE" IN THREE COUNTRIES

Analysis of British efficacy experiment wave 1 data in ESS
Data ng=3 ni=6 no=885 ma=cm
Km
*
1.00
–.317 1.00
–.381 .301 1.00
 .538 –.328 –.373 1.00
–.330 .646 .303 –.362 1.00
–.390 .310 .489 –.401 .358   1.00

me
*
3.21 2.32 3.13 3.19 2.36 3.15
sd
*
1.122 1.355 1.069 1.049 1.121 .983
label
complex1 active1 understand1 complex2 active2 understand2
model ny=6 ne=3 nk=1 ly=fu,fi te=di,fr ps=di,fr be=fu,fi ga=fu,fi ph=sy,fr
ty=fr ka=fr al=fi
value 1 ly 4 1 ly 5 2 ly 6 3
free ly 1 1 ly 2 2 ly 3 3
value 1 ga 3 1
free   ga 2 1 ga 1 1

fixed ty 1 ty 2 ty 3
value 1 ty 1 ty 2 ty 3
free al 1 al 3
start 1 ly 1 1
out
Analysis of Dutch efficacy experiment wave 1 data in ESS
Data   ni=6 no=885 ma=cm
Km
1.00
–.334 1.00
–.465 .389 1.00
 .624 –.296 –.399 1.00
–.355   .671   .370 –.349 1.00
–.454   .341   .586 –.424   .420   1.00

me
*
3.00   2.18 2.96   2.89 2.25 2.98
sd
*
1.09   1.30 .986   1.07 1.133 1.055

label
complex1 active1 understand1 complex2 active2 understand2
model ny = 6 ne = 3 nk = 1 ly = fu,fi te = di,fr ps = di,fr be = fu,fi ga = in ph = sy,fr ty = sp
ka = fr al = in
value 1 ly 4 1 ly 5 2 ly 6 3
free ly 1 1 ly 2 2 ly 3 3
fixed ty 1 ty 2 ty 3
value 1 ty 1 ty 2 ty 3
start 1 ly 1 1
out

Analysis of Spanish efficacy experiment wave 1 data in ESS
Data ni = 6 no = 280 ma = cm
Km
1.00
−.306 1.00
−.508   .403 1.00
 .562 −.289 −.551 1.00
−.301   .643 .379 −.289 1.00
−.457 .311 .530 −.486   .297   1.00

me
*
3.44 1.65 2.64 3.31 1.83 2.83
sd
*
1.200 1.044 1.175 1.094 1.084 1.177
label
complex1 active1 understand1 complex2 active2 understand2
model ny = 6 ne = 3 nk = 1 ly = fu,fi te = di,fr ps = di,fr be = fu,fi ga = in ph = sy,fr ty = sp
ka = fr al = in
value 1 ly 4 1 ly 5 2 ly 6 3
free ly 1 1 ly 2 2 ly 3 3
value 1 ty 1 ty 2 ty 3
fixed ps 3 3
start 1 te 6 6 te 5 5
start 1 ly 1 1
out   adm = off

APPENDIX 16.5   LISREL INPUT TO ESTIMATE THE NULL MODEL FOR ESTIMATION OF THE RELATIONSHIP BETWEEN "SUBJECTIVE COMPETENCE" AND "POLITICAL TRUST"

```
estimation of relations between pol eff and pol trust
!first group UK
data ng=3 ni=6 no=880 ma=cm
km
1.00
-.317 1.00
-.382 .301  1.00
-.153 .104  .093 1.00
-.093 -.001  .027   .456 1.00
-.027 -.034 -.063 .353 .529 1.00
me
3.21  2.32  3.13  4.75  5.17  6.17
sd
1.122  1.355  1.069  2.33 9 2.357  2.374
label
complex active understand parliament juridical police

model ny=6 ne=6 nk=2 ly=fu,fi te=sy,fi ga=fu,fi ph=fu,fr ps=sy,fi
value 1 ga 3 1 ga 4 2
free ga 1 1 ga 2 1 ga 5 2 ga 6 2

!later corrections
!free ga 4 1
!free ga 1 2
!free ga 5 1

free ps 1 1 ps 2 2 ps 3 3 ps 4 4 ps 5 5 ps 6 6

value 1.15 ly 1 1
value 1.15 ly 2 2
value 1.02 ly 3 3
value 1.17 ly 4 4
value 1.03 ly 5 5
value 1.06 ly 6 6

value .56 te 1 1
value .61 te 2 2
value .46 te 3 3
value .73 te 4 4
value 1.18 te 5 5
value .40 te 6 6
```

start .42 ph 1 1
start 1.14 ph 2 2
fr ph 2 1
out   adm = of ns

Netherlands
data ni = 6 no = 1150 ma = cm
km
1.00
−.334 1.00
−.465 .389 1.00
−.264 .132 .052 1.00
−.299 .170 .112 .597 1.00
−.122 .057 −.001 .445 .606 1.00
me
3.00 2.18 2.98 5.18 5.34 5.84
sd
1.093 1.295 .986 2.025 2.217 1.934
label
complex active understand parliament juridical police
model ny = 6 ne = 6 nk = 2 ly = fu,fi te = sy,fi ga = in ph = sp ps = sp
value 1 ga 3 1 ga 6 2

!free ga 4 1
!free ga 1 2
!free ga 5 1

value 1.06 ly 1 1
value 1.02 ly 2 2
value 0.93 ly 3 3
value 1.15 ly 4 4
value 1.17 ly 5 5
value 1.11 ly 6 6

value .40 te 1 1
value .67 te 2 2
value .41 te 3 3
value .79 te 4 4
value .79 te 5 5
value .32 te 6 6
start .51 ph 1 1
start 1.32 ph 2 2

out

Spain
data ni=6 no=281 ma=cm
km
1.00
−.306 1.00
−.508 .403 1.00
−.147 .076 .158 1.00
−.064 .005 .121 .617 1.00
−.030 .080 .068 .526 .561 1.00
me
3.44 1.65 2.64 4.96 4.48 5.69
sd
1.200 1.044 1.175 2.265 2.328 2.359
label
complex active understand parliament juridical police
model ny=6 ne=6 nk=2 ly=fu,fi te=sy,fi ga=in ph=sp ps=sp
value 1 ga 3 1 ga 6 2
free ga 2 1 ga 5 2

value 0.96 ly 1 1
value 1.00 ly 2 2
value 1.13 ly 3 3
value 1.20 ly 4 4
value 1.14 ly 5 5
value 1.12 ly 6 6

value .70 te 1 1
value .36 te 2 2
value .52 te 3 3
value .17 te 4 4
value .56 te 5 5
value .55 te 6 6
start .67 ph 1 1
start 2.05 ph 2 2

out

APPENDIX 16.6   DERIVATION OF THE COVARIANCE BETWEEN THE COMPOSITE SCORES

The covariance "$\sigma_{CP_1CP_2}$" between the variables of interest CP_1 and CP_2 as expressed in deviation from their mean is defined for the population as follows:

$$\sigma_{CP_1CP_2} = \left(\frac{1}{N}\right)\sum(CP_1CP_2) \text{ summed over all people (N) for the population}$$

$$(16.6A.1)$$

When the indicators for each latent variable are expressed in deviation from their means, the relationships between the latent variables of interest and the indicators corrected for measurement error can be formulated as

$$F_{11} = c_{11}CP_1 + u_{11} \qquad\qquad (16.6A.2a)$$

$$F_{21} = c_{21}CP_1 + u_{21} \qquad\qquad (16.6A.2b)$$

$$F_{31} = c_{31}CP_1 + u_{31} \qquad\qquad (16.6A.2c)$$

$$F_{12} = c_{12}CP_2 + u_{12} \qquad\qquad (16.6A.2d)$$

$$F_{22} = c_{22}CP_2 + u_{22} \qquad\qquad (16.6A.2e)$$

$$F_{32} = c_{22}CP_2 + u_{32} \qquad\qquad (16.6A.2f)$$

Assuming cov $(CP_i, u_j) = 0$ for all i and j

$$\text{and cov}(u_i, u_j) = 0 \quad \text{for all } i \neq j \qquad\qquad (16.6A.2g)$$

and that the means of all disturbances (u) are equal to 0.

Based on the scores from the respondents on the indicators, we can calculate composite scores of C_1 for CP_1 and C_2 for CP_2. The covariance between the composite scores is not necessarily the same as the covariance between the latent variables CP_1 and CP_2 assuming that the model is correct.[6] In order to proceed, we take a simple unweighted sum for the composite scores. This means that

$$C_1 = F_{11} + F_{21} + F_{31} \quad \text{and} \quad C_2 = F_{12} + F_{22} + F_{32} \qquad\qquad (16.6A.3a)$$

By substituting (16.6A.2a)–(16.6A.2f) into (16.6A.3a), we get

$$C_1 = c_{11}CP_1 + u_{11} + c_{21}CP_1 + u_{21} + c_{31}CP_1 + u_{31} \qquad\qquad (16.6A.3b)$$

$$C_2 = c_{12}CP_2 + u_{12} + c_{22}CP_2 + u_{22} + c_{32}CP_2 + u_{32} \qquad\qquad (16.6A.3c)$$

This can be rewritten as

$$C_1 = (c_{11} + c_{21} + c_{31})CP_1 + (u_{11} + u_{21} + u_{31}) \qquad\qquad (16.6A.3d)$$

$$C_2 = (c_{12} + c_{22} + c_{32})CP_2 + (u_{12} + u_{22} + u_{32}) \qquad\qquad (16.6A.3e)$$

---

[6] Note that this is not tested and it can lead to biased estimates.

Given that the means of CP_1 and CP_2 and all disturbance terms (u) are 0, the means of C_1 and C_2 are also equal to 0; therefore, the covariance of C_1 and C_2 is defined as

$$\sigma_{C_1,C_2} = \left(\frac{1}{N}\right)\sum (C_1 \cdot C_2) \qquad (16.6A.4a)$$

$$\sigma_{C_1,C_2} = \left(\frac{1}{N}\right)\sum \left[(c_{11}+c_{21}+c_{31})CP_1 + (u_{11}+u_{21}+u_{31})\right]$$
$$\left[(c_{12}+c_{22}+c_{32})CP_2 + (u_{12}+u_{22}+u_{32})\right] \qquad (16.6A.4b)$$

However, (16.6A.4b) can be simplified after multiplying it out using (16.6A.2g) to

$$\sigma_{C_1,C_2} = (c_{11}+c_{21}+c_{31})(c_{12}+c_{22}+c_{32})\left(\frac{1}{N}\right)\sum CP_1 CP_2 \qquad (16.6A.4c)$$

or

$$\sigma_{C_1,C_2} = (c_{11}+c_{21}+c_{31})(c_{12}+c_{22}+c_{32})\sigma_{CP_1 CP_2} \qquad (16.6A.4d)$$

REFERENCES

Abelson R. P., D. R. Kinder, M. D. Peters, and S. T. Fiske 1982. Affective and semantic components in political person perception. *Journal of Personality and Social Psychology*, 42, 619–630.

Ajzen I. 1989. Attitude structure and behavior. In A. R. Pratkanis, S. J. Breckler, and A. G. Greenwald (eds.), *Attitude Structure and Function*. Hillsdale: Erlbaum, 241–247.

Ajzen I. 1991. The theory of planned behavior. *Organizational Behavior and Human Decision Processes*, 50, 179–211.

Ajzen I., and M. Fishbein 1980. Understanding attitudes and predicting social behaviour. The expectancy-value model. *Actes du Congrès de l'AFM (Poitiers)*, 681–695.

Allison P. D. 1987. Estimation of linear models with incomplete data. In C. C. Clogg (ed.), *Sociological Methodology*. Washington, DC: American Sociological Association, 71–103.

Althauser R. P., T. A. Heberlein, and R. A. Scott 1971. A causal assessment of validity: The augmented mulitrait-multimethod matrix. In H. M. Blalock Jr. (ed.), *Causal Models in the Social Sciences*. Chicago: Aldine, 151–169.

Alwin D. F. 1974. An analytic comparison of four approaches to the interpretation of relationships in the multitrait-multimehod matrix. In H. L. Costner (ed.), *Sociological Methodology*. San Francisco: Jossey Bass, 79–105.

Alwin D. F. 1997. Feeling thermometers versus 7-point scales: Which are better? *Sociological Methods and Research*, 25, 318–341.

Alwin D. 2007. *Margin of Error. A Study of Reliability in Survey Measurement*. Hoboken: Wiley.

Design, Evaluation, and Analysis of Questionnaires for Survey Research, Second Edition.
Willem E. Saris and Irmtraud N. Gallhofer.
© 2014 John Wiley & Sons, Inc. Published 2014 by John Wiley & Sons, Inc.

Alwin D. F., and J. A. Krosnick 1991. The reliability of survey attitude measurement. The influence of question and respondent attributes. *Sociological Methods and Research*, 20, 139–181.

Andrews F. M. 1984. Construct validity and error components of survey measures: A structural equation approach. *Public Opinion Quarterly*, 48, 409–442.

Andrews F. M., and S. B. Withey 1974. Developing measures of perceived life quality: Results from several surveys. *Social Indicators Research*, 1, 1–26.

Aquilino W. S. 1993. Effects of spouse presence during the interview on survey responses concerning marriage. *Public Opinion Quarterly*, 57, 358–376.

Aquilino W. S. 1994. Interview mode effects in surveys of drug and alcohol use. *Public Opinion Quarterly*, 58, 210–240.

Aquilino W. S., and L. LoSciuto 1990. Effect of interview mode on self-reported drug use. *Public Opinion Quarterly*, 54, 362–395.

Arminger G., and M. E. Sobel 1991. Pseudo-maximum likelihood estimation of mean and covariance structures with missing data. *Journal of the American Statistical Association*, 85, 195–203.

Bagozzi R. P. 1989. An investigation in the role of affective and moral evaluations in the purposeful behavior model of attitude. *British Journal of Social Psychology*, 28, 97–113.

Bagozzi R. P., and Y. Yi 1991. Multitrait-multimethod matrices in consumer research. *Journal of Consumer Research*, 17, 426–439.

Barker M. 1981. *The New Racism: Conservatives and the Ideology of the Tribe*. London: Junction Books.

Bartelds J. F., E. P Jansen, and Th. H. Joosten 1994. *Enquêteren: Het opstellen en gebruiken van vragenlijsten*. Groningen: Wolters-Noordhoff.

Belson W. 1981. *The Design and Understanding of Survey Questions*. London: Gower.

Bemelmans-Spork M., and D. Sikkel 1986. Data collection with handheld computers. In *Proceedings of the International Statistical Institute*, 3, Voorburg: International Statistical Institute.

Billiet J. G., and J. McClendon 2000. Modelling acquiescence in measurement models for two balanced sets of items. *Structural Equation Modeling*, 7, 608–629.

Billiet J., G. Loosveldt, and L. Waterplas 1986. *Het Survey-interview onderzocht: Effecten van het onderwerp en gebruik van vragenlijsten op de kwaliteit van antwoorden*. Leuven: Sociologisch Onderzoeksinstituut KU Leuven.

Blalock H. M. Jr. 1964. *Causal Inferences in Nonexperimental Research*. Chapel Hill: University of North Carolina Press.

Blalock H. M. Jr. 1968. The measurement problem: A gap between languages of theory and research. In H. M. Blalock and A. B. Blalock (eds.), *Methodology in the Social Sciences*. London: Sage, 5–27.

Blalock H. M. Jr. 1990. Auxiliary measurement theories revisited. In J. J. Hox and J. de Jong-Gierveld (eds.), *Operationalization and Research Strategy*. Amsterdam: Swets and Zeitlinger, 33–49.

Blauner R. 1966. Work satisfaction and industrial trends in modern society. In R. Bendix and S. M. Lipset (eds.) *Class, Status and Power*. New York: The Free Press, 473–487.

Bobo L., J. R. Kluegel, and R. A. Smith 1997. Laissez faire racism: The crystallization of a kinder, gentler anti-black ideology. In S. A. Tuch and J. K. Martin (eds.), *Racial Attitudes in the 1990's: Continuity and Change*. Westport: Praeger, 76–90.

Bollen K. A. 1989. *Structural Equations with Latent Variables*. New York: Wiley.

Bollen K. A., and R. Lennox 1991. Conventional wisdom on measurement: A structural equation perspective. *Psychological Bulletin*, 110, 305–314.

Bradburn N. M., and S. Sudman 1988. *Polls and Surveys. Understanding What They Tell Us.* San Francisco: Jossey Bass.

Breiman L. 2001. Random forests. *Machine Learning*, 45, 5–32.

Browne M. W. 1984. The decomposition of multitrait-multimethod matrices. *British Journal of Mathematical and Statistical Psychology*, 37, 1–21.

Bunting, B., G. Adamson, and P.K. Mulhall 2002. A Monte Carlo examination of a MTMM model with planned incomplete data structures. *Structural Equation modeling*, 9, 369–389.

Bütschi D. 1997. *Informations et opinions: Promesses et limites du questionnaire de choix.* PhD thesis, University of Genève.

Byrne B. M., R. J. Shavelson, and B. O. Muthen 1989. Testing for the equivalence of factor covariance and mean structures: The issue of partial measurement invariance. *Psychological Bulletin*, 105, 456–466.

Campbell D. T., and D. W. Fiske 1959. Convergent and discriminant validation by the multi-trait-multimethod matrices. *Psychological Bulletin*, 56, 81–105.

Campbell D. T., and E. J. O'Connell 1967. Method factors in multitrait-multimethod matrices: Multiplicative rather than additive? *Multivariate Behavioral Research*, 2, 409–426.

Campbell A., P. E. Converse, W. E. Miller, and D. E. Stokes 1960. *The American Voter.* New York: Wiley.

Carpenter E. H., and M. Just 1975. Sentence comprehension: A psychological model of verification. *Psychological Review*, 82, 45–73.

Chisholm W. E., L. T. Milic, and J. A. Grepin 1984. *Interrogativity: A Colloquium on the Grammar Typology and Pragmatics of Questions in Seven Diverse Languages*. Amsterdam: J. Benjamins.

Clark A. F. 1998. Measurement of job satisfaction; What makes a good job? Evidence from OECD countries. *OECD Labour Market and Social Policy Occasional Papers*, no. 34.

Clark H. H., and E. V. Clark 1977. *Psychology and Language*. New York: Harcourt.

Cochran W. G. 1977. *Sampling Techniques*. New York: Wiley.

Coenders G., and W. E. Saris 1998. Relationship between a restricted correlated uniqueness model and a direct product model for multitrait-multimethod data. In A. Ferligoj (ed.), *Advances in Methodology, Data Analysis and Statistic., Metodološki Zvezki*, 14. Ljubljana: FDV, 151–172.

Coenders, G., and W. E. Saris 2000. Testing nested additive, multiplicative and general multitrait-multimethod models. *Structural Equation Modeling*, 7, 219–250.

Coenders, G., W. E. Saris, J. M. Batista-Foguet, and A. Andreenkova 1999. Stability of three-wave simplex estimates of reliability. *Structural Equation Modeling*, 6, 135–157.

Coleman J. S. 1988. Social capital in the creation of human capital. *American Journal of Sociology*, 94, Supplement S95–S120.

Coleman J. S. 1990. *Foundations of Social Theory*. Cambridge, MA: Belknap Press of Harvard University.

Converse P. 1964. The nature of belief systems in mass publics. In D. A. Apter (ed.), *Ideology and Discontent*. New York: Free Press, 206–261.

Converse J. M., and H. Schuman 1984. The manner of inquiry. An analysis of survey questions from across organizations and over time. In C. F. Turner, and E. Martins (eds.), *Surveying Subjective Phenomena*, Vol. 2. New York: Russel Sage Foundation, 283–316.

Converse J. M., and S. Presser 1986. *Survey Questions: Handcrafting the Standardized Questionnaire.* Beverly Hills: Sage.

Coombs C. H. 1964. *A Theory of Data.* New York: Wiley.

Cornelius R. R. 1996. *The Sciences of Emotion. Research and Tradition in the Psychology of Emotions.* Upper Saddle River, NJ: Prentice Hall.

Coromina L., Saris, W. E. and Oberski, D. (2008) The quality of the measurement of interest in the political issues presented in the media in the European Social Survey. *Ask, Research & Methods,* 17, 7–30.

Corten I., W. E. Saris, G. Coenders, W. van der Veld, C. Albers, and C. Cornelis 2002. The fit of different models for multitrait-multimethod experiments. *Structural Equation Modeling,* 9, 213–232.

Couper M. P. 2000. Web surveys. A review of issues and approaches. *Public Opinion Quarterly,* 64, 464–494.

Couper M. P., S. E. Hansen, and S. A. Sadovski 1997. Evaluating interviewer use of CAPI. In L. Lyberg, P. Biemer, M. Collins, E. de Leeuw, C. Dippo, N. Schwarz, and D. Trewind (eds.), *Survey Measurement and Process Quality.* New York: Wiley, 267–285.

Couper M. P., R. P. Baker, J. Bethlehem, C. Z. F. Clark, J. Martin, W. L. Nicholls II and J. M. O'Reilly (eds.) 1998. *Computer Assisted Survey Information Collection.* New York: Wiley.

Cox E. P. 1980. The optimal number of response alternatives for a scale. *Journal of Marketing Research,* 17, 407–422.

Cronbach L. J. 1951. Coefficient alpha and the internal structure of tests. *Psychometrika,* 16, 294–334.

Cudeck, R. 1988. Multiplicative models and MTMM matrices. *Journal of Educational Statistics,* 13, 131–147.

Daniel F. 2000. *QUAID Question Taxonomy.* www.psyc.memphis.edu/quaid.html (Accessed on).

Danielson L., and P. A. Maarstad 1982. *Statistical Data, Collection with Hand-held Computers: A Consumer Price Index.* Orebro: Statistics Sweden.

Davidov E. 2008. A cross-country and cross-time comparison of the human values measurement with the second round of the European Social Survey. *Survey Research Methods,* 2(1), 33–46.

De Groot A. D., and F. L. Medendorp 1986. *Term, Begrip, Theorie: Inleiding tot Significhe Begripsanalyse.* Meppel: Boom.

De Heer W. 1999. International response trends: Results of an international survey. *Journal of Official Statistics,* 15, 129–142.

De Pijper W. M., and W. E. Saris 1986. *The Formulation of Interviews Using the Program Interv.* Amsterdam: Sociometric Research Foundation.

Dijkstra W., and J. van der Zouwen 1982. *Response Behaviour in the Survey-Interview.* London: Academic Press.

Dillman D. A. 1991. The design and administration of mail surveys. *Annual Review of Sociology,* 17, 225–249.

Dillman D. A. 2000. *Mail and Internet Surveys. The Tailored Design Method.* New York: Wiley.

Duncan O. D. 1975. *Introduction to Structural Equation Models.* New York: Academic Press.

Eagly A. H., and S. Chaiken 1993. *The Psychology of Attitudes*. New York: Harcourt-Brace-Jovanovich.

Edwards J. R., and R. P. Bagozzi 2000. On the nature and direction of relationships between constructs and measures. *Psychological Methods*, 5, 155–174.

Eid M. 2000. Multitrait-multimethod model with minimal assumptions. *Psychometrika*, 65, 241–261.

Emans B. 1990. *Interviewen, Theorie, Techniek en Training*. Groningen: Stenfert Kroese.

Ericsson K. A., and H. A. Simon 1984. *Protocol Analysis: Verbal Reports as Data*. Cambridge, MA: MIT Press.

Esposito J. P., and J. M. Rothgeb 1997. Evaluating survey data: Making the transition from pretesting to quality assessment. In P. Lyberg, P. Biemer, L. Collins, E. de Leeuw, C. Dippo, N. Schwarz, and D. Trewin (eds.). *Survey Measurement and Process Quality*. New York: Wiley, 541–571.

Esposito J, P. C. Campanelli, J. Rothgeb, and A. E. Polivka 1991. Determining which questions are best: Methodologies for evaluating survey questions. In *Proceedings of the Section on Survey Methods Research of the American Statistical Association*, Washington, DC: American Statistical Association, 46–55.

Essed P. 1984. *Alledaags Racisme*. Amsterdam: Feministische Uitgeverij SARA.

European Social Survey (ESS) 2002. *European Social Survey Round 1: Report of the First Year*. London: NatCen.

Ferligoj A., and V. Hlebec 1999. Evaluation of social network measurement instruments. *Social Networks*, 21, 111–130.

Fishbein M., and I. Ajzen 1975. *Belief, Attitude, Intention and Behavior: An Introduction to Theory and Research*. Reading: Addison Wesley.

Forsyth B. H., J. T. Lessler, and M. L.Hubbard 1992. Cognitive evaluation of the questionnaire. In C. F. Tanur and R. Tourangeau (eds.), *Cognition and Survey Research*. New York: Wiley, 183–198.

Fowler F. J., and T. W. Mangione 1990. *Standardized Survey Interviewing. Minimizing Interview-Related Error*. Newbury Park: Sage.

Gaertner S. L., and J. F. Dovidio (eds.) 1986. *The Aversive Form of Racism. Prejudice, Discrimination and Racism*. New York: Academic Press.

Gallhofer I. N., and W. E. Saris 1995. *Foreign Policy Decision-Making. A Qualitative and Quantitative Analysis of Political Argumentation*. Westport: Praeger.

Gfröerer J. C., and A. L. Hughes 1992. Collecting data on illicit drug use by phone. In C. F. Turner, J. T. Lessler, and J. C. Gförer (eds.), *Survey Measurement of Drug Use: Methodological Studies*. Rockville: National Institute on Drug Abuse, 277–295.

Ginzburg J. 1996. Interrogatives: Questions, facts and dialogue. In S. Lappin (ed.), *The Handbook of Contemporary Semantic Theory*. Cambridge, MA: Blackwell, 385–421.

Givon T. 1984. Syntax. *A Functional–Typological Introduction* Vol. I–II. Amsterdam: J. Benjamin.

Graesser A. C., C. L. McMahen, and B. K. Johnson 1994. Question asking and answering. In M. Gernsbacher (ed.), *Handbook of Psycholinguistics*. San Diego: Academic Press, 517–538.

Graesser, A. C., P. K. Wiemer-Hastings, R. Kreuz, and P. Wiemer-Hastings, 2000a. QUAID: A questionnaire evaluation aid for survey methodologists. *Behavior Research Methods, Instruments, and Computers*, 32, 254–262.

Graesser, A. C., K. Wiemer-Hastings, P. Wiemer-Hastings, and R. Kreuz 2000b. The gold standard of question quality on surveys: Experts, computer tools, versus statistical indices. In *Proceedings of the Section on Survey Research Methods of the American Statistical Association*. Washington, DC: American Statistical Association, 459–464.

Groenendijk J., and M. Stokhof 1997. Questions. In J. van Benthem and A. ter Meulen (eds.), *Handbook of Logic and Language*. Amsterdam: Elsevier, 1055–1124.

Grouzet F. M. E, N. Otis, and L. G. Pelletier 2006. Longitudinal cross-gender factorial invariance of the academic motivation. *Structural Equation Modeling*, 13, 73–98.

Groves, R. M. 1989. *Survey Errors and Survey Costs*. New York: Wiley.

Groves R. M., and M. P. Couper 1998. *Nonresponse in Household Interview Surveys*. New York: Wiley.

Guttman L. 1954. A new approach to factor analysis: The Radex. In P. F. Lazersfeld (ed.), *Mathematical Thinking in the Social Sciences*. Glencoe: The Free Press, 258–348.

Guttman L. 1981. Definitions and notations for the facet theory of questions. In I. Borg (ed.), *Multidimensional Data Representations: When and Why*. Ann Arbor: Mathesis Press, 95–125.

Guttman L. 1986. Science and empirical scientific generalizations. In H. Gratch (ed.), *Research Report 1984–1985*. Jerusalem: Israel Institute of Applied Research.

Halpern D. 2005. *Social Capital*. Malden: Polity Press.

Hambleton R. K., and H. Swaminathan 1985. *Item Response Theory. Principles and Applications*. Boston: Kluwer–Nijhoff.

Hamblin R. L. 1974. Social attitudes: Magnitude measurement and theories. In H. M. Blalock (ed.), *Measurement in the Social Sciences*. Chicago: Aldine, 61–120.

Harary F. 1971. *Graph Theory*. London: Addison–Wesley.

Harkness J. A., F. J. R. Van de Vijver, and P. Ph. Mohler (eds.) 2003. *Cross-Cultural Survey Methods*. Hoboken: Wiley.

Harrel L., and R. Clayton 1991. *Voice Recognition Technology in Survey Data Collection: Results of the First Field Tests*. Paper presented at the National Field Technologies Conference. San Diego.

Harris Z. 1978. The interrogative in a syntactic framework. In H. Hiz (ed.), *Questions*. Dordrecht: Reidel, 37–89.

Hartman H., and W. E. Saris 1991. *Data Collection on Expenditures*. Paper presented at the Workshop on Diary Surveys, Stockholm, Sweden.

Heise D. R. 1969. Separating reliability and stability in test-retest-correlation. *American Sociological Review*, 34, 93–101.

Heise D. R., and G. W. Bohrnstedt 1970. Validity, invalidity and reliability. *Sociological Methodology*, 2, 104–129.

Helmers H. M., R. J. Mokken, R. C. Plijter, and F. N. Stokman 1975. *Graven naar Macht. Op zoek naar de Kern van de Nederlandse Economie*. Amsterdam: Van Gennep.

Hippler H. J., and N. Schwarz 1987. Response effects in surveys. In H. J. Hippler, N. Schwarz, and S. Sudman (eds.), *Social Information Processing and Survey Methodology*. New York: Springer Verlag, 102–122.

Holsti O. R. 1996. *Public Opinion and American Foreign Policy*. Ann Arbor: The University of Michigan Press.

Homans G. C. 1965. *The Human Group*. London: Routledge–Kegan.

Horn J. L., J. McArdle, and R. Mason 1983. When is invariance not invariant: A practical scientist's look at the ethereal concept of factor invariance. *The Southern Psychologist*, 1, 179–188.

Hox J. J. 1997. From theoretical concept to survey questions. In L. Lyberg, P. Biemer, M. Collins, E. de Leeuw, C. Dippo, N. Schwarz, and D. Trewin (eds.), *Survey Measurement and Process Quality*. New York: Wiley, 47–70.

Huddleston R. 1994. The contrast between interrogatives and questions. *Journal of Linguistics*, 30, 411–439.

Huddleston R. 1988. *English Grammar: An Outline*. Cambridge, UK: Cambridge University Press.

Hyman H. H., and P. B. Sheatsley 1950. The current status of American public opinion. In J. C. Payne (ed.), *The Teaching of Contemporary Affairs, Twenty-first Yearbook of the National Council of Social Studies*. Princeton: Princeton University Press, 11–34.

Jobe J. B., W. F. Pratt, R. Tourangeau, A. Baldwin, and K. Rasinski 1997. Effects of interview mode on sensitive questions in a fertility survey. In L. Lyberg, P. Biemer, M. Collins, E. de Leeuw, C. Dippo, N. Schwarz, and D. Trewin (eds.), *Survey Measurement and Process Quality*. New York: Wiley, 322–329.

Jöreskog K. G. 1971. Simultaneous factor analysis in several populations, *Psychometrika*, 34, 409–426.

Kahn R. L. 1972. The meaning of work: Interpretation and proposals for measurement. In A. Campbell and P. E. Converse (eds.) *The Human Meaning of Social Change*. New York: Russel Sage Foundation, 159–203.

Kalfs P. 1993. *Hour by Hour. Effects of the Data Collection Mode in Time Use Research*. PhD thesis, University of Amsterdam.

Kallenberg A. L. 1974. A causal approach to the measurement of job satisfaction. *Social Science Research*, 3, 299–322.

Kallenberg A. L. 1975. *Work Values, Job Rewards and Job Satisfaction: A Theory of the Quality of Work Experience*. Unpublished PhD dissertation, University of Wisconsin.

Kallenberg A. L. 1977. Work values and job rewards: A theory of job satisfaction. *American Sociological Review*, 42, 124–143.

Kalton G. 1983. *Introduction to Survey Sampling*. Newbury Park: Sage.

Kaper E. 1999. *Panel Effects in Consumer Research. Statistical Models for Underreporting*. Amsterdam: Tinbergen Institute Research Series.

Kaplan D. 2000. *Structural Equation Modeling: Foundations and Extensions*. London: Sage.

Kay A. F. 1998. *Locating Consensus for Democracy. A Ten-Year U.S. Experiment*. St. Augustine: American Talks Issues.

Kelley H. H., and J. L. Michela 1980. Attribution theory and research. *Annual Review of Psychology*, 31, 475–501.

Kenny D. A. 1976. An empirical application of confirmatory factor analysis to the multitrait-multimethod matrix. *Journal of Experimental Social Psychology*, 12, 247–252.

Kenny D. A., and D. A. Kashy 1992. Analysis of the multitrait-multimethod matrix by confirmatory factor analysis. *Psychological Bulletin*, 112, 165–172.

Kiesler S., and L. S. Sproull 1986. Response effects in the electronic survey. *Public Opinion Quarterly*, 50, 402–413.

Kinder D. R., and D. O. Sears 1981. White opposition to busing: On conceptualizing and operationalizing group conflict. *Journal of Personal and Social Psychology*, 40, 414–431.

Kish L. 1965. *Survey Sampling*. New York: Wiley.

Klingemann H. D. 1997. The left-right self-placement question in face to face and telephone surveys. In W. E. Saris and M. Kaase (eds.), *Eurbarometer Measurement Instruments for Opinions in Europe, Zuma–Nachrichten Spezial*. Mannheim: ZUMA, 2, 113–123.

Knoke D., and J. H. Kuklinski 1982. *Network Analysis*. Beverly Hills: Sage.

Kogovšek T., A. Ferligoj, G. Coenders, and W. E. Saris 2001. Estimating reliability and validity of personal support measurements: Full information ML estimation with planned missing data. *Social Networks*, 24, 4–20.

Költringer R. 1993. *Gültigkeit von Umfragedaten*. Wien: Bohlau.

Költringer R. 1995. Measurement quality in Austrian personal interview surveys. In W. E. Saris and A. Münnich (eds.), *The Multitrait-Multimethod Approach to Evaluate Measurement Instruments*. Budapest: Eötvös University Press, 207–225.

Koning P. L., and P. J. van der Voort 1997. *Sentence Analysis*. Groningen: Wolters-Noordhoff.

Kovel J. 1971. *White Racism: A Psychohistory*. London: Allen-Lane.

Krech D., and R. Crutchfield 1948. *Theories and Problems in Social Psychology*. New York: McGraw Hill.

Krech D., Crutchfield R., and E. Ballachey 1962. *Individual in Society*. New York: McGraw Hill.

Krosnick J. A. 1991. Response strategies for coping with cognitive demands of attitude measures in surveys. *Applied Cognitive Psychology*, 5, 201–219.

Krosnick J. A., and H. Schuman 1988. Attitude intensity, importance and certainty and susceptibility to response effects. *Journal of Personality and Social Pschology*, 54, 940–952.

Krosnick J. A., and R. P. Abelson 1991. The case for measuring attitude strength in surveys. In J. M. Tanur (ed.), *Questions about Questions. Inquiries into the Cognitive Bases of Surveys*. New York: Russel Sage Foundation, 177–203.

Krosnick J. A., and L. R. Fabrigar 1997. Designing rating scales for effective measurement in surveys. In L. Lyberg, P. Biemer, M. Collins, E. de Leeuw, C. Dippo, N. Schwarz, and D. Trewin (eds.), *Survey Measurement and Process Quality*. New York: Wiley, 141–164.

Krosnick J. A., and L. R. Fabrigar (forthcoming). Designing good questionnaires; insights from cognitive psychology.

Lambrecht K. 1995. *Information Structure and Sentence Form. Topic, Focus and the Mental Representations of Discourse Referents*. Cambridge, UK: Cambridge University Press.

Lance C. E., B. Dawson, D. Birkelbach, and B. J. Hoffman 2010. Method effects, measurement error and substantive conclusions. *Organizational Research Methods*, 13, 435–455.

Lass J., W. E. Saris, and M. Kaase 1997. Sizes of the different effects: Coverage, mode and non-response. In W. E. Saris, and M. Kaase (eds.), *Eurbarometer Measurement Instruments for Opinions in Europe*, ZUMA–Nachrichten Spezial 2. Mannheim: ZUMA, 73–86.

Lawley D. N., and A. E. Maxwell 1971. *Factor Analysis as a Statistical Method*. London: Butterworth.

Lehnert W. G. 1977. Human and computational question asking. *Cognitive Science*, 1, 47–73.

Lessler J. T., and B. H. Forsyth 1966. A coding system for appraising questionnaires. In N. Schwarz and S. Sudman (eds.), *Answering Questions: Methodology for Determining Cognitive and Communicative Processes in Survey Research*. San Francisco: Jossey Bass, 259–292.

Liaw A., and M. Wiener 2002. Classification and regression by Random Forests. *R.News*, 2, 18–22.

Little T. D. 1997. Mean and covariance structures (MACS) analyses of cross-cultural data: Practical and theoretical issues. *Multivariate Behavioral Research*, 32, 53–76.

Lodge M. 1981. *Magnitude Scaling. Quantitative Measurement of Opinions*. Beverly Hills: Sage.

Lodge M., J. Tannenhaus, D. Cross, B. Tursky, M. A. Foley, and M. Foley 1976. The calibration and cross model validation of ratio scales of political opinion in survey research. *Social Science Research*, 5, 325–347.

Lord F., and M. R. Novick 1968. *Statistical Theories of Mental Test Scores*. Reading: Addison-Wesly.

Marsh H. W. 1989. Confirmatory factor analysis of multitrait-multimethod data: Many problems and few solutions. *Applied Psychological Measurement*, 13, 335–361.

Marsh, H. W., and L. Bailey 1991. Confirmatory factor analyses of multitrait-multimethod data: A comparison of alternative models. *Applied Psychological Measurement*, 15, 47–70.

Martini M. C. 2001. *Assessment of Erroneous but Compatible Answers in Social Surveys*. PhD thesis, University of Padua.

McConahay J. B., and J. C. Hough Jr. 1976. Symbolic racism. *Journal of Social Issues*, 32, 23–45.

Meredith W. 1993. Measurement invariance, factor analysis and factorial invariance. *Psychometrika*, 58, 525–543.

Messick L. 1989. Validity. In R. L. Linn (ed.), *Educational Measurement*. New York: Macmillan, 13–103.

Miethe T. D. 1985. The validity and reliability of value measurements. *Journal of Psychology*, 119, 441–453.

Miller G. A. 1956. The magical number seven plus or minus two—some limits on our capacity for processing information. *Psychological Review*, 63, 81–97.

Mokken R. J. 1971. *A Theory and Procedure of Scale Analysis with Applications in Political Research*. New York: Walter-Gruyter-Mouton.

Molenaar. N. J. 1986. *Formuleringseffecten in Survey-Interviews*. PhD thesis. Amsterdam: Free University.

Münnich A. 2004. *Judgement and Choice*. PhD thesis, University of Amsterdam.

Muthen B., D. Kaplan, and M. Hollis 1987. On structural equation modeling with data that are not missing completely at random. *Psychometrika*, 52, 431–462.

Neijens P. 1987. *The Choice Questionnaire. Design and Evaluation of an Instrument for Collecting Informed Opinions of a Population*. PhD thesis. Amsterdam: Free University.

Newton K. 1997. Social capital and democracy. *American Behavioral Scientist*, 40, 575–586.

Newton K. 1999. Social trust and political trust in established democracies. In P. Norris (ed.), *Critical Citizens*. Oxford: Oxford University Press, 169–187.

Newton K. 2001. *Social Trust and Political Disaffection: Social Capital and Democracy*. Paper presented at the EURESCO Conference on Social Capital: Interdisciplinary Perspectives, Exeter.

Nisbett R. E., and T. D. Wilson 1977. Telling more than we know: Verbal reports on mental processes. *Psychological Review*, 84, 231–259.

Northrop F. S. C. 1947. *The Logic of the Sciences and the Humanities*. New York: World Publishing Company.

Nunnally J. C, and L. H. Bernstein 1994. *Psychometric Theory*. New York: McGraw Hill.

Oppenheim A. N. 1966. *Questionnaire Design and Attitude Measurement*. London: Heinemann.

Oskamp S. 1991. *Attitudes and Opinions*. Englewoods Cliffs: Prentice Hall.

Parsons T. 1951. *The Social System*. Glencoe: Free Press.

Payne S. 1951. *The Art of Survey Questions*. Princeton: Princeton University Press.

Pettigrew T. F., and R. W. Meertens 1995. Subtle and blatant prejudice in Western Europe. *European Journal of Social Psychology*, 25, 57–75.

Phipps P. A., and A. R. Tupek 1991. Assessing measurement errors in a touchtone recognition survey. *Survey Methodology*, 17, 15–26.

Piazza T., and P. M. Sniderman 1991. Incorporating experiments into computer-assisted surveys. In M. P. Couper, R. P. Bakker, J. Bethlehem, C. Z. F. Clark, J. Martin, W. L. Nicholls II., and J. M. O'Reilly (eds.), *Computer Assisted Survey Information Collection*. Hoboken: Wiley, 167–185.

Poulton E. C. 1968. The new psychophysics: Six models for magnitude estimation judgments. *Psychological Bulletin*, 69, 1–19.

Presser S. 1984. The use of survey data in basic research in the social sciences. In. C. F. Turner, and E. Martin (eds.), *Surveying Subjective Phenomena*. New York: Russell Sage Foundation, 110–132.

Presser S., and J. Blair 1994. Survey pretesting: Do different methods produce different results? In P. V. Marsden (ed.), *Sociological Methodology*. Oxford: Basil Blackwell, 73–104.

Pruchno R. A., and J. M. Hayden 2000. Interview modality: Effects on costs and data quality in a sample of older women. *Journal of Aging and Health*, 12, 3–24.

Putnam R. D. 2000. *Bowling Alone*. New York: Simon and Schuster.

Quirk R., S. Greenbaum, G. Leech, and J. Svartvik 1985. *A Comprehensive Grammar of the English Language*. London: Longman.

R Development Core Team 2011. R: A language and environment for statistical computating. *R Foundation for Statistical Computing*, Vienna, Austria. www.R-project.org (Accessed on October 5, 2013).

Rabinowitz G., S. E. Macdonald, and O. Lishuag 1991. New players in an old game: Party strategy in multiparty systems. *Comparative Political Studies*, 24, 147–185.

Rasch G. 1960. *Probabilistic Models for some Intelligence and Attainment Tests*. Copenhagen: Danish Institute for Education Research.

Raykov T. 1997. Scale reliability. Cronbach's coefficient alpha and violations of essential tau-equivalence with fixed congeneric components. *Multivariate Behavioral Research*, 32, 329– 353.

Raykov T. 2001. Bias of Cronbach's coefficient alpha for fixed congeneric measures with correlated errors. *Applied Psychological Measurement*, 25, 69–76.

Revilla M., and W. E. Saris 2011. The problem and solutions of the analysis of the MTMM experiments. In W. Saris, D. Oberski, M. Revilla, D. Zavala, L. Lilleoja, I. Gallhofer, and T. Gruner *The Development of the Program SQP 2.0 for the Prediction of the Quality of Survey Questions*, RECSM Working paper 24.

Revilla M., and W. E. Saris 2013. The split-ballot multitrait-multimethod approach: Implementation and problems, *Structural Equation Modeling: A Multidisciplinary Journal*, 20(1), 27–46.

Revilla M., W. E. Saris, and J. A. Krosnick (forthcoming). Choosing the number of categories in agree-disagree scales. *Sociological Methods and Research*.

Richards J., C. J. Platt, and H. Platt 1993. *Dictionary of Language Teaching and Applied Linguistics*. Harlow: Longman.

Robinson J. P, R. Athanadiou, and K. B. Head 1969. *Measures of Occupational Attitudes and Occupational Characteristics*. Ann Arbor: Institute for Social Science.

Rodgers W. L., F. M. Andrews, and A. R. Herzog 1992. Quality of survey measures: A structural modelling approach. *Journal of Official Statistics*, 8, 251–275.

Rokeach M. 1973. *The Nature of Human Values*. New York: Free Press.

Saris W. E. 1981. Different questions, different variables. In C. Fornell (ed.), *A Second Generation of Multivariate Analysis*, Vol. 2. New York: Praeger, 78–96.

Saris W. E. (ed.) 1988a. *Variations in Response Functions: A Source of Measurement Error in Attitude Research*. Amsterdam: Sociometric Research Foundation.

Saris W. E. 1988b. A measurement model for psychophysical scaling. *Quality and Quantity*, 22, 417–483.

Saris W. E. 1990. The choice of a model for evaluation of measurement instruments. In W. E. Saris and A. van Meurs (eds.), *Evaluation of Measurement Instruments by Meta-analysis of Multitrait-Multimethod studies*. Amsterdam: North Holland, 118–133.

Saris W. E. 1991. *Computer Assisted Interviewing*. Newbury Park: Sage.

Saris W. E. 1996. A facet design to describe measures for ethnocentrism. In *Proceedings of the IRMCS conference in Preddvor*, Slovenia.

Saris W. E. 1997. Comparability across mode and country. In W. E. Saris, and M. Kaase (eds.), *Eurobarometer Measurement for Opinions in Europe. ZUMA-Nachrichten Spezial 2*. Mannheim: ZUMA, 125–139.

Saris W. E. 1998a. Ten years of interviewing without interviewers: The telepanel. In M. P. Couper, R. P. Baker, J. Bethlehem, C. Clark, J. Martin, W. L. Nicholls II, and J. M. O'Reilly (eds.), *Computer-Assisted Survey Information Collection*. New York: Wiley, 409–431.

Saris W. E. 1998b. Words are sometimes not enough to express the existing information. In M. Fenema, C. van der Eyck, and H. Schijf (eds.), *In Search of Structure: Essays in Social Science and Methodology*. Amsterdam: Het Spinhuis, 98–115.

Saris W. E. 1998c. A new approach for evaluation of measurement instruments: The split-ballot MTMM design. Paper presented at the International Conference on Methodology and Statistics. Preddvor, Slovenia, September 20–22.

Saris W. E., and H. Stronkhorst 1984. *Causal Modeling in Nonexperimental Research: An Introduction to the LISREL Approach*. Amsterdam: Sociometric Research Foundation.

Saris W. E., and W. M. de Pijper 1986. Computer assisted interviewing using home computers. *European Research*, 14, 144–152.

Saris W. E., and K. de Rooy 1988. What kind of terms should be used for reference points. In W. E. Saris (ed.), *Variations in Response Functions: A Source of Measurement Error in Attitude Research*. Amsterdam: Sociometric Research Foundation, 199–219.

Saris W. E., and F. M. Andrews 1991. Evaluation of measurement instruments using a structural modeling approach. In P. P. Biemer, R. M. Groves, L. E. Lyberg, N. Mathiowetz, and S. Sudman (eds.), *Measurement Errors in Surveys*. New York: Wiley, 575–599.

Saris W. E., and M. Kaase 1997 (eds.). *Eurobarometer Measurement for Opinions in Europe. ZUMA-Nachrichten Spezial 2.* Mannheim: ZUMA, 125–139.

Saris W. E., and I. N. Gallhofer 2002. *Cross-Cultural Research Comparability: The Effects of Random and Systematic Errors. Report for the ESS.* London: ESS.

Saris W. E., and C. Aalberts 2003. Different explanantions for correlated errors in MTMM studies. *Structural Equation Modeling*, 10, 193–214.

Saris W. E., and I. N. Gallhofer 2004. Operationalization of social science concepts by intuition. *Quality and Quantity*, 38, 235–258.

Saris W. E., and I. N. Gallhofer 2007a. Estimation of the effects of measurement characteristics on the quality of survey questions. *Survey Research Methods*, 1, 31–46.

Saris W. E., and I. N. Gallhofer 2007b. *Design, Evaluation and Analysis of Questionnaires for Survey Research.* Hoboken: Wiley-Interscience.

Saris W. E., C. Bruinsma, W. Schoots, and C. Vermeulen 1977. The use of magnitude estimation in large scale survey research. *Mens en Maatschappij*, 52, 369–359.

Saris W. E., W. M. de Pijper, and J. Mulder 1978. Optimal procedures for estimation of factor scores. *Sociological Methods and Research*, 7, 85–105.

Saris W. E., M. de Pijper, and P. Neijens 1982. Some notes on the computer steered interview. In C. P. Middendorp (ed.), *Handelingen van het Congres gehouden in Rotterdam.* Rotterdam: Dutch Sociometric Society, 95–104.

Saris W. E., P. Neijens, and J. A. de Ridder 1984. Resultaten van de keuze-enquête in het kader van de B. M. D. *In Stuurgroep Maatschappelijke Diskussie Engergiebeleid: Het Eindrapport, Appendix.* Leiden: Stenfert Kroese.

Saris W. E., A. Satorra, and D. Sörbom 1987. Detection and correction of structural equation models. *Sociological Methodology*, 17, 105–131.

Saris W. E., A. Satorra, and G. Coenders 2004. A new approach for evaluating quality of measurement instruments. *Sociological Methodology*, 3, 311–347.

Saris W. E., M. Revilla, J. Krosnick, and E. M. Schaefer 2009a. Comparing questions with agree/disagree response options to questions with item specific response options. *Survey Research Methods*, 16, 561–582.

Saris W. E., Satorra A., and W. M. van der Veld 2009b. Testing structural equation models or detection of misspecifications? *Structural Equation Modeling: A Multidisciplinary Journal*, 16(4), 561–582.

Saris W. E., D. Oberski, M. Revilla, D. Zavala, L. Lilleoja, I. Gallhofer, and T. Gruner 2011. The development of the program SQP 2.0 for the prediction of the quality of survey questions, RECSM Working paper 24.

Satorra A. 1990. Robustness issues in structural equation modelling: A review of recent developments. *Quality and Quantity*, 24, 367–387.

Satorra A. 1992. Asymptotic robust inferences in the analysis of mean and covariance structures. In P. V. Marsden (ed.), *Sociological Methodology 1992.* Oxford: Basil Blackwell, 249–278.

Satorra, A. 1993. Asymptotic robust inferences in multi-sample analysis of augmented-moment matrices. In C. R. Rao and C. M. Cuadras (eds.), Amsterdam: North Holland, 211–229.

Satorra A. 2000. Goodness of fit testing of structural equation models with multiple group data and nonnormality. In R. Cudeck, S. du Toit, and D. Sörbom (eds.), *Structural Equation Modeling: Present and Future.* Lincolnwood: SSI, 231–257.

Scherpenzeel A. C. 1995. *A Question of Quality. Evaluating Survey Questions by Multitrait-Multimethod Studies.* Leidschendam: KPN Research.

Scherpenzeel A. C., and W. E. Saris 1993. The quality of indicators of satisfaction across Europe. A meta-analysis of multitrait-multimethod studies. *Bulletin de Methodologie Sociologique,* 39, 3–19.

Scherpenzeel A., and W. E. Saris 1996. Causal direction in a model of life satisfaction: The top-down/bottom- up controversy. *Social Indicators,* 38, 161–180.

Scherpenzeel A. C., and W. E. Saris 1997. The validity and reliability of survey questions: A meta-analysis of MTMM studies. *Sociological Methods and Research,* 25, 341–383.

Scherpenzeel A. C., and W. E. Saris 2006. Multitrait-multimethod models for longitudinal research. In K.van Montford, H. Oud, and A. Satorra (eds.), *Longitudinal Models in Behavioral and Related Sciences.* London: Lawrence Erlbaum, 381–403.

Schober M. F., and F. G. Conrad 1997. Does conversational interviewing reduce survey measurement error? *Public Opinion Quarterly,* 60, 576–602.

Schuman H., and S. Presser 1981. *Questions and Answers in Attitude Survey: Experiments on Question Form, Wording and Context.* New York: Academic Press.

Schwartz S. H. 1997. Values and culture. In D. Muno, S. Carr, and J. Schumaker (eds.), *Motivation and Culture.* New York: Routledge, 69–84.

Schwarz N., and H. J. Hippler 1987. What response scales may tell your respondents: Information functions of response alternatives. In H. J. Hippler, N. Schwarz, and S. Sudman (eds.), *Social Information Processing and Survey Methodology.* New York: Springer, 163–178.

Schwarz N., and S. Sudman (eds.) 1996. *Answering Questions: Methodology for Determining Cognitive and Communicative Processes in Survey Research.* San Francisco: Jossey-Bass.

Schwartz, S. H., and A. Bardi 2001. Value hierarchies across cultures: Taking similarities perspective. *Journal of Cross Cultural Psychology,* 32, 268–290.

Silberstein A. S., and S. Scott 1991. Expenditure diary surveys and their associated errors. In P. Biemer, R. M. Groves, L. E. Lyberg, N. A. Mathiowetz, and S. Sudman (eds.), *Measurement Errors in Surveys.* New York: Wiley, 303–327.

Skinner B. F. 1953. *Science and Human Behavior.* New York: Macmillan.

Smith T. W. 1987. The art of asking questions 1936–1985. *Public Opinion Quarterly,* 51, 95–108.

Sniderman P. M., and Ph. E. Tetlock 1986. Reflections on American racism. *Journal of Social Issues,* 42, 173–187.

Sniderman P. M., and S. Theriault 2004. The structure of political argument and the logic of issue framing. In W. E. Saris, and P. M. Sniderman (eds.), *Studies in Public Opinion: Gauging Attitudes, Nonattitudes, Measurement Error and Change.* Princeton: Princeton University Press, 133–166.

Sniderman P. M., R. A. Brody, and P. E. Tetlock 1991. *Reasoning and Choice. Explorations in Political Psychology.* Cambridge, MA: Cambridge University Press.

Snijkers G. 2002. *Cognitive Laboratory Experiences: On Pretesting, Computerized Questionnaires and Data Quality.* PhD thesis, University of Utrecht.

Sorokin P. 1928. *Contemporary Sociological Theories.* New York: Harper.

Steenkamp J., and H. Baumgartner 1998. Assessing measurement invariance in Cross-national consumer research. *Journal of Consumer Research,* 25, 78–90.

Stevens S. S. 1975. *Psychophysics: Introduction to its Perceptual, Neural and Social Prospects.* New York: Wiley.

Stokes D. E. 1963. Spatial models of party competition. *American Political Science Review*, 57, 368–377.

Stoop I. 2005. *The Hunt for the Last Respondent: Nonresponse in Sample Surveys*. The Hague: SCP.

Sudman S., and N. M. Bradburn 1983. *Asking Questions: A Practical Guide to Questionnaire Design*. San Francisco: Jossey Bass.

Sudman S., N. M. Bradburn, and N. Schwarz 1996. *Thinking about Answers: The Application of Cognitive Processes to Survey Methodology*. San Francisco: Jossey-Bass.

Swan M. 1995. *Practical English Usage*. Oxford: Oxford University Press.

Tesser A., and L. Martin 1996. The psychology of evaluation. In E. T. Higgins and A. W. Kruglinski (eds.), *Social Psychology. Handbook of Basic Principles*. New York: Guilford Press, 400–432.

Thomassen J. 2002. http://www.europeansocialsurvey.org/docs/methodology/core_ess_ questionnaire/ESS_core_questionnaire_political_issues.pdf (Accessed on October 19, 2013).

Tönnies F. 1887. *Gemeinschaft in der Gesellschaft. Grundbegriffe der reinen Sociologie*. Berlin: Curtius.

Torcal Loriente M. L., M. Diez de Ulzurrun, and S. Perez-Nievas Montiel (eds.) 2005. *España; Sociedad y Politica en Perspectiva Comparada*. Valencia: Tirant Lo Blanch.

Torgerson W. S. 1958. *Theory and Methods of Scaling*. New York: Wiley.

Tortora R. D. 1985. Cati in an agricultural statistical agency. *Journal of Official Statistics*, 1, 301–314.

Tourangeau R., and T. W. Smith 1996. Asking sensitive questions: The impact of data collection mode, question format and question context. *Public Opinion Quarterly*, 60, 275–304.

Tourangeau R., K. Rasinski, J. B. Jobe, T. W. Smith, and W. Pratt 1997. Sources of error in a survey of sexual behavior. *Journal of Official Statistics*, 13, 342–365.

Tourangeau R., L. J. Rips, and K. Rasinski 2000. *The Psychology of Survey Response*. Cambridge, MA: Cambridge University Press.

Trabasso T., H. Rollins, and E. Shaughnessey 1971. Storage and verification stages is processing concepts. *Cognitive Psychology*, 2, 239–289.

Troldahl V. C., and R. E. Carter 1964. Random selection of respondents within households in phone surveys. *Journal of Marketing Research*, 1, 71–76.

Turner C. F., L. Ku, S. M. Rogers, L. D. Lindberg, J. H. Pleck, and F. L. Sonenstein, 1998. Adolescent sexual behavior, drug use, and violence: Increased reporting with computer survey technology. *Science*, 280, 867–873.

Van der Pligt J., and N. K. de Vries 1995. *Opinies en Attitudes. Meting, Modellen en Theorie*. Amsterdam-Meppel: Boom.

Van der Veld W. 2006. *The Survey Response Dissected: A New Theory about the Survey Response Process*. PhD thesis, University of Amsterdam.

Van der Veld W., and W. E. Saris 2003. Separation of error, method effects, instability and attitude strength. In W. E. Saris and P. Sniderman (eds.), *The Issue of Belief: Essays in the Intersection of Nonattitudes and Attitude Change*. Princeton: Princeton University Press, 37–63.

Van der Zouwen J. 2000. An assesment of the difficulty of questions used in the ISSP questionnaires, the clarity of their wording and the comparability of the responses. *ZA-information*, 45, 96–114.

Van der Zouwen J., and W. Dijkstra 1996. Trivial and non-trivial question-answer sequences, types determinants and effects on data quality. In *Proceedings of the International Conference on Survey Measurement and Process Quality*, Bristol, April 1–4, 1995. Alexandria: American Statistical Association (ASA), 81–86.

Van der Zouwen J., W. Dijkstra, and J. H. Smit 1991. Studying respondent-interviewer interaction: The relationship between interviewing style, interviewer behavior and response behavior. In P. Biemer, R. M. Groves, L. E. Lyberg, N. A. Mathiowetz, and S. Sudman (eds.), *Measurement Errors in Surveys*. New York: Wiley, 419–437.

Van Meurs A., and W. E. Saris 1990. Memory effects in MTMM studies. In W. E. Saris and A. van Meurs (eds.), *Evaluations of Measurement Instruments by Metaanalysis of Multitrait-Multimethod Studies*. Amsterdam: North Holland, 134–146.

Van Schuur W. H. 1988. Stochastic unfolding. In W. E. Saris and I. Gallhofer (eds.), *Sociometric Research, Vol. I, Data collection and scaling*. London: McMillan Press, 137–159.

Vetter A. 1997. Political efficacy: Alte und neue Meßmodelle im Vergleich. *Kölner Zeitschrift für Soziologie und Sozialpsychologie*, 49, 53–73.

Von Winterfeldt D., and W. Edwards 1986. *Decision Analysis and Behavioral Research*. Cambridge, MA: Cambridge University Press.

Voogt R. 2003. *"I am interested". Nonresponse Bias, Response Bias and Stimulus Effects in Election Research*. PhD thesis, University of Amsterdam.

Voogt R., and W. E. Saris 2003. To participate or not participate: The link between survey participation, electoral participation and political interest. *Political Analysis*, 11, 164–179.

Wanous J. P., and E. E. Lawler 1972. Measurement and meaning of job satisfaction. *Journal of Applied Psychology*, 56, 95–105.

Weber E. G. 1993. *Varieties of Questions in English Conversation*. Amsterdam: J. Benjamin.

Wegener B. (ed.) 1982. *Social Attitudes and Psychophysical Measurement of Opinions*. Hillsdale: Erlbaum.

Werts C. E., and R. L. Linn 1970. Path analysis. Psychological examples. *Psychological Bulletin*, 74, 193–212.

Whitman D. S., D. L. Van Rooy, and C. Viswesvaran 2010. Satisfaction, citizenship behaviors and performance in work units: A meta analysis of collective construct relations. *Personel Psychology*, 63, 41–81.

Wilensky H. L. 1964. Varieties in work experience. In H. Borow (ed.) *Man in the World at Work*. Boston: Houghton-Mifflin, 125–154.

Wilensky H. L. 1966. Work as a social problem. In H. S. Becker (ed.) *Social Problems: A Moderm Approach*. New York: Wiley, 117–166.

Wiley D. E., and J. A. Wiley 1970. The estimation of measurement error in panel data. *American Sociological Review*, 35, 112–117.

Wilson T. D., and D. Dunn 1986. Effects of introspection on attitude-behavior consistency: Analyzing reasons versus focusing on feelings. *Journal of Experimental Social Psychology*, 22, 249–263.

Wothke W. 1996. Models for multitrait-multimethod matrix analysis. In G. C. Marcoulides, and R. E. Schumacker (eds.), *Advanced Structural Equation Modeling: Issues and Techniques*. Mahwah: L. Erlbaum, 7–56.

Wouters M. 2001. *A Design for the Evaluation of the Quality of Open-ended Questions*. Paper presented at the IRMCS Meeting, Gent, May 25–27.

Yule G. 1998. *Explaining English Grammar*. Oxford: Oxford University Press.

Zaller J. R. 1992. *The Nature and Origins of Mass Opinion*. Cambridge, UK: Cambridge University Press.

Zanna M. P., and J. K. Rempel 1988. Attitudes: A new look at an old concept. In D. Bar-Tal, and A. Kruglanski (eds.), *The Social Psychology of Knowledge*. Cambridge, UK: Cambridge University Press, 210–245.

INDEX

Design, Evaluation, and Analysis of Questionnaires for Survey Research, Second Edition.
Willem E. Saris and Irmtraud N. Gallhofer.
© 2014 John Wiley & Sons, Inc. Published 2014 by John Wiley & Sons, Inc.

WILEY SERIES IN SURVEY METHODOLOGY
Established in Part by WALTER A. SHEWHART AND SAMUEL S. WILKS

Editors: *Mick P. Couper, Graham Kalton, J. N. K. Rao, Norbert Schwarz, Christopher Skinner*
Editor Emeritus: *Robert M. Groves*

The *Wiley Series in Survey Methodology* covers topics of current research and practical interests in survey methodology and sampling. While the emphasis is on application, theoretical discussion is encouraged when it supports a broader understanding of the subject matter.

The authors are leading academics and researchers in survey methodology and sampling. The readership includes professionals in, and students of, the fields of applied statistics, biostatistics, public policy, and government and corporate enterprises.

*Now available in a lower priced paperback edition in the Wiley Classics Library.

HARKNESS, VAN DE VIJVER, and MOHLER (editors) · Cross-Cultural Survey Methods

KALTON and HEERINGA · Leslie Kish Selected Papers

KISH · Statistical Design for Research

*KISH · Survey Sampling

KORN and GRAUBARD · Analysis of Health Surveys

KREUTER (editor) · Improving Surveys with Paradata: Analytic Uses of Process Information

LEPKOWSKI, TUCKER, BRICK, DE LEEUW, JAPEC, LAVRAKAS, LINK, and SANGSTER (editors) · Advances in Telephone Survey Methodology

LESSLER and KALSBEEK · Nonsampling Error in Surveys

LEVY and LEMESHOW · Sampling of Populations: Methods and Applications, *Fourth Edition*

LUMLEY · Complex Surveys: A Guide to Analysis Using R

LYBERG, BIEMER, COLLINS, de LEEUW, DIPPO, SCHWARZ, TREWIN (editors) · Survey Measurement and Process Quality

LYNN · Methodology of Longitudinal Surveys

MADANS, MILLER, and MAITLAND (editors) · Question Evaluation Methods: Contributing to the Science of Data Quality

MAYNARD, HOUTKOOP-STEENSTRA, SCHAEFFER, and VAN DER ZOUWEN · Standardization and Tacit Knowledge: Interaction and Practice in the Survey Interview

PORTER (editor) · Overcoming Survey Research Problems: New Directions for Institutional Research, No. 121

PRESSER, ROTHGEB, COUPER, LESSLER, MARTIN, MARTIN, and SINGER (editors) · Methods for Testing and Evaluating Survey Questionnaires

RAO · Small Area Estimation

REA and PARKER · Designing and Conducting Survey Research: A Comprehensive Guide, *Third Edition*

SARIS and GALLHOFER · Design, Evaluation, and Analysis of Questionnaires for Survey Research, *Second Edition*

SÄRNDAL and LUNDSTRÖM · Estimation in Surveys with Nonresponse

SCHWARZ and SUDMAN (editors) · Answering Questions: Methodology for Determining Cognitive and Communicative Processes in Survey Research

SIRKEN, HERRMANN, SCHECHTER, SCHWARZ, TANUR, and TOURANGEAU (editors) · Cognition and Survey Research

SNIJKERS, HARALDSEN, JONES, and WILLIMACK · Designing and Conducting Business Surveys

STOOP, BILLIET, KOCH and FITZGERALD · Improving Survey Response: Lessons Learned from the European Social Survey

SUDMAN, BRADBURN, and SCHWARZ · Thinking about Answers: The Application of Cognitive Processes to Survey Methodology

UMBACH (editor) · Survey Research Emerging Issues: New Directions for Institutional Research No. 127

VALLIANT, DORFMAN, and ROYALL · Finite Population Sampling and Inference: A Prediction Approach

WALLGREN and WALLGREN · Register-based Statistics: Administrative Data for Statistical Purposes

Printed and bound by CPI Group (UK) Ltd, Croydon, CR0 4YY

27/10/2024

14580268-0004